Digital Image Processing Techniques

This is Volume 2 in

COMPUTATIONAL TECHNIQUES

Edited by BERNI J. ALDER and SIDNEY FERNBACH

Digital Image
Processing Techniques

Edited by

Michael P. Ekstrom

Schlumberger-Doll Research
Ridgefield, Connecticut

1984

ACADEMIC PRESS, INC.
(Harcourt Brace Jovanovich, Publishers)
Orlando   San Diego   San Francisco   New York   London
Toronto   Montreal   Sydney   Tokyo

COPYRIGHT © 1984, BY ACADEMIC PRESS, INC.
ALL RIGHTS RESERVED.
NO PART OF THIS PUBLICATION MAY BE REPRODUCED OR
TRANSMITTED IN ANY FORM OR BY ANY MEANS, ELECTRONIC
OR MECHANICAL, INCLUDING PHOTOCOPY, RECORDING, OR ANY
INFORMATION STORAGE AND RETRIEVAL SYSTEM, WITHOUT
PERMISSION IN WRITING FROM THE PUBLISHER.

ACADEMIC PRESS, INC.
Orlando, Florida 32887

United Kingdom Edition published by
ACADEMIC PRESS, INC. (LONDON) LTD.
24/28 Oval Road, London NW1 7DX

Library of Congress Cataloging in Publication Data

Main entry under title:

Digital image processing techniques.

    (Computational techniques ; v.   )
    Bibliography: p.
    Includes index.
    1. Image processing--Digital techniques. I. Ekstrom,
Michael P. II. Series.
TA1632.D496 1983   621.36'7   83-22321
ISBN 0-12-236760-X (alk. paper)

PRINTED IN THE UNITED STATES OF AMERICA

84 85 86 87   9 8 7 6 5 4 3 2 1

Contents

Contributors

JOHN R. ADAMS (289), International Imaging Systems, Milpitas, California 95035

V. RALPH ALGAZI (171), Signal and Image Processing Laboratory, Department of Electrical and Computer Engineering, University of California, Davis, California 95616

EDWARD C. DRISCOLL, JR. (289), International Imaging Systems, Milpitas, California 95035

PAUL M. FARRELLE (171), British Telecom Research Labs, Martlesham Heath, Ipswich, IP5 7RE, England

B. R. HUNT (53), Department of Electrical Engineering, University of Arizona, Tucson, Arizona 85721

ANIL K. JAIN (171), Signal and Image Processing Laboratory, Department of Electrical and Computer Engineering, University of California, Davis, California 95616

A. C. KAK (111), School of Electrical Engineering, Purdue University, West Lafayette, Indiana 47907

S. W. LANG (227), Schlumberger-Doll Research, Ridgefield, Connecticut 06877

JAE S. LIM (1), Research Laboratory of Electronics, Department of Electrical Engineering and Computer Science, Massachusetts Institute of Technology, Cambridge, Massachusetts 02139

T. L. MARZETTA (227), Schlumberger-Doll Research, Ridgefield, Connecticut 06877

CLIFF READER (289), International Imaging Systems, Milpitas, California 95035

AZRIEL ROSENFELD (257), Center for Automation Research, University of Maryland, College Park, Maryland 20742

JOHN W. WOODS (77), Electrical, Computer, and Systems Engineering Department, Rensselaer Polytechnic Institute, Troy, New York 12181

ix

Preface

Over the past decade the impact of computational techniques on all fields of information sciences has been both profound and widespread. This is perhaps nowhere more true than in the field of digital image processing, where digital approaches to the manipulation of imagery and other two-dimensional data sets have been employed almost exclusively in a broad range of scientific applications. Such processing has become an important, integral part of applications like computerized tomography, geophysical exploration, nondestructive testing, x-ray and radio astronomy, remote sensing, medical ultrasound, and industrial robotics. Indeed, it is difficult to conceptualize many of these applications without their image processing and display components.

What has caused this remarkable adoption of digital image processing? The availability of computing machinery capable of handling large-scale image processing tasks has been a primary element. The analog-to-digital conversion of image fields typically involves massive amounts of information. Sampling an image on a 1024 × 1024 grid, for example, with each picture element quantized to 10 bits, results in a data array of over 10 million bits. Although it remains true that the size of such arrays places severe demands on digital image processing systems, numerous systems do exist that have sufficient speed and memory to routinely process these large image arrays. In common with other digital signal processors, they are highly reliable, modular, and allow control of precision.

The evolution of increasingly sophisticated digital image processing algorithms has been a concurrent and complementary element. In this regard, a sort of "push–pull" relationship has existed between the processor hardware and algorithm development components: As processor capabilities have expanded, new algorithms have been developed to exploit their capabilities (and vice versa). The variety of algorithms considered is extensive, reflecting the variety of applications mentioned above. Contributions have been made from many disciplines including digital signal processing, computer science, statistical communications theory, control systems, and applied physics.

Our objective in preparing this volume is to provide a state-of-the-art review of digital image processing techniques, with an emphasis on the processing approaches and their associated algorithms. We have approached this by considering a canonical set of image processing problems. While this set is not all-inclusive, it does represent the class of functions typically required in most image processing applications. It also represents a reasonably accurate cross section of the literature presently being published in this field.

The volume is organized into two major sections. The first and principal section (Chapters 1–7) deals directly with processing techniques associated with the following tasks:

Image Enhancement—This processing involves subjectively improving the quality of various types of image information (e.g., edges), often in a perceptual context.

Image Restoration—Correcting for deblurring and/or other distorting effects introduced in the image forming process, usually performed with regard to an objective fidelity measure.

Image Detection and Estimation—Deciding on the presence or absence of an object or class of objects in an image scene and estimating certain of the object's features or parameters.

Image Reconstruction—Reconstructing a two-dimensional field from its one-dimensional projections, where the projections are taken at various angles relative to the object.

Image Data Compression—Encoding the imagery, based on information redundancy, to reduce its data storage/transmission requirements while maintaining its overall fidelity.

Image Spectral Estimation—Developing two-dimensional power spectral density estimates for image fields (used in many of the above tasks).

Image Analysis—Representing image information in a form compatible with automated, machine-based processing, usually in an interpretative context.

The emphasis in these contributions is on problem definition and technique description. Each chapter broadly addresses the following questions: What problem is being considered? What are the best techniques for this particular problem? How do they work? What are their strengths and limitations? How are the techniques actually implemented and what are their computational aspects?

The second section (Chapter 8) describes hardware and software systems for digital image processing. Aspects of commercially available systems that combine both processing and display functions are described, as are future prospects for their technological and architectural evolution. Specifics of system design trade-offs are explicitly presented in detail.

Some care has been taken in the preparation of all chapters to make the material as accessible as possible to the reader. To the extent practical, standard notation has been adopted, and each chapter concludes with an annotated bibliography, noting particularly important contributions, and an extensive list of references. This is intended to provide a useful entry point into the rich yet diverse literature of the field. For their consideration and cooperation in preparing these materials, I wish to publicly acknowledge the contributing authors.

1 Image Enhancement

Jae S. Lim

Research Laboratory of Electronics
Department of Electrical Engineering and Computer Science
Massachusetts Institute of Technology
Cambridge, Massachusetts

I. Introduction

Image enhancement is the processing of images to increase their usefulness. Methods and objectives vary with the application. When images are
enhanced for human viewers, as in television, the objective may be to improve perceptual aspects: image quality, intelligibility, or visual appearance. In other applications, such as object identification by machine, an
image may be preprocessed to aid machine performance. Because the
objective of image enhancement is dependent on the application context,
and the criteria for enhancement are often subjective or too complex to be
easily converted to useful objective measures, image enhancement algorithms tend to be simple, qualitative, and ad hoc. In addition, in any given

DIGITAL IMAGE PROCESSING TECHNIQUES

1

application, an image enhancement algorithm that performs well for one class of images may not perform as well for other classes.

Image enhancement is closely related to image restoration. When an image is degraded, restoration of the original image often results in enhancement. There are, however, some important differences between restoration and enhancement. In image restoration, an ideal image has been degraded, and the objective is to make the processed image resemble the original as much as possible. In image enhancement, the objective is to make the processed image better in some sense than the unprocessed image. In this case, the ideal image depends on the problem context and often is not well defined. To illustrate this difference, note that an original, undegraded image cannot be further restored but can be enhanced by increasing sharpness through high-pass filtering.

Image enhancement is desirable in a number of contexts. In one important class of problems, an image is enhanced by modifying its contrast and/or dynamic range. For example, a typical image, even if undegraded, will often appear better when the edges are sharpened. Also, if an image with large dynamic range is recorded on a medium with small dynamic range, such as film or paper, the contrast and, therefore, the details of the image are reduced, particularly in the very bright and dark regions (see Fig. 1). The contrast of an image taken from an airplane is reduced when the scenery is covered by cloud or mist (see Fig. 2); increasing the local contrast

Fig. 1. Image with little detail in dark regions.

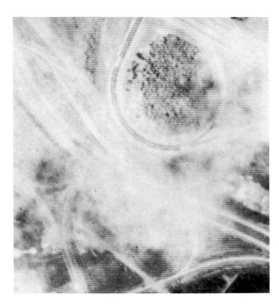

Fig. 2. Image taken from an airplane through varying amounts of cloud cover.

and reducing the overall dynamic range can significantly enhance the quality of such an image.

In another class of enhancement problems, a degraded image may be enhanced by reducing the degradation. When an image is quantized for the purpose of bit rate reduction, it may be degraded by random noise or signal-dependent false contours [31,32,80]. An example of an image with false contours is shown in Fig. 3. When a coherent light source is used, as in infrared radar imaging, the image may be degraded by a speckly effect [17,59]. An example of an image degraded by speckle noise is given in Fig. 4. An image recorded on film is degraded by film-grain noise. The speckle noise and film-grain noise, in some instances, can be approximately modeled by multiplicative noise [5,17,37]. When an image is coded and transmitted over a noisy channel or degraded by electrical sensor noise, as in a vidicon TV camera, degradation appears as salt-and-pepper noise (see Fig. 5). An image may also be degraded by blurring (convolutional noise) due to misfocus of lenses, to motion, or to atmospheric turbulence. In this case, high-frequency details of the image are often reduced, and the image appears blurred. An image degraded by one or more of these factors can be enhanced by reducing the degradation.

Another important class of image enhancement problems is the display of two-dimensional data that may or may not represent the intensities of an actual image. In two-dimensional spectral estimation, the spectral esti-

Fig. 3.   Image with false contours due to signal-dependent quantization noise in a 2-bit PCM image coding system.

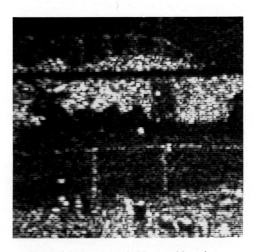

Fig. 4.   Image degraded by speckle noise.

Fig. 5. Image degraded by salt-and-pepper noise.

mates have traditionally been displayed as contour plots. Even though such two-dimensional data displays are not images in the conventional sense, their appearance may be improved and information more clearly conveyed when enhanced with gray scale and/or color. In other applications, such as infrared radar imaging, range information as well as image intensities may be available. By displaying the range information with color, one can highlight relative distances of objects in an image. Even typical images may be enhanced by certain types of distortion. When an object in an image is displayed with false color, the object may stand out more clearly to a human viewer.

Thus, an image can often be enhanced when one or more of the following objectives is accomplished: modification of contrast or dynamic range; edge enhancement; reduction of additive, multiplicative, and salt-and-pepper noise; reduction of blurring; and display of nonimage data. In the next section, we shall discuss these methods of enhancement.

II. Enhancement Techniques

A. Gray-Scale Modification

Gray-scale modification is a simple and effective way of modifying an image's dynamic range or contrast [28,29,34,35,46,80,100]. In this method the gray scale of an input image is changed to a different gray scale according to a specific transformation. The transformation $n_y = T[n_x]$ relates an input intensity n_x to an output intensity n_y and is often represented by a plot or a table. Consider a simple illustration of this method. Figure 6(a) shows an image of 4 × 4 pixels, with each pixel represented by three bits, so that there are eight reconstruction levels, that is, $n_x = 0, 1, 2, \ldots , 7$. The transformation that relates the input intensity to the output intensity is shown in Fig. 6(b) as a plot and in Fig. 6(c) as a table. For each pixel in the input image, the corresponding output intensity is obtained from the plot in Fig. 6(b) or from the table in Fig. 6(c). The result is shown in Fig. 6(d). By properly choosing the specific transformation, one can modify the contrast or dynamic range.

When the transformation is specified, gray-scale modification requires one table look-up per output point, and the table has 2^M entries, where M is the number of bits used for the gray scale. The specific transformation desired depends on the application. In some applications, physical considerations determine the transformation selected. For example, when a display system has nonlinear characteristics, the objective of the modification may be to precompensate for the nonlinearities. In such a case, the most

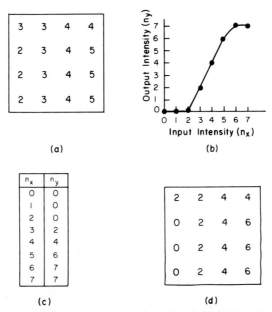

Fig. 6. (a) Image of 4 × 4 pixels, with each pixel represented by 3 bits. (b) Gray-scale transformation function represented by a plot. (c) Gray-scale transformation function represented by a table. (d) Result of modifying the image in (a) using the gray-scale transformation function of (b) or (c).

suitable transformation can be determined from the nonlinearity of the display system.

A good transformation in typical applications can be identified by computing the histogram of the input image and studying its characteristics. The histogram of an image, denoted by $p(n_x)$, represents the number of pixels that have a specific intensity as a function of the intensity variable n_x. For example, the 4×4 pixel image shown in Fig. 6(a) has the histogram shown in Fig. 7(a). The histogram obtained in this way displays some important image characteristics that help determine which particular gray-scale transformation is desirable. In Fig. 7(a), the image's intensities are clustered in a small region, and the available dynamic range is not very well utilized. In such a case, a transformation of the type shown in Fig. 6(b) would increase the overall dynamic range, and the resulting image would appear to have greater contrast. This is evidenced by Fig. 7(b), which is the histogram of the processed image shown in Fig. 6(d).

Because computing the histogram of an image and modifying its gray scale for a given gray-scale transformation requires little computation (on the order of one table look-up and one arithmetic operation per output pixel), the desirable gray-scale transformation can be determined by a human operator in real time. Based on the initial histogram computation, the operator chooses a gray-scale transformation to produce a processed image. By looking at the processed image and its histogram, the operator can choose another gray-scale transformation, obtaining a new processed image. These steps can be repeated until the output image satisfies the operator.

When there are too many images for individual attention by a human operator, the gray-scale transformation must be chosen automatically. Histogram modification is useful in these circumstances. In this method, the gray-scale transformation that produces a desired histogram is chosen for each individual image. The desired histogram of the output image, denoted by $P_d(n_y)$, that is useful for typical pictures has a maximum around

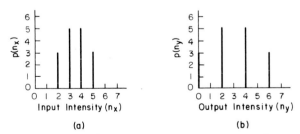

Fig. 7. Histogram of the 4×4 pixel image in Fig. 6(a). (b) Histogram of the 4×4 pixel image in Fig. 6(d).

the middle of the dynamic range and decays slowly as the intensity increases or decreases. To obtain the transformation $n_y = T[n_x]$ so that the processed image has approximately the same histogram as $p_d(n_y)$, the cumulative histograms $p'(n_x)$ and $p'_d(n_y)$ are first computed from $p(n_x)$ and $p_d(n_y)$ as follows:

$$p'(n_x) = \sum_{k=0}^{n_x} p(k) = p'(n_x - 1) + p(n_x)$$

$$p'_d(n_y) = \sum_{k=0}^{n_y} p_d(k) = p'_d(n_y - 1) + p_d(n_y)$$

(1)

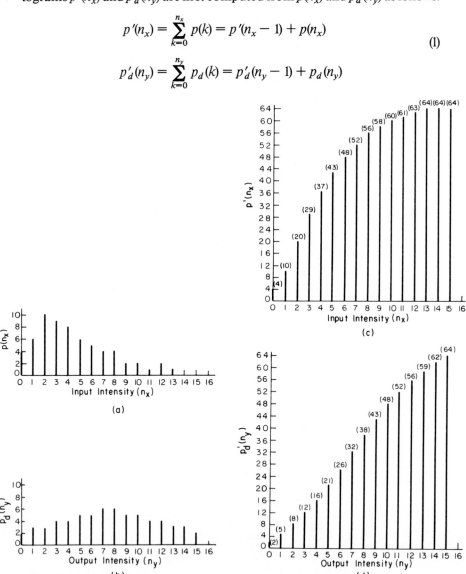

Fig. 8. (a) Histogram of an 8×8 pixel image represented by 4 bits/pixel. (b) Desired histogram for 8×8 pixel images represented by 4 bits/pixel. (c) Cumulative histogram derived from (a). (d) Cumulative histogram derived from (b).

An example of the cumulative histograms is shown in Fig. 8. Figures 8(a) and (b) show an example of $p(n_x)$ and $p_d(n_y)$, and Figs. 8(c) and (d) show $p'(n_x)$ and $p'_d(n_y)$ obtained using Eq. (1). From $p'(n_x)$ and $p'_d(n_y)$, the gray-scale transformation $n_y = T[n_x]$ can be obtained by choosing n_y for each n_x such that $p'_d(n_y)$ is closest to $p'(n_x)$. This can be accomplished by exploiting the fact that $T[n_x]$ obtained in this way is a monotonically nondecreasing function of n_x. The gray-scale transformation obtained for Fig. 8 is shown in Fig. 9(a), and the histogram of the resulting image obtained by using this transformation is shown in Fig. 9(b). The amount of computation involved in the procedure is quite small because $p'_d(n_y)$ is computed only once for all images unless the desired histogram $p_d(n_y)$ is different for different images, and the transformation $T[n_x]$ has only 2^M entries, where M is the number of bits used to represent a pixel.

Even though gray-scale modification is conceptually and computationally simple, it can often afford significant improvement in image quality or intelligibility. This is illustrated in the following two examples. Figure 10(a) shows an original undegraded image of size 256×256 pixels, with each

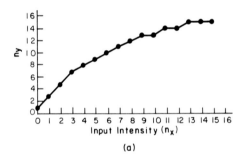

(a)

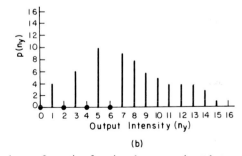

(b)

Fig. 9. (a) Gray-scale transformation function that approximately transforms the histogram in Fig. 8(a) to the desired histogram in Fig. 8(b). (b) Histogram of gray-scale–transformed image obtained by applying the transformation function in Fig. 9(a) to an image with the histogram shown in Fig. 8(a).

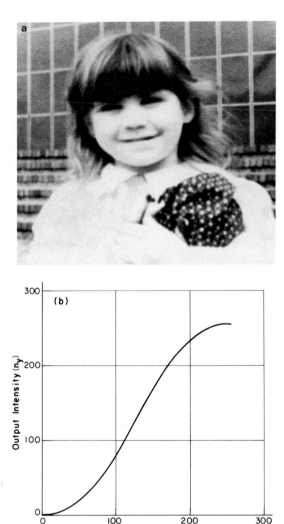

Fig. 10. (a) Original image. (b) Gray-scale transformation function used to modify the gray scale of the image in (a). (c) Processed image.

pixel represented by eight bits. Figure 10(b) shows the gray-scale transformation used to obtain the processed image in Fig. 10(c). The processed image appears sharper and thus is more visually pleasant than the unprocessed image. Figure 11(a) shows an image of size 256 × 256 pixels with eight bits per pixel, which has large shadow areas where the details are not visi-

Fig. 10. *(Continued)*

ble. By increasing the contrast in the shadow regions, the details can be made more visible. This can be accomplished by using the transformation shown in Fig. 11(b). The processed image based on this transformation is shown in Fig. 11(c).

Gray-scale modification has several limitations. One limitation is that it is global: The same operation is applied to all the pixels of the same intensity in an image. As in all global operations, this method cannot be used differently in different regions of an image, unless the regions are distinguishable by image intensity. Applying different amounts of contrast modification in several regions of the same intensity, for example, would be impossible. Adaptive methods to overcome this limitation are discussed in Section II.F. Another limitation is that a gray-scale transformation that is good for one image may not be good for others. Finding a good gray-scale transformation of any given image often requires human intervention.

B. Low-Pass Filtering

The energy of a typical image is concentrated primarily in low-frequency components. This is due to the high spatial correlation among neighboring pixels. The energy of image degradation, for instance, wideband random noise, is typically more spread out in the frequency domain. By reducing

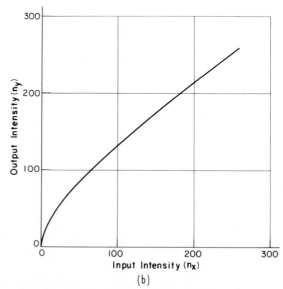

(b)

Fig. 11. (a) Image with large shadow regions. (b) Gray-scale transformation function used to modify the gray scale of the image in (a). (c) Processed image.

Fig. 11. *(Continued)*

the high-frequency components while preserving the low-frequency components, low-pass filtering reduces a large amount of noise at the expense of reducing a small amount of signal.

In addition to reducing additive random noise, low-pass filtering can also be used to reduce multiplicative noise. Consider an image $x(n_1,n_2)$ degraded by multiplicative noise $d(n_1,n_2)$, so that the degraded image $y(n_1,n_2)$ is given by

$$y(n_1,n_2) = x(n_1,n_2) \cdot d(n_1,n_2). \tag{2}$$

By taking the logarithmic operation in both sides of Eq. (2), we obtain

$$\log y(n_1,n_2) = \log x(n_1,n_2) + \log d(n_1,n_2). \tag{3}$$

If we denote $\log y(n_1,n_2)$ by $y'(n_1,n_2)$ and denote $\log x(n_1,n_2)$ and $\log d(n_1,n_2)$ in a similar way, Eq. (3) becomes

$$y'(n_1,n_2) = x'(n_1,n_2) + d'(n_1,n_2). \tag{4}$$

The multiplicative noise $d(n_1,n_2)$ now has been transformed to additive noise $d'(n_1,n_2)$, and low-pass filtering may be used to reduce $d'(n_1,n_2)$. The resulting image is exponentiated to compensate for the logarithmic operation. This transformation of a multiplicative component into an additive component by a logarithmic operation and subsequent linear filtering to reduce the additive component is referred to as homomorphic filtering [14,15,76,95].

Low-pass filtering reduces the high-frequency components of noise, but at the same time it reduces the high-frequency components of the signal. Because the edges or details of an image usually contribute to the high-frequency components, low-pass filtered images often look blurred. Thus, when using low-pass filtering, the tradeoff between noise reduction and image blurring should be considered.

Low-pass filtering is a linear shift invariant operation and can be represented by using the convolution operator $*$. The low-pass filtered image $z(n_1,n_2)$ can thus be represented by

$$z(n_1,n_2) = y(n_1,n_2) * h(n_1,n_2)$$
$$= \sum_{(k_1,k_2) \in A} \sum h(k_1,k_2)y(n_1 - k_1,n_2 - k_2)$$
$$= \sum_{(n_1-k_1,n_2-k_2) \in A} \sum h(n_1 - k_1,n_2 - k_2)y(k_1,k_2), \qquad (5)$$

where $h(n_1,n_2)$ represents the impulse response of the low-pass filter, and the region A represents the support of $h(n_1,n_2)$. If the size of A is finite, then the filter is called a finite impulse response (FIR) filter and can be implemented directly by use of Eq. (5). If the size of A is infinite, the filter is called an infinite impulse response (IIR) filter, and its implementation using Eq. (5) is computationally inefficient. For an IIR filter, the impulse response is constrained to have a rational z-transform, so that it can be represented by a difference equation. Even though an IIR filter typically requires less computation than an FIR filter to achieve the same magnitude response, the FIR filter predominates in image enhancement for several reasons. It is quite simple to generate a linear phase FIR filter, and an FIR filter is stable regardless of coefficient quantization. Generating a linear phase IIR filter is more involved and checking the stability of a two-dimensional IIR filter is quite difficult [18,39,63,64].

There are various ways of designing (determining $h(n_1,n_2)$ of) an FIR filter. If the desired frequency response $H_d(\omega_1,\omega_2)$ is given, then one simple but effective method of determining $h(n_1,n_2)$ is the window method. In this method, the desired frequency response $H_d(\omega_1,\omega_2)$ is inverse Fourier transformed and windowed as

$$h(n_1,n_2) = \mathcal{F}^{-1}[H_d(\omega_1,\omega_2)] \cdot w(n_1,n_2), \qquad (6)$$

where $\mathcal{F}^{-1}$ represents the inverse Fourier transform [18,75] and $w(n_1,n_2)$ is some smooth window such as a separable Kaiser window [18,39,75]. Details on two-dimensional windows for designing two-dimensional filters can be found in references [18] and [39].

The inverse Fourier transform operation $\mathcal{F}^{-1}$ can be computed analytically if $H_d(\omega_1,\omega_2)$ is a simple function by

$$\mathcal{F}^{-1}[H_d(\omega_1,\omega_2)] = \frac{1}{(2\pi)^2} \int_{\omega_1=-\pi}^{\pi} \int_{\omega_2=-\pi}^{\pi} H_d(\omega_1,\omega_2)$$

$$\times \exp(j\omega_1 n_1) \exp(j\omega_2 n_2) \, d\omega_1 \, d\omega_2. \tag{7}$$

Alternatively, the inverse Fourier transform operation can be performed by sampling $H_d(\omega_1,\omega_2)$ on a very dense grid and then performing the inverse discrete Fourier transform (IDFT) operation as follows:

$$\mathcal{F}^{-1}[H_d(\omega_1,\omega_2)] \simeq \frac{1}{L^2} \sum_{k_1=0}^{L-1} \sum_{k_2=0}^{L-1} H_d(\omega_1,\omega_2)$$

$$\times \exp\left(j\frac{2\pi}{L}k_1 n_1\right) \exp\left(j\frac{2\pi}{L}k_2 n_2\right), \tag{8}$$

where $\omega_1 = (2\pi/L)k_1$, $\omega_2 = (2\pi/L)k_2$, and $L \times L$ is much larger (typically 512×512) than the support size. The two-dimensional IDFT in Eq. (8) can be computed by using a two-dimensional fast Fourier transform (FFT) algorithm. In a typical two-dimensional FFT algorithm, a one-dimensional FFT is used with respect to the variable k_1 for each k_2 and then with respect to the variable k_2 for each k_1. Details on two-dimensional FFT algorithms, including their efficient implementation in a minicomputer environment with small on-line memory, can be found in [18], [19], and [39].

The desired frequency response $H_d(\omega_1,\omega_2)$ depends on the specific application. For example, in an application to reduce additive background noise, the cutoff frequency of the desired low-pass filter depends on the frequency contents of the image and the background noise. If the image and the background noise can be modeled by samples of random processes with known power spectra, then the Wiener filter [4] will minimize the mean square error between the original noise-free image and the processed image. The frequency response of the Wiener filter is a function of the image and noise power spectra.

When the optimal frequency response is not known, a simple yet reasonable low-pass filter $h(n_1,n_2)$ is a smooth window function. In this case, several different shapes and sizes may be used to obtain several processed images, from which the best can be chosen. Some examples of $h(n_1,n_2)$ are shown in Fig. 12.

In general, a low-pass filter with larger support region has better characteristics in terms of smaller transition bandwidth and smaller deviation from the desired frequency response, but it requires more computation. For a square-shaped region with support size of $M \times M$, from Eq. (5), the

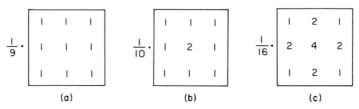

Fig. 12.   Examples of low-pass filters useful for image enhancement.

number of computations required per output pixel is on the order of M^2 multiplications and M^2 additions.

When the filter support size $M \times M$ is small, the most efficient way to low-pass filter an image is by using Eq. (5).   When the support size $M \times M$ is large, it may be more efficient to filter the image in the frequency domain by

$$z(n_1,n_2) = y(n_1,n_2) * h(n_1,n_2) = \text{IDFT}[Y(k_1,k_2) \cdot H(k_1,k_2)]. \qquad (9)$$

The sequences $Y(k_1,k_2)$ and $H(k_1,k_2)$ represent the discrete Fourier transforms (DFTs) of $y(n_1,n_2)$ and $h(n_1,n_2)$.   In Eq. (9), the size of the DFT and IDFT is at least $(N + M - 1) \times (N + M - 1)$ when the image size is $N \times N$ and the filter size is $M \times M$.   If the DFT size is less than $(N + M - 1) \times (N + M - 1)$, then the result of $\text{IDFT}[Y(k_1,k_2) \cdot H(k_1,k_2)]$ is not identical to the convolution near the boundaries of the processed image due to aliasing effects [5,18].   In most cases, however, adequate results can be obtained when the size of the DFT and IDFT used is $N \times N$.   When an image is low-pass filtered by use of Eq. (9) with the DFT and IDFT computed by an FFT algorithm, the number of computations required per output point is on the order of $2 \log_2 N$.   For a sufficiently large M, M^2 is larger than $2 \log_2 N$.   An additional advantage of using Eq. (9) arises when the desired frequency response $H_d(\omega_1,\omega_2)$ is known.   In this case, a reasonable $H(k_1,k_2)$ in Eq. (9) may be obtained by sampling the desired frequency response $H_d(\omega_1,\omega_2)$ as follows:

$$H(k_1,k_2) = H_d(\omega_1,\omega_2)|_{\omega_1 = (2\pi/L)k_1, \ \omega_2 = (2\pi/L)k_2} \qquad (10)$$

where $L \times L$ is the DFT and IDFT size.   Further details of the design and implementation of a two-dimensional FIR filter can be found in [18] and [39].

To illustrate the performance of low-pass filtering for image enhancement, two examples are considered.   Figure 13(a) shows an original noise-free image of 256×256 pixels with eight bits per pixel, and Fig. 13(b) shows the image degraded by wideband Gaussian random noise at signal-to-noise ratio (SNR)† of 15 dB.   Figure 13(c) shows the result of low-pass filtering the

† Signal-to-noise ratio is defined as SNR $= 10 \log_{10}$ (image variance/noise variance).

Fig. 13. (a) Original image. (b) Image in (a) degraded by wideband Gaussian random noise at SNR of 15 dB. (c) Result of processing the image in (b) with a low-pass filter.

Fig. 13. *(Continued)*

degraded image. The low-pass filter used is shown in Fig. 12(b). The second example is of an image degraded by multiplicative noise. Figure 14(a) shows the original image of 256 × 256 pixel with eight bits per pixel, and Fig. 14(b) shows the image degraded by multiplicative noise. The multiplicative noise was generated by using the exponential density function to approximate speckle noise [17,59]. The result of low-pass filtering in the image density (log intensity) domain is shown in Fig. 14(c). The low-pass filter used is shown in Fig. 12(b).

In both Figs. 13 and 14, low-pass filtering clearly reduces the additive or multiplicative random noise, but at the same time it blurs the image. Blurring is a primary limitation of low-pass filtering. The performance of a low-pass filter can be significantly improved by adapting it to the local characteristics of the image. The adaptive low-pass filter, which typically requires considerably more computation than the nonadaptive low-pass filter, will be discussed in Section II.F.

C. High-Pass Filtering

High-pass filtering emphasizes the high-frequency components of a signal while reducing the low-frequency components. Because the high-fre-

Fig. 14. (a) Original image. (b) Image in (a) degraded by multiplicative noise generated from an exponential density function. (c) Result of processing the image in (b) with a low-pass filter in the density domain.

Fig. 14. *(Continued)*

quency components of a signal generally correspond to edges or fine details of an image, high-pass filtering often increases the local contrast and thus sharpens the image. Images with sharper edges are often more pleasant visually.

High-pass filtering is also useful in reducing image blur. When an image $x(n_1,n_2)$ is degraded by blurring due to misfocus of a lens, motion, or atmospheric turbulence, the blurred image $y(n_1,n_2)$ can be represented by

$$y(n_1,n_2) = x(n_1,n_2) * b(n_1,n_2)$$

or   (11)

$$Y(\omega_1,\omega_2) = X(\omega_1,\omega_2)B(\omega_1,\omega_2),$$

where $b(n_1,n_2)$ is the point spread function of the blur, and $B(\omega_1,\omega_2)$ is its Fourier transform. Because $B(\omega_1,\omega_2)$ is typically of low-pass character, one approach to reducing the blurring is to pass $y(n_1,n_2)$ through a high-pass filter.

High-pass filtering can also be used in preprocessing an image prior to its degradation by noise. In applications such as image coding, an original undegraded image is available for processing prior to its degradation by noise such as quantization noise. In such applications, the undegraded image can be high-pass filtered prior to its degradation and then low-pass filtered after degradation. The usual effect of this is an improvement in the quality or intelligibility of the resulting image. For example, when the

0	-1	0
-1	5	-1
0	-1	0

(a)

1	-2	1
-2	5	-2
1	-2	1

(b)

$\frac{1}{7}\cdot$

-1	-2	-1
-2	19	-2
-1	-2	-1

(c)

Fig. 15. Examples of high-pass filters useful for image enhancement.

degradation is due to wideband random noise, the effective SNR of the degraded image is much lower in the high-frequency components than in the low-frequency components, due to the low-pass character of a typical image. High-pass filtering prior to the degradation significantly improves the SNR in the high-frequency components at the expense of a small SNR decrease in the low-frequency components. This process typically results in quality or intelligibility improvement of the resulting image.

Because high-pass filtering is a linear, space-invariant operation, our previous discussions on filter design, implementation, and required computation for low-pass filters apply equally well to high-pass filters. Some typical examples of the impulse response of a high-pass filter used for image enhancement are shown in Fig. 15.

To illustrate the performance of high-pass filtering for image enhancement, three examples are given. Figure 16(a) shows an original noise-free image of 256 × 256 pixels with 8 bits/pixel, and Fig. 16(b) shows the result of

Fig. 16. (a) Original image. (b) Image in (a) processed by a high-pass filter.

Fig. 16. *(Continued)*

Fig. 17. (a) Original image. (b) Image in (a) degraded by blurring. (c) Image in (b) processed by a high-pass filter.

Fig. 17. *(Continued)*

high-pass filtering. Even though the original image has not been degraded, some high-pass filtering increases the local contrast of an image and thus gives a sharper visual appearance. The second example is the case in which an image is degraded by blurring. Figure 17(a) shows an original image of 256 × 256 pixels with 8 bits/pixel, and Fig. 17(b) shows the image degraded by blurring. The point spread function used for blurring the image is shown in Fig. 12(b). Figure 17(c) shows the result of applying a high-pass filter to the image in Fig. 17(b). The third example is high-pass filtering an image prior to its degradation by quantization noise in pulse code modulation (PCM) image coding. Figure 18(a) shows an original image of 256 × 256 pixels with 8 bits/pixel. Figure 18(b) shows the image coded by a PCM system with Roberts' pseudonoise technique [85] at 2 bits/pixel. Figure 18(c) shows the result of high-pass filtering before coding and low-pass filtering after coding. The high-pass filter shown in Fig. 15(c) has been used in all three of the above cases.

Because a high-pass filter emphasizes the high-frequency components, and because background noise typically has significant high-frequency components, high-pass filtering tends to increase background noise power when it is applied to a degraded image. This is a major limitation of high-pass filtering for image enhancement and should be noted when considering a high-pass filter to enhance an image.

Fig. 18. (a) Original image. (b) Image in (a) coded by a PCM system with Roberts' pseudonoise technique at 2 bits/pixel. (c) Result of high-pass filtering the image in (a) before coding and low-pass filtering after coding with a PCM system with Roberts' pseudonoise technique at 2 bits/pixel.

Fig. 18. *(Continued)*

D. Median Filtering

Median filtering [1,9,39,41,80] is a nonlinear process useful in reducing impulsive or salt-and-pepper noise. It is also useful in preserving edges in an image while reducing random noise. In a median filter, a window slides along the image, and the median intensity value of the pixels within the window replaces the intensity of the pixel being processed. For example, when the pixel values within a window are 5, 6, 35, 10, and 15, and the pixel being processed has a value of 35, its value is changed to 10, which is the median of the five values.

Like low-pass filtering, median filtering smoothes the image and, there-fore, is useful in reducing noise. Unlike low-pass filtering, median filtering can preserve discontinuities in a step function and can smooth a few pixels whose values differ from their surroundings without affecting the other pixels. Figure 19(a) shows a one-dimensional step sequence degraded by a small amount of random noise. Figure 19(b) shows the result of a low-pass filter whose impulse response is a five-point rectangular window. Fig-ure 19(c) shows the result of a five-point median filter. It is clear from the figure that the step discontinuity is better preserved by the median filter. Figure 20(a) shows a one-dimensional sequence with two pixel values that are significantly different from the surrounding pixels. Figures 20(b) and 20(c) show the result of a low-pass filter and a median filter, respectively. The filters used in Fig. 20 are the same as those used in Fig. 19. If the

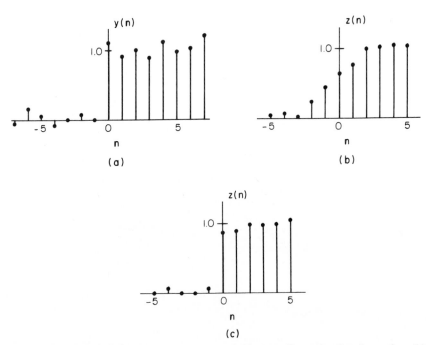

Fig. 19. (a) One-dimensional step sequence degraded by a small amount of random noise. (b) Result of low-pass filtering the sequence in (a) with a 5-point rectangular impulse response. (c) Result of applying a 5-point median filter to the sequence in (a).

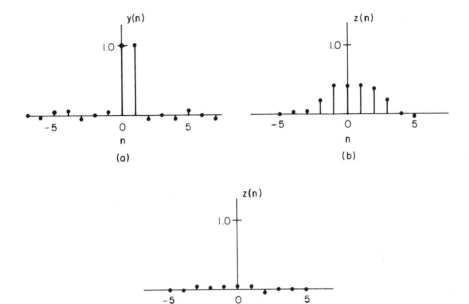

Fig. 20.   (a) One-dimensional sequence with two consecutive samples significantly different from surrounding samples.   (b) Result of low-pass filtering the sequence in (a) with a 5-point rectangular impulse response.   (c) Result of applying a 5-point median filter to the sequence in (a).

impulsive values of the two pixels are due to noise, the result of using a median filter will be to reduce the noise.   If the two pixel values are truly part of the signal, however, the use of the median filter will distort the signal.

An important design parameter in using a median filter is the size of the window used.   Figure 21 illustrates the result of median filtering of the signal in Fig. 20(a) as a function of window size.   If the window size is smaller than 5, the two pixels with impulsive values are not significantly affected.   For a larger window size, they are.   Thus, the choice of the window size depends on the context.   Because it is difficult to choose the optimum window size in advance, it may be useful to try several median filters of different window sizes and choose the best of the resulting images.

Because median filtering is a nonlinear operation and is defined in the spatial domain, it can be implemented much more easily in the spatial domain.   For a window size of $M \times M$, the largest pixel value can be found by $M^2 - 1$ arithmetic comparisons, and the second largest pixel value can be found by $M^2 - 2$ arithmetic comparisons.   For odd M, therefore, the

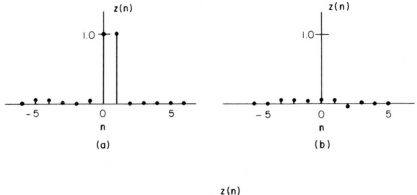

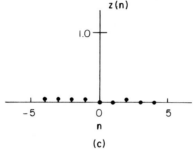

Fig. 21. Result of applying a median filter to the sequence in Fig. 20(a) as a function of window size. (a) Window size = 3. (b) Window size = 5. (c) Window size = 7.

median value of M^2 values can be obtained by $\sum_{k=(M^2-1)/2}^{M^2-1} k$ arithmetic comparisons. For a window size of 3×3, for example, each pixel can be processed by median filtering with 30 arithmetic comparisons. When M is large, it is possible to compute the median value with the number of arithmetic comparisons proportional to $M^2 \log_2 M$ [1]. In addition, if running medians are computed, the median value can be computed with the number of arithmetic operations proportional to M [41].

To illustrate the performance of median filtering, two examples are given. In the first, the original image of 256×256 pixels with 8 bits/pixel, shown in Fig. 22(a), is degraded by wideband Gaussian random noise at SNR of 15 dB. The degraded image is shown in Fig. 22(b). Figure 22(c) shows the image processed by median filtering with window size of 3×3. In the second example, the original image from Fig. 22(a) is degraded by salt-and-pepper noise. The degraded image is shown in Fig. 23(a), and the image processed by median filtering with window size of 3×3 is shown in Fig. 23(b).

Fig. 22.   (a) Original image.   (b) Image in (a) degraded by wideband Gaussian random noise.
(c) Result of applying a 3 × 3 median filter to the image in (b).

Fig. 22. *(Continued)*

Fig. 23. (a) Image in Fig. 22(a) degraded by salt-and-pepper noise. (b) Result of applying a 3×3 median filter to the image in Fig. 23(a).

Fig. 23. *(Continued)*

E. Out-Range Pixel Smoothing

Out-range pixel smoothing is useful in reducing salt-and-pepper noise. In this method, a window slides along the image, and the average of the pixel values, excluding the pixel being processed, is obtained. If the difference between the average and the value of the pixel being processed is above some threshold, then the current pixel value is replaced by the average. Otherwise, the value is unaffected.

Like median filtering, out-range pixel smoothing is a nonlinear operation and is more easily implemented in the spatial domain. For a window size of $M \times M$, each pixel can be processed by $M^2 - 2$ additions, one division, and one arithmetic comparison. The threshold value and window size can be varied. Because the best parameter values are difficult to determine in advance, it may be useful to process an image using several different threshold values and window sizes and select the best result.

To illustrate the performance of out-range pixel smoothing, an original image of 256×256 pixels with 8 bits/pixel is shown in Fig. 24(a), and the original image degraded by salt-and-pepper noise is shown in Fig. 24(b). Figure 24(c) shows the image processed by out-range pixel smoothing with threshold value of 64 and window size of 3×3. Figure 24(d) shows the processed image with threshold value of 48 and window size of 5×5.

Fig. 24. (a) Original image. (b) Image in (a) degraded by salt-and-pepper noise. (c) Result of applying out-range pixel smoothing to the image in (b) with threshold value of 64 and window size of 3 × 3. (d) Same as (c) with threshold value of 48 and window size of 5 × 5.

Fig. 24. *(Continued)*

F. Adaptive Filtering

In the preceding methods, a single operation is applied in all parts of the image.   For example, in low-pass filtering, the impulse response of the filter remains unchanged throughout the image.   In many image enhancement problems, however, results can be significantly improved by adapting the processing to local characteristics of the image and the degradation [24,57,58,67,78,101]. Details of image characteristics typically differ considerably between one image region and another. For example, walls and skies have approximately uniform background intensities, whereas buildings and trees have large, detailed variations in intensity. Degradation may also vary from one region to another. For example, an image of scenery taken from an airplane may be degraded by varying amounts of cloud cover. It is quite reasonable, then, to adapt the processing to the changing characteristics of the image and the degradation.

Two approaches to adaptive image enhancement exist. In one, the image is divided into many subimages, and each subimage is processed separately and then combined with the others [57]. The size of the subimage is typically between 8 $\times$ 8 and 32 $\times$ 32 pixels. For each subimage, a space-invariant operation appropriate to the subimage is chosen, based on the local image and degradation characteristics. Because the processing that is applied to a subimage is a space-invariant operation, our discussions in previous sections apply to this approach. For example, a low-pass filter can be implemented in both the spatial domain and the frequency domain. Because a different type of processing is applied to each subimage in this approach, the boundaries of the adjacent subimages may have discontinuities in the intensities, which may show blocking effect in the processed image. This blocking effect can be reduced in some cases by overlapping the subimages. Specifically, to obtain a subimage, a window $w(n_1,n_2)$ is applied to $y(n_1,n_2)$, the input image to be processed. The window $w(n_1,n_2)$ is chosen to satisfy two conditions. The first condition can be expressed as

$$\sum_i \sum_j w_{ij}(n_1,n_2) = 1 \qquad \text{for all } n_1,n_2 \text{ of interest.} \qquad (12)$$

This condition guarantees that simple addition of the unprocessed subimages results in the original image. The second condition requires $w_{ij}(n_1,n_2)$ to be a smooth function that falls to zero near the window boundaries. This tends to reduce possible discontinuities or degradations that may appear at the subimage boundaries in the processed image.

One way to find a smooth two-dimensional window function that satisfies Eq. (12) is to form a separable two-dimensional window from two one-dimensional windows that satisfy similar conditions:

$$w_{ij}(n_1,n_2) = w_i(n_1) \cdot w_j(n_2). \qquad (13)$$

Two such window functions are two-dimensional separable triangular or Hamming windows overlapped with the neighboring window by half the window duration in each dimension. The case of a two-dimensional sepa-

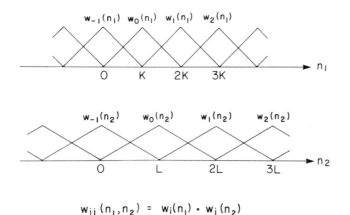

$$w_{ij}\,(n_1,n_2) \;=\; w_i(n_1) \cdot w_j\,(n_2)$$

Fig. 25.   Example of a two-dimensional separable triangular window.

rable triangular window is shown in Fig. 25.   When an image is windowed to form a subimage, processing on the subimage must take window shape and size into account.

Another approach to adaptive image enhancement is to adapt the processing method to each pixel.   This approach does not suffer from boundary discontinuities but can be considerably more expensive computationally than the approach previously described.   The amount of additional computation depends on the complexity of decisions as to which type of processing to perform for each pixel.

A general adaptive image enhancement system is represented in Fig. 26. The processing to be done on each subimage or pixel is determined by the local image and degradation characteristics.   Information on these characteristics can be obtained from two sources.   One is some prior knowledge of the problem at hand.   We may know, for example, what class of images to expect in a given application, or we may be able to deduce the degradation

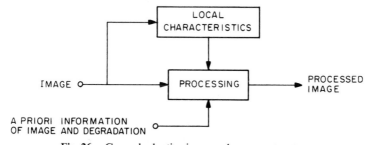

Fig. 26.   General adaptive image enhancement system.

characteristics from our knowledge of the degrading source. Another source of information is the image to be processed. By measuring factors such as local variance, one may be able to determine the presence of significant high-frequency details.

Determining which specific type of processing to use based on the nature of the image and degradation depends on a number of factors, for example, the type of knowledge we have about the image and degradation and how this knowledge can be exploited in estimating the parameters of a processing method, such as low-pass filter cutoff frequency. Without a specific application context, only general statements can be made. In general, the more the available knowledge is used in choosing the processing type, the more an image enhancement system's performance can be improved. If the knowledge is inaccurate, however, system performance may be degraded.

The following two examples illustrate how an adaptive image enhancement system can be developed. In the first example, a low-pass filter is adapted to local image characteristics. When an image is degraded by wideband random noise, the local variance of the degraded image consists of two components, one due to signal variance and the other due to noise variance. Large local variance indicates high-frequency image components such as edges. Edge preservation is important, and noise is less visible in the region that contains high-frequency components. A reasonable approach to image enhancement, therefore, is to determine the local image variance and apply less smoothing when the local image variance is large. Such treatment will preserve edges while reducing noise and is schematically illustrated in Fig. 27. Figures 28(a) through 28(d) illustrate the performance of this type of system. Figure 28(a) shows an original image of 256 × 256 pixels with 8 bits/pixel, and Fig. 28(b) shows the image degraded by wideband Gaussian random noise at SNR of 15 dB. Figure 28(c) shows the image that has been processed by the system shown in Fig. 27. In generating this image, the local variance was obtained at each pixel with a window size of 9 × 9 pixels. Depending on the local variance, the size of the low-pass filter impulse response, whose shape is Gaussian, was changed at each pixel with a shorter impulse response corresponding to the larger local image variance. The number of computations required per pixel is on

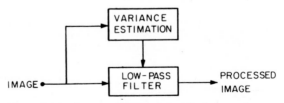

Fig. 27. Adaptive low-pass filtering system for image enhancement.

Fig. 28. (a) Original image. (b) Image in (a) degraded by wideband Gaussian random noise. (c) Result of processing the image in (a) with the system in Fig. 27. (d) Result of processing the image in (b) with a fixed low-pass filter.

Fig. 28. *(Continued)*

the order of 80 multiplications and 160 additions to compute the variance and on the order of 25 multiplications and additions to perform the filtering operation. Figure 28(d) shows the image that was low-pass filtered with a fixed impulse response throughout the image.

In the second example, we consider the problem of enhancing an image that was taken from an airplane and was degraded by varying amounts of

cloud cover [78].   According to one simplified model of image degradation
due to cloud cover, regions of an image covered by cloud have increased local
luminance mean and decreased local contrast, with the amount of change
determined by the amount of cloud cover.   One approach to the enhancement of images degraded by cloud cover, then, is to increase the local
contrast and decrease the local luminance mean whenever the local luminance is high.   This can be accomplished by using the system shown in Fig.
29, which modifies the local contrast and local luminance mean as a function of the local luminance mean.   In this figure, $x(n_1,n_2)$ denotes the unprocessed image, and $x_L(n_1,n_2)$ which denotes the local luminance mean of
$x(n_1,n_2)$, is obtained by low-pass filtering $x(n_1,n_2)$.   The sequence $x_H(n_1,n_2)$,
which denotes the local contrast, is obtained by subtracting $x_L(n_1,n_2)$ from
$x(n_1,n_2)$.   The local contrast is modified by multiplying $x_H(n_1,n_2)$ with $k(x_L)$,
a scalar that is a function of $x_L(n_1,n_2)$.   The modified contrast is denoted by
$x'_H(n_1,n_2)$.   If $k(x_L)$ is greater than one, the local contrast is increased,
whereas $k(x_L)$ less than one represents local contrast decrease.   The local
luminance mean is modified by a point nonlinearity, and the modified local
luminance mean is denoted by $x'_L(n_1,n_2)$.   The modified local contrast and
local luminance mean are then combined to obtain the processed image,
$g(n_1,n_2)$.   To increase the local contrast and decrease the local luminance
mean when the local luminance is high, a larger $k(x_L)$ is chosen for a larger,
x_L, and the nonlinearity is chosen taking into account the local luminance
change and the contrast increase.   Figure 30 shows the result of using the
system in Fig. 29 to enhance an image degraded by cloud cover.   Figure 30(a)
shows an image of 256 × 256 pixels with 8 bits/pixel degraded by differing amounts of cloud cover in different regions of the image.   Figure 30(b)
shows the processed image.   The function $k(x_L)$ and the nonlinearity used
are shown in Figs. 30(c) and 30(d).   The low-pass filtering operation was
performed using a filter whose impulse response is an 8 × 8 rectangular
window.

Even though an adaptive image enhancement system often requires considerably more computation than a nonadaptive system, its performance is
often considerably better, and it is worthwhile to explore adaptive systems in
any image enhancement problem that requires high performance.

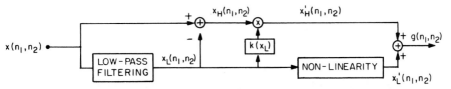

Fig. 29.   System for the modification of local contrast and local luminance mean as a function
of local luminance mean.

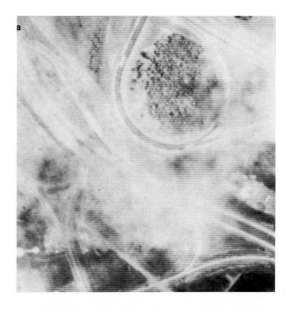

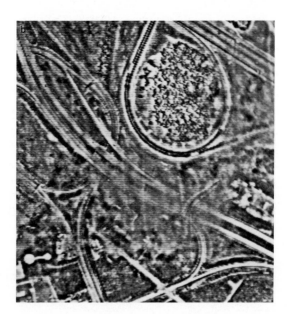

Fig. 30.   (a) Image taken from an airplane through varying amounts of cloud cover.   (b) Result of processing the image in (a) with the system in Fig. 29.   (c) Function $k(x_L)$ used in obtaining the image in Fig. 30(b).   (d) Nonlinearity used in obtaining the image in Fig. 30(b).

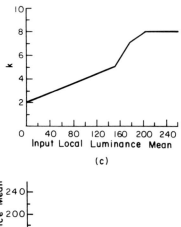

(c)

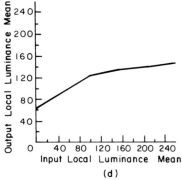

(d)

Fig. 30. *(Continued)*

G. *Image Processing in the Density Domain*

The enhancement techniques discussed in previous sections can also be applied in the image density (log intensity) domain. There is some justification for the view that processing an image in the density domain may be better than processing an image in the intensity domain. At the peripheral level of the human visual system, the image intensity appears to be subject to some form of nonlinearity, such as logarithmic operation [80,95]. This is evidenced in part by the approximate validity of Weber's law, which states that the just noticeable difference (jnd) in image intensity is proportional to the intensity or, alternatively, that the jnd in image density is independent of the density. Thus, the image density domain is, in a sense, closer to the center of the human visual system than is the intensity domain.

In one method, known as *homomorphic image enhancement* [76] or *unsharp masking* [91,92], processing an image in the density domain has led to successful results. Specifically, by applying some form of high-pass filtering in the image density domain, the contrast of an image has been in-

Fig. 31. (a) Original image. (b) Result of processing the image in Fig. 30(a) by applying a high-pass filter in the density domain, after Oppenheim *et al.* [76].

creased, and the dynamic range of an image has been reduced. One example is shown in Fig. 31. Figure 31(a) shows an original image of 256×256 pixels with 8 bits/pixel, and Fig. 31(b) shows an image processed by applying a high-pass filter to the image in the density domain and then exponentiating the resulting image.

Even though processing an image in its density domain has some appeal from point of view of the human visual system, has led to some successful results, and is a natural thing to do in reducing multiplicative noise (as discussed in Section II.B), it is not always better than processing an image in the intensity domain. Processing in the density domain requires more computations due to the logarithmic and exponential operations and may not perform as well as intensity domain processing in a given application. In addition, image degradations such as additive noise or convolutional noise do not remain as additive or convolutional noise in the density domain, and this has to be taken into account in processing an image in the density domain.

H. False Color, Pseudocolor, Display of Nonimage Data

It is well known that the human visual system is quite sensitive to color [65]. The number of distinguishable intensities, for example, is much smaller than the number of distinguishable colors and intensities. In addition, color images are generally much more pleasant to look at than black and white images. The aesthetic aspect of color can be used for image enhancement [22,25,54,93,96]. In some applications, such as television commercials, false color can be used to emphasize a particular object in an image. For example, a red banana in a surrounding of other fruits of natural color will receive more of a viewer's attention. In other applications, data that do not represent an image in the conventional sense can be represented by a black and white or color image. For example, a speech spectrogram showing speech energy as a function of time and frequency can be represented by a color image, with silence, voiced segments, and unvoiced segments distinguished by different colors and energy represented by color brightness.

Nature generates color by filtering out, or subtracting, some wavelengths and reflecting others. The wavelengths that are reflected determine the color that is perceived. This process of wavelength subtraction is accomplished by molecules called pigments, which absorb particular parts of the spectrum. Most colors in nature can be reproduced by mixing a small selection of pigments. One such set is yellow, cyan, and magenta, known as the three primary colors. When yellow and magenta are mixed appropriately, for example, red is generated. Because wavelengths are subtracted

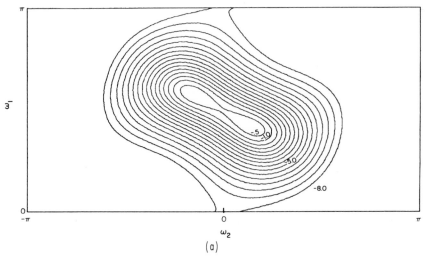

Fig. 32. (a) Two-dimensional maximum-likelihood spectral estimate represented by a contour plot. (b) Spectral estimate in (a) represented by a color image (see opposite page).

Fig. 33. (a) Intensity image from an infrared radar imaging system with range information discarded. (b) Image in (a) with range information displayed with color (see opposite page).

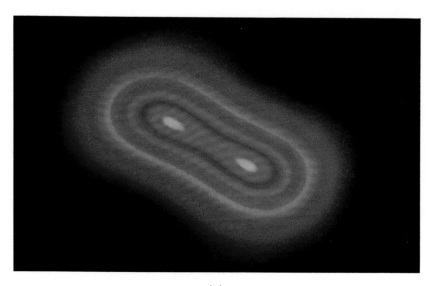

(b)

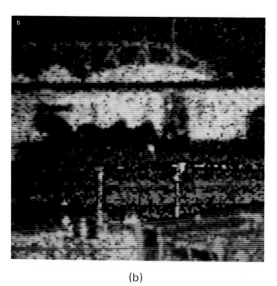

(b)

each time different color pigments are added, this is called the subtractive color system. Different colors can also be generated by adding light sources with different wavelengths. In this case, the wavelengths are added, so it is called the additive color system. For example, the lighted screen of a color television tube is covered with small, glowing phosphor dots arranged in groups of three. Each of these groups contains one red, one green, and one blue dot. These three colors are used because they can produce the widest range of colors; they are the primary colors of the additive color system.

The use of color in image enhancement is limited only by artistic imaginations, and there are no simple guidelines or rules to follow. In this section, therefore, we shall concentrate on two specific examples that illustrate the type of image enhancement that can be achieved by using color. In the first example, we consider the display of a two-dimensional power spectrum estimate on a display system with a television tube. The two-dimensional spectral estimate, represented by $f(\omega_1,\omega_2)$ in dB, is typically displayed using a contour plot. The variables ω_1 and ω_2 represent the two-dimensional frequencies. An example of a two-dimensional maximum-likelihood spectral estimate for the data of two sinusoids in white noise is shown in Fig. 32(a). The maximum corresponds to 0 dB and the contours are in increments of 0.5 dB downward from the maximum point. In applications such as detection of low-flying aircraft by an array of microphone sensors, we wish to determine the number of sinusoids present and their frequencies. An alternative way of representing the spectral estimate is to represent the spectral estimate using pseudocolor. An example is shown in Fig. 32(b). Comparison of the two figures shows that the presence of two peaks and their locations in the spectral estimate stand out more clearly in Fig. 32(b).

The second example is the display of range information using color [96]. In applications such as infrared radar imaging systems, range information and image intensity are available. Figure 33(a) shows an intensity image of several buildings located 2 to 4 km away from the radar; the range information has been discarded. Figure 33(b) shows an image that uses color to display range information. The range value determines the display hue, and the corresponding intensity value determines the brightness level of the chosen hue. The most striking aspect of this technique is demonstrated by the observation that a horizontal line seen at close range (actually a telephone wire) is visible in Fig. 33(b) but is completely obscured in Fig. 33(a).

III. Further Comments

In Section II we presented a number of different image enhancement techniques and discussed their capabilities and limitations. It should be noted that this discussion does not provide off-the-shelf algorithms that can be used in specific application problems. Instead, our discussion should be used as a guide to developing the image enhancement algorithm most suitable to a given application. The factors that have to be considered in developing an image enhancement algorithm differ considerably in different applications, and thus one cannot usually expect to find an off-the-shelf image enhancement algorithm ideally suited to a particular application.

An important step in developing an image enhancement system in a practical application environment is to identify clearly the overall objective. In applications in which images are enhanced for human viewers, the properties of human visual perception have to be taken into account. In applications in which an image is processed to improve machine performance, the characteristics of the machine are important considerations. Thus, approaches to developing an image enhancement system vary considerably, depending on the overall objective in a given application.

Another important step is to identify the constraints imposed in a given application environment. In some applications, system performance is the overriding factor, and any reasonable cost may be justified, whereas in other applications, cost may be an important factor. In some applications, real-time image processing is necessary, whereas in others, the real-time aspect may not be as critical. Clearly, the approach to developing an image enhancement system is influenced by the constraints imposed by the application context.

A third important step is to gather information about the images to be processed. This information can be exploited in developing an image enhancement system. For example, if the class of images to be processed consists mostly of buildings, which have a large number of horizontal and vertical lines, this information can be used in designing an adaptive low-pass filter to reduce background noise.

Given the overall objective, the constraints imposed, and information about the class of images to be processed, a reasonable approach to developing an image enhancement algorithm is to determine if one or more existing methods, such as the ones discussed in the previous section, are applicable to the given problem. In general, significant developmental work is needed to adapt existing methods to a given problem. If existing methods do not apply, or if the performance achieved by existing methods does not meet the requirements of the problem, new approaches have to be developed. Digital image processing has a relatively short history, and there is considerable room for new approaches and methods of image enhancement.

Developing an image enhancement algorithm suitable for a given application can be a tedious process, sometimes requiring many iterations and considerable human interaction. This chapter is intended to be an aid to this difficult process.

IV. Bibliographic Notes

For general information on image enhancement and quality, see Budrikis [13], Graham [32], Huang [38], Levi [56], Pratt [80], and Rosenfeld and Kak [89]. For general information on digital signal processing, see Oppenheim and Schafer [75] and Rabiner and Gold [84]. For general information on two-dimensional digital signal processing, see Dudgeon and Mersereau [18], Huang [39], and Mersereau and Dudgeon [63].

For gray-scale modification, see Gonzalez and Fittes [29], Hall [34], Hall *et al.* [35], and Troy *et al.* [100]. For image enhancement by low-pass and high-pass filtering, see O'Handley and Green [74] and Prewitt [81]. For algorithms on median filtering, see Aho *et al.* [1], Ataman *et al.* [9], Huang *et al.* [37], and Huang [39]. For image enhancement by homomorphic filtering or unsharp masking, see Oppenheim *et al.* [76], Schreiber [91,92], and Stockham [95]. For adaptive methods of image enhancement, see Lim [57], Peli and Lim [78], and Trussel and Hunt [101]. For color and human perception of color, see Mueller *et al.* [65].

Acknowledgments

The author would like to thank Professor William F. Schreiber of the Massachusetts Institute of Technology (M.I.T.) for his valuable comments, Michael McIlrath and Farid Dowla for generating most of the examples used in this chapter, Jean-Pierre Schott for his help in generating the references, and many others who contributed to this chapter. The author would also like to thank the M.I.T. Lincoln Laboratory and Cognitive Information Processing Group of the M.I.T. Research Laboratory of Electronics for the use of their facilities.

References*

1. A. V. Aho, J. E. Hopcroft, and J. D. Ullman, "The Design and Analysis of Computer Algorithms," Addison-Wesley, Reading, Mass., 1974.
2. G. B. Anderson and T. S. Huang, Frequency-domain image errors, *Pattern Recognition* **3**, 1961, 185–196.
3. H. C. Andrews, "Computer Techniques in Image Processing." Academic Press, New York, 1970.
4. H. C. Andrews, Digital image restoration: A survey, *Computer* **7**, 1974, 88–94.

* Some references listed here are not cited in the text.

5. H. C. Andrews and B. R. Hunt, "Digital Image Restoration." Prentice-Hall, Englewood Cliffs, N.J., 1977.
6. H. C. Andrews and W. K. Pratt, "Digital image transform processing," *Proceedings of the Applications of Walsh Functions,* 1970, 183–194.
7. H. C. Andrews, A. G. Tescher, and R. P. Kruger, Image processing by digital computers, *IEEE Spectrum* **9,** 1972, 20–32.
8. R. J. Arguello, H. R. Sellner, and J. Z. Stuller, Transfer function compensation of sampled imagery, *IEEE Trans. Comput.* **C-21,** 1972, 812–818.
9. F. Ataman, V. K. Aatre, and K. M. Wong, A fast method for real-time median filtering, *IEEE Trans. Acoust. Speech, Signal Process.* **ASSP-28,** 1980, 415–421.
10. L. M. Biberman, (ed.), "Perception of Displayed Information." Plenum, New York, 1973.
11. F. C. Billingsley, Applications of digital image processing, *Appl. Opt.* **9,** 1970, 289–299.
12. F. C. Billingsley, A. F. H. Goetz, and J. N. Lindsley, Color differentiation by computer image processing. *Photogr. Sci. Eng.* **14,** 1970, 28–35.
13. Z. L. Budrikis, Visual fidelity criterion and modeling, *Proc. IEEE* **60,** 1972, 771–779.
14. T. M. Cannon, "Digital Image Deblurring by Nonlinear Homomorphic Filtering," Ph.D. Thesis, University of Utah, 1974.
15. E. R. Cole, "The Removal of Unknown Image Blurs by Homomorphic Filtering," Ph.D. Thesis, University of Utah, 1973.
16. *Computer,* **7,** Special Issue on "Digital Picture Processing," 1974.
17. J. C. Dainty, The statistics of speckle patterns, *in* "Progress in Optics" (E. Wolf, ed.), Vol. XIV. North-Holland, Amsterdam, 1976.
18. D. E. Dudgeon, and R. M. Mersereau, "Multi-Dimensional Signal Processing." Prentice-Hall, Englewood Cliffs, N.J., 1983.
19. J. O. Eklundh, A fast computer method for matrix transposing, *IEEE Trans. Comput.* **C-21,** 1972, 801–803.
20. D. G. Falconer, Image enhancement and film-grain noise, *Opt. Acta* **17,** 1970, 693–705.
21. J. Feinberg, Real-time edge enhancement using the photorefractive effect, *Optics Letters* **5,** 1980, 330–332.
22. W. Fink, Image coloration as an interpretation aid, *Proc. SPIE/OSA Conf. Image Process.* **74,** 1976, 209–215.
23. B. R. Frieden and H. H. Barrett, Image noise reduction using Stein's procedure, *Annual Meeting of the Optical Society of America,* 1977.
24. V. S. Frost *et al.* An adaptive filter for smoothing noisy radar images, *Proc. IEEE* **69,** 1981, 133–135.
25. C. Gazley, J. E. Reiber, and R. H. Stratton, Computer works a new trick in seeing pseudo color processing, *Aeronaut. Astronaut.* **4,** 1967, 56.
26. B. Gold and C. M. Rader, "Digital Processing of Signals." McGraw-Hill, New York, 1969.
27. P. C. Goldmark and J. M. Hollywood, "A new technique for improving the sharpness of television pictures, *Proc. IRE* **39,** 1951, 1314–1322.
28. R. C. Gonzalez and B. A. Fittes, Gray-level transformation for interactive image enhancement, *Proc. Second Conference on Remotely Manned Systems,* 1975.
29. R. C. Gonzalez and B. A. Fittes, Gray-level transformation for interactive image enhancement, *Mech. Mach. Theory* **12,** 1977, 111–122.
30. R. C. Gonzalez and P. Wintz, "Digital Image Processing." Addison-Wesley, Reading, Mass., 1977.
31. W. M. Goodall, Television transmission by pulse code modulation, *Bell System Technical J.* **30,** 1951, 33–49.
32. C. H. Graham (ed.), "Vision and Visual Perception." Wiley, New York, 1965.
33. R. E. Graham, Snow removal—A noise-stripping process for picture signals, *IRE Trans. Inf. Theory* **IT-8,** 1962, 129–144.

34. E. L. Hall, Almost uniform distributions for computer image enhancement, *IEEE Trans. Comput.* **C-23**, 1974, 207–208.
35. E. L. Hall, R. P. Kruger, S. J. Dwyer, III, D. L. Hall, R. W. McLaren, and G. S. Lodwick, A survey of preprocessing and feature extraction techniques for radiographic images, *IEEE Trans. Comput.* **C-20**, 1971, 1032–1044.
36. J. E. Hall, and J. D. Awtrey, Real-time image enhancement using 3 × 3 pixel neighborhood operator functions, *Opt. Eng.* **19**, 1980, 421–424.
37. T. S. Huang, Some notes on film-grain-noise, *in* "Restoration of Atmospherically Degraded Images," Appendix 14, pp. 105–109. Woods Hole Oceanographic Institution, Woods Hole, Mass., 1966.
38. T. S. Huang, Image enhancement: A review, *Opto-Electronics* **1**, 1969, 49–59.
39. T. S. Huang (ed.), Two-dimensional digital signal processing I & II, *in* "Topics in Applied Physics," Vols. 42–43. Springer-Verlag, Berlin, 1981.
40. T. S. Huang, W. F. Schreiber, and O. J. Tretiak, Image processing, *Proc. IEEE* **59**, 1971, 1586–1609.
41. T. S. Huang, G. J. Yang, and G. Y. Tang, A fast two-dimensional median filtering algorithm, *IEEE Trans. Acoust. Speech, Signal Process.* **ASSP-27**, 1979, 13–18.
42. R. A. Hummel, "Histogram Modification Techniques," Technical Report TR-329 F-44620-726-0062. Computer Science Center, University of Maryland, College Park (1974).
43. B. R. Hunt, Block-mode digital filtering of pictures, *Math. Biosci.* **11**, 1971, 343–354.
44. B. R. Hunt, Data structures and computational organization in digital image enhancement, *Proc. IEEE* **60**, 1972, 884–887.
45. B. R. Hunt and J. Breedlove, Scan and display consideration in processing images by digital computer, *IEEE Trans. Comput.* **C-24**, 1975, 848–853.
46. B. R. Hunt, D. H. Janney, and R. K. Zeigler, "Introduction to Restoration and Enhancement of Radiographic Images," Report LA 4305. Los Alamos Scientific Laboratory, Los Alamos, N.M., 1970.
47. P. M. Joseph, Image noise and smoothing in computed tomography (CT) scanners, *Opt. Eng.* **17**, 1978, 396–399.
48. M. A. Karim and H. Liu, Linear versus logarithmic spatial filtering in the removal of multiplicative noise, *Opt. Lett.* **6**, 1981, 207–209.
49. H. R. Keshavan and M. V. Srinath, Interpolative models in restoration and enhancement of noisy images, *IEEE Trans. Acoust. Speech, Signal Process.* **ASSP-25**, 1977, 525–534.
50. H. R. Keshavan and M. V. Srinath, Sequential estimation technique for enhancement of noisy images, *IEEE Trans. Comput.* **C-26**, 1977, 971–987.
51. D. J. Ketcham, Real time image enhancement technique, *Proc. SPIE/OSA Conf. Image Process.* **74**, 1976, 120–125.
52. R. J. Kohler and H. K. Howell, Photographic image enhancement by superimposition of multiple images, *Photogr. Science Eng.* **7**, 1963, 241–245.
53. L. S. G. Kovasznay and H. M. Joseph, Image processing, *Proc. IRE* **43**, 1955, 560–570.
54. E. R. Kreins and L. J. Allison, Color enhancement of nimbus high resolution infrared radiometer data, *Appl. Opt.* **9**, 1970, 681.
55. M. J. Lahart, Local segmentation of noisy images, *Opt. Eng.* **18**, 1979, 76–78.
56. L. Levi, On image evaluation and enhancement, *Opt. Acta* **17**, 1970, 59–76.
57. Jae S. Lim, Image restoration by short space spectral subtraction, *IEEE Trans. Acoust. Speech, Signal Process.* **ASSP-28**, 1980, 191–197.
58. Jae S. Lim, "Short Space Implementation of Wiener Filtering for Image Restoration," Technical Note 1980-11. M.I.T. Lincoln Laboratory, Lexington, Mass., 1980.
59. Jae S. Lim and H. Nawab, Techniques for speckle noise removal, *Opt. Eng.* **20**, 1981, 472–480.

60. C. S. McCamy, The evaluation and manipulation of photographic images., *in "Picture Processing and Psychopictorics"* (A. Rosenfeld, ed.). Academic Press, New York, 1970.

61. A. Martelli and U. Montanari, Optimal smoothing in picture processing: An application to fingerprints, *Proc. IFIP Congr.* **71,** 86–90. **(Booklet TA-2)**

62. W. Matuska, Jr. *et al.,* Enhancement of solar corona and comet details, *Optical Engineering* **17,** 1978, 661–665.

63. R. Mersereau and D. Dudgeon, Two-dimensional digital filter, *Proc. IEEE* **63,** 1975, 610–623.

64. S. K. Mitra and M. P. Ekstrom (eds.), "Two-Dimensional Digital Signal Processing." Dowden, Hutchinson, and Ross, Stroudsburg, Pa., 1978.

65. C. G. Mueller, M. Rudolph, and the editors of LIFE, Light and vision, *Time,* New York, 1966.

66. N. E. Nahi, Role of recursive estimation in statistical image enhancement, *Proc. IEEE* **60,** 1972, 872–877.

67. N. E. Nahi, "Nonlinear Adaptive Recursive Image Enhancement," USCEE Report 459. University of Southern California, Los Angeles, 1973.

68. R. Nathan, Picture enhancement for the moon, Mars, and man, *in* "Pictorial Pattern Recognition" (G. C. Cheng, ed.), pp. 239–266. Thompson, Washington, D.C., 1968.

69. R. Nathan, Spatial frequency filtering, *in* "Picture Processing and Psychopictorics" (A. Rosenfeld, ed.), pp. 151–164. Academic Press, New York, 1970.

70. A. N. Netravali, Noise removal from chrominance components of a color television signal, *IEEE Trans. Commun.* **COM-26,** 1978, 1318–1321.

71. T. G. Newman and H. Dirilten, A nonlinear transformation for digital picture processing, *IEEE Trans. Comput.* **C-22,** 1973, 869–873.

72. K. N. Ngan and R. Steel, Enhancement of PCM and DPCM images corrupted by transmission errors, *IEEE Trans. Commun.* **COM-30,** 1982, 257–269.

73. L. W. Nichols and J. Lamar, Conversion of infrared images to visible in color, *Appl. Opt.* **7,** 1968, 1757.

74. D. A. O'Handley and W. B. Green, Recent developments in digital image processing at the image processing laboratory at the Jet Propulsion Laboratory, *Proc. IEEE* **60,** 1972, 821–828.

75. A. V. Oppenheim and R. W. Schafer, "Digital Signal Processing." Prentice-Hall, Englewood Cliffs, N.J., 1975.

76. A. V. Oppenheim, R. W. Schafer, and T. G. Stockham, Jr., Nonlinear filtering of multiplied and convolved signals, *Proc. IEEE* **56,** 1968, 1264–1291.

77. W. A. Pearlman Visual-error criterion for color-image processing, *J. Opt. Soc. Am.* **72,** 1982, 1001–1007.

78. T. Peli and J. S. Lim, Adaptive filtering for image enhancement, *J. Opt. Eng.* **21,** 1982, 108–112.

79. S. C. Pohlig *et al.,* New technique for blind deconvolution, *Opt. Eng.* **20,** 1981, 281–284.

80. W. K. Pratt, "Digital Image Processing." Wiley, New York, 1978.

81. J. M. S. Prewitt, Object enhancement and extraction, *in* "Picture Processing and Psychopictorics" (B. S. Lipkin and A. Rosenfeld, eds.). Academic Press, New York, 1970.

82. *Proc. IEEE,* **60,** Special Issue on "Digital Picture Processing," 1972.

83. P. R. Prucnal and B. E. A. Saleh, Transformation of image-signal-dependent noise into image-signal-independent noise, *Opt. Lett.* **6,** 1981, 316–318.

84. L. R. Rabiner and B. Gold, "Theory and Applications of Digital Signal Processing." Prentice-Hall, Englewood Cliffs, N.J., 1975.

85. L. G. Roberts, Picture coding using pseudo-random noise, *IRE Trans. Inf. Theory* **IT-8,** 1962, 145–154.

86. G. S. Robinson and W. Frei, *"Final Research Report on Computer Processing of ERTS Images,"* Report USCIPI 640. University of Southern California, Image Processing Institute, Los Angeles, 1975.
87. P. G. Roetling, Image Enhancement by Noise Suppression, *J. Opt. Soc. Am.* **60,** 1970, 867–869.
88. A. Rosenfeld, "Picture Processing by Computer." Academic Press, New York, 1969.
89. A. Rosenfeld and A. C. Kak, "Digital Picture Processing." Academic Press, New York, 1976.
90. A. Rosenfeld, C. M. Park, and J. P. Strong, Noise cleaning in digital pictures, *Proc. EASCON Convention Record,* 1969, 264–273.
91. W. F. Schreiber, "Image processing for quality improvement," *Proc. IEEE* **66,** 1978, 1640–1651.
92. W. F. Schreiber, Wirephoto quality improvement by unsharp masking, *J. Pattern Recognition* **2,** 1970, 171–121.
93. J. J. Sheppard, Jr., Pseudocolor as a means of image enhancement, *Am. J. Ophthalmol. Arch. Am. Acad. Optom.* **46,** 1969, 735–754.
94. L. Sica, Image-sharpness criterion for space-variant imaging, *J. Opt. Soc. Am.* **71,** 1981, 1172–1175.
95. T. G. Stockham, Jr., Image processing in the context of a visual model, *Proc. IEEE* **60,** 1972, 828–842.
96. D. R. Sullivan, R. C. Harney, and J. S. Martin, Real-time quasi-three-dimensional display of infrared radar images, *SPIE* **180,** 1979, 56–64.
97. A. G. Tescher and H. C. Andrews, Data compression and enhancement of sampled images, *Applied Optics* **11,** 1972, 919–925.
98. H. Thiry, Some qualitative and quantitative results on spatial filtering of granularity, *Applied Optics* **3,** 1964, 39–43.
99. D. E. Troxel and others, Image enhancement/coding systems using pseudorandom noise processing, *IEEE Proc.* **67,** 1979, 972–973.
100. E. B. Troy, E. S. Deutsch, and A. Rosenfeld, Gray-level manipulation experiments for texture analysis, *IEEE Trans. Syst. Man Cybernet.* **SMC-3,** 1973, 91–98.
101. H. J. Trussel and B. R. Hunt, Sectioned methods for image restoration *IEEE Trans. Acoustics, Speech, and Signal Processing* **ASSP-26,** 1978, 157.
102. C. S. Vikram, Simple approach to process speckle-photography data, *Optics Letters* **7,** 1982, 374–375.
103. J. F. Walkup and R. F. Choens, Image processing in signal-dependent noise, *Optical Engineering* **13,** 1974, 258–266.
104. R. H. Wallis, An approach for the space variant restoration and enhancement of images, *Proc. Symposium on Current Mathematical Problems in Image Science,* 1976.
105. G. W. Wecksung and K. Campbell, Digital image processing at EG&G, *Computer* **7,** 1974, 63–71.
106. R. E. Woods and R. C. Gonzalez, Real-time digital image enhancement, *Proc. IEEE* **69,** 1981, 643–654.
107. H. J. Zweig, E. B. Barett, and P. C. Hu, Noise-cheating image enhancement, *J. Optical Society of America* **65,** 1975.

2 Image Restoration

B. R. Hunt

Department of Electrical Engineering
University of Arizona
Tucson, Arizona

Science Applications Inc.
5151 East Broadway
Suite 1100
Tucson, Arizona

I. Statement of the Problem

A completely accurate statement of the problem of image restoration requires a discussion of the concepts of image formation and the description of image formation by concepts of linear systems and Fourier transforms. Because such a discussion is beyond the scope and purpose of this book, we shall present the relevant results and processes without justification. The reader is referred to other books for detailed exposition of this theory; see, for example, [1,3,5,7,14,15].

An *object* is defined as a two-dimensional entity that emits or reflects

53

DIGITAL IMAGE PROCESSING TECHNIQUES

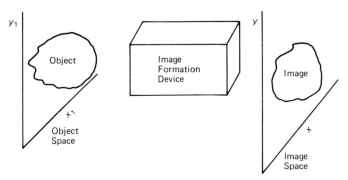

Fig. 1. Image formation schematic.

radiant energy.† The energy may be of a variety of forms, for example, optical energy, acoustic energy, or nuclear particles. An *imaging system* is a "black box" that is capable of intercepting some or all of the radiant energy reflected or emitted by the object. The imaging system has the function of bringing to focus the intercepted energies and making an *image,* that is, a representation of the original object that emitted or reflected the energy. Schematically, we signify the image formation process as in Fig. 1, in which we show a set of coordinates for the object space and the image space, and a generic box capable of image formation.

The description of the image formation process in the simplest mathematical terms requires that we assume spatial invariance and linearity in the process of image formation; that is, the image formed from two objects simultaneously present in the object space is equal to the sum of the images formed when each object is present by itself in the object space and does not change with position. Under such assumptions, it is possible to describe image formation by the following equation:

$$g(x,y) = \int_{-\infty}^{\infty} \int_{-\infty}^{\infty} h(x - x_1, y - y_1) f(x_1, y_1) \, dx_1 \, dy_1. \tag{1}$$

The quantities in this equation have the following meanings. The function f represents the two-dimensional distribution of energy corresponding to the original object. The function g is the representation of f that is created in the image space by the image-formation process. The function h is most important, because it embodies all the important behavior of the image formation system; it is known as the *point spread function* (PSF). It is entirely analogous to the impulse response function used in the analysis of linear

---

† By taking the object to be two-dimensional, we are, in effect, discarding imaging situations in which the depth of field of the imaging system is not small compared with the third-dimensional extent of the object.

temporal systems but is extended to two dimensions for linear spatial systems.

It is important to recognize that, because image formation is a process of capturing and focusing radiant energy, the image $g(x,y)$ in Eq. (1) is a transient phenomenon; that is, it fluctuates in response to fluctuations in the energy radiated by the object. Because the image is literally a flow of energy, it is necessary to use some recording mechanism, which captures the energy flow at a particular time and makes a permanent record for later usage. Thus, into Eq. (1) we must introduce a function $S(\cdot)$ that accounts for the sensing and recording of the energy flow. Common recording mechanisms abound: photographic film, electronic video, xerographic copy, and so on. Finally, the mechanism of recording is always imperfect, that is, there are distortions in the recorded image. These distortions can be either predictable or random. The predictable or systematic distortions we encompass in the function $S(\cdot)$. The random distortions we account for by a noise component. The result of adding these effects is the equation

$$g(x,y) = S\left[\int_{-\infty}^{\infty} \int_{-\infty}^{\infty} h(x - x_1, y - y_1) f(x_1, y_1) \, dx_1 \, dy_1 \right] + n(x,y). \quad (2)$$

We should note that, as formulated, the noise or random component in this equation is more general than just that due to the image recording process. Images are formed from radiant energy, and because quantum mechanics demonstrates that all energy is quantized, there is a random component to an image even in the absence of any randomness introduced by sensing and recording of the image. For example, the number of photons exposing a fixed area of photographic film fluctuates from instant to instant. There is no loss of generality in the model of Eq. (2) in lumping all random distortions of the image into $n(x,y)$, whether they are induced by the image formation and recording or are inherent in the quantized nature of the radiant energy used in image formation.

The statement of the image restoration problem is now a direct consequence of Eq. (2): Given a recorded image $g(x,y)$, make an estimate of the original object $f(x,y)$. This estimate must be made with or without a priori knowledge of the point spread function h (*without* is more difficult) and with only a statistical description of the nature of the noise process n. The source of the terminology "image restoration" is obvious; we seek to restore, from the degrading mechanism of the point spread function, the original radiant energy distribution of the object. Often in literature and discussions on image restoration the term "ideal image" can be seen in referring to the original object energy distribution, the source of the term being the image formed under ideal conditions of image formation.

It is obvious from Eq. (2) that image restoration requires the solution of an integral equation. As we shall see, it is also the most difficult type of integral equation to solve. The difference among the various methods of image restoration can be reduced to three different classes: solution methods that differ in how the difficulty of solving the associated integral equation is treated, solution methods that differ in how various assumptions for sensitometry $S(\cdot)$ and noise n are incorporated, and solution methods that require minimal a priori knowledge of $h, S(\cdot)$, or n and infer these quantities from the underlying data relationships in the image.

What are the applications of image restoration? That is, what real-life situations give rise to a necessity to employ image restoration techniques. The following is a brief (and incomplete) discussion of applications of image restoration.

There are occasions when the imperfections or degradations inherent in an image formation system cannot be avoided. For example, it may be impossible to improve the image formation, due to engineering constraints such as weight, power consumption, or cost. An example is the imaging systems launched on NASA satellites. NASA designs the best imaging possible, consistent with the space vehicle constraints. But even this may not be adequate in all situations, and image restoration is used to produce better images. Such processing has been performed by NASA from the first pictures of the moon's surface obtained by the first remote lander [6].

There are occasions when adverse exposure conditions occur, such as when the individual holding the camera jumps or twitches during the time when the camera shutter is open. The result is a motion blur of the image. If the image is sufficiently unique and cannot be retaken, then restoration could be employed. As an example, the assassination of President John F. Kennedy was recorded on film by a number of amateur photographers. Some of the most important images were blurred by motions of the photographer resulting from the terror and excitement of the occasion. Thus, when the U. S. Congress reopened the investigation of the Kennedy assassination, one of the tasks taken on by the congressional consultants was to apply image restoration to the available photographs, so as to clarify the photographs and, hopefully, the issues about the assassination [13].

What types of degradations commonly arise in image formation, and for which types is image restoration chosen? In compensating for expected deficiencies in image formation, the degradation is that associated with the composite point spread function (or optical transfer function) of the entire imaging system. In dealing with unexpected errors in image formation, the most common degradations are due to focus error or to relative motion between the optical system and the object during the exposure interval. The processing of space probe images by NASA is an ideal example of the application of image restoration for the compensation of expected deficiencies in an imaging device. Likewise, the processing of amateur photographs of the Kennedy assassination for focus and motion blurs is exemplary of compensating for unexpected image degradations.

II. Direct Techniques of Image Restoration

Image restoration has been the subject of extensive research over the past 15 years. The extent of the research is visible in the number of published papers on this topic. The research worker who enters the field of image restoration for the first time, and who has a practical problem that must be solved, can be intimidated by the large number of techniques published. The one important criterion that we shall apply in the following discussion of techniques of image restoration is what works and has been proven in a variety of situations. This criterion is a powerful filter of the available techniques. Although interesting methods for image restoration have been proposed, only a very small number have been implemented in *realistic* situations. The emphasis on the realistic is a subjective assessment by the author, who believes that realism in the application of image restoration is measured by the following dimensions.

First, an image restoration technique should be computable at an average computer facility; that is, successful implementation should not depend on a super-computer with megabytes of storage and special high-speed array-processing arithmetic units. All the algorithms to be discussed in this chapter can be (and have been) implemented on a minicomputer of modest capability, for example, a PDP-11/34.

Second, the computational requirements of the restoration technique must be applicable to pictures of substantial size. When processed by computer, images are sampled as matrices of picture elements, or *pixels,* by a suitable scanning device [9]. The quality of a reconstructed digital image is partially a function of how many pixels are presented to the eye within the image frame. For example, a high-quality video image can display about 500×500 pixels in a frame of approximately 30 cm square, and the eye – brain combination views this as a superb image. On the other hand, a display of 64×64 pixels within the same dimensions is objectionable, the eye being very sensitive to the aliasing and spatial quantization that is usually associated with coarse sampling. Consequently, it is our position that, for realism, an image restoration technique must be readily applicable to images sampled at least 200×200 pixels over the image frame *displayed after processing.*

Third, a realistic image restoration technique must be successful in the presence of the information that is available a priori or must be a technique that is relatively insensitive to errors in the state of knowledge of the a priori information. By this we mean that a method is of little use if it requires perfect knowledge of an unavailable quantity, for example, the noise power spectrum. Conversely, a method can be considered workable if it produces useful results with a reasonable estimate of a quantity such as the noise power spectrum.

Applying the preceding dimensions of realism and the criterion of discussing only methods that have been shown to work in a variety of situations, we can greatly reduce the number of methods that must be discussed.

A. Ill-Conditioned Behavior and Image Restoration

The basic difficulty of image restoration is that it is an example of an ill-conditioned integral equation. To be specific, let us assume the image sensing and recording processes are linear, so that in Eq. (2) the function $S(\cdot) = I(\cdot)$, the identity operator. The resulting equation has the convolution property; and if we calculate the two-dimensional Fourier transform of both sides of the equation, we have

$$G(\omega_1,\omega_2) = H(\omega_1,\omega_2)F(\omega_1,\omega_2) + N(\omega_1,\omega_2). \tag{3}$$

Given that F is the Fourier transform of the desired quantity, the object energy distribution, we are tempted to compute a quantity F', the Fourier transform of the restored image by a simple division,

$$F'(\omega_1,\omega_2) = \frac{G(\omega_1,\omega_2)}{H(\omega_1,\omega_2)} = F(\omega_1,\omega_2) + \frac{N(\omega_1,\omega_2)}{H(\omega_1,\omega_2)}. \tag{4}$$

This expression contains the desired quantity F plus a term due to noise. The real difficulty is the quantity H in the denominator of the term containing N. If H is small or zero in any region of the Fourier spectrum where N is not correspondingly smaller or where N is nonzero, then the contribution of the noise will be amplified. This is the source of the term "ill-conditioned." A small perturbation in the data (the noise in the recorded image) can produce a large effect in the solution (the restored image).

The quantity $H(\omega_1,\omega_2)$ in Eq. (3) is the *optical transfer function* of the image formation system, that is, the Fourier transform of the optical point spread function. As can be seen from simple optical analysis (e.g., see Goodman [1] or Gaskill [5]), a number of optical transfer functions become small or identically zero. (Indeed, every optical transfer function is zero beyond the diffraction limit of the optical system.) Thus, we expect to encounter the ill-conditioned behavior in most situations in which image restoration will be required. One of the major facets of any method of image restoration, therefore, is how the method controls the solution perturbations caused by noise in the data when ill-conditioned behavior is present.

B. Image Restoration and Spatial Filtering

The above discussion indicates a simple basis for approaching image restoration. However, it is dependent on the assumption that the character-istic function of image sensing and recording, $S(\cdot)$, is a linear function. For many new image acquisition devices, linearity of the sensing and recording mechanism can be assured; for example, the CCD imaging devices are exemplary of a class of semiconductor imaging technologies that has very good linearity characteristics. But many image acquisition devices are nonlinear. The most common and oldest image sensing and recording device, photographic film, can be extremely nonlinear. How do we treat this nonlinearity in the relation between recorded data and the quantity we seek to estimate, $f(x,y)$?

There are three basic approaches to the problem of the nonlinearity in Eq. (2). First is the most pragmatic approach: Ignore the nonlinearity and process the data as though $S(\cdot) = I(\cdot)$. If the variations in the image signal are of small magnitude, this is adequate, because a small enough perturba-tion allows a nonlinear function to be expressed as a linear Taylor series representation [1]. This is known as the "low-contrast" assumption, be-cause small signal variations are equivalent to low pictorial contrast. Sec-ond is the sensitometric inversion approach. If the sensitometry of the image acquisition device is known and calibrated, then a function may be constructed that is the inverse to the function $S(\cdot)$. If we apply this to both sides of Eq. (2), then

$$g_c(x,y) = S^{-1}(g(x,y))$$

$$= S^{-1}\left[S\left[\int_{-\infty}^{\infty}\int_{-\infty}^{\infty} h(x-x_1,y-y_1)f(x_1,y_1)f(x_1,y_1)\,dx_1\,dy_1\right] + n(x,y)\right].$$

Because noise perturbations usually are small compared with the signal term, the magnitude of the integral in this equation is usually much greater than the term $n(x,y)$, and we can approximate the nonlinearity in $S^{-1}(\cdot)$ with a Taylor series expansion. For such a case we can make the approximation

$$g_c(x,y) \cong S^{-1}\left[S\left[\int_{-\infty}^{\infty}\int_{-\infty}^{\infty} h(x-x_1,y-y_1)f(x_1,y_1)\,dx_1\,dy_1\right]\right] + S^{-1}\left[n(x,y)\right]$$

$$= \int_{-\infty}^{\infty}\int_{-\infty}^{\infty} h(x-x_1,y-y_1)f(x_1,y_1)\,dx_1\,dy_1 + S^{-1}[n(x,y)], \tag{5}$$

which again shows a linear relation between the corrected or calibrated image and the process of image formation. The amplitude statistics of the noise process are changed, of course, by the nonlinear action of $S^{-1}(\cdot)$. The usual course is to assume the noise is uncorrelated with the signal, so that the change in amplitude statistics is not reflected in any portion of the solution. The noise is also assumed to be uncorrelated with itself, so that the point transformation $S^{-1}(\cdot)$ does not change the actual shape of the noise power spectrum in any fashion.

The third course to pursue in dealing with nonlinearity is to construct a restoration procedure that can estimate $f(x,y)$ in the presence of a nonlinearity. We shall defer any further discussion of this until later sections of this chapter.

The resolution of the issue of nonlinearity is important, because of the nature of spatial filters. As is shown from optics (for example, see [1–6]), the relation between image and object is governed by the optical transfer function operating as a two-dimensional spatial filter on the Fourier spectrum of the object. For a nonlinear sensing and recording of the image, the data that are available for processing does not have this linear (spatial filtering) relationship preserved. If the linear relationship is preserved, however, then we can conceive of image restoration in terms of a spatial filter. That is what we see suggested in Eq. (4) and which we shall now elaborate.

Spatial filters for image restoration are important in the context of rapid, simple computation. In spite of the power of today's computers, full-scale image restoration would still be impractical in most cases had it not been for the discovery of the FFT algorithm, which makes implementation of large or complicated spatial filters a reality on even modest minicomputers. Because spatial filtering is equivalent to convolution with a specific point spread function, that is

$$f * h \overset{F}{\Leftrightarrow} FH, \tag{6}$$

where $*$ denotes convolution and the arrow notation means Fourier transform, the computer implementation of spatial filters makes it possible to conceive of image restoration schemes in terms of implementable convolutions. The use of this duality — of creating image restoration schemes in either their convolution–space domain form or their spatial filter–Fourier domain form — has expanded the insight into image restoration techniques [1].

The practical and realistic image restoration schemes, which we shall discuss in the following sections, all rest upon implementation by spatial filters. Thus, the single most important factor that will distinguish one technique from another is how the spatial filter is specified or the criteria used in constructing the filter. In the following we shall not discuss the actual details by which a spatial filter is to be implemented, because this encompasses programming questions that are of greater detail than can be addressed by this book. The reader interested in spatial filter implementations is referred to the literature (Hunt [8].)

C. Inverse Filter Restoration

The simplest restoration technique is embodied within Eq. (4), and it is known as the inverse filter, for the obvious reason that it consists of nothing more than a product of the Fourier transforms of the image and the algebraic inverse of the optical transfer function. The preceding discussion alerts us to a basic concern with the inverse filter: noise amplification. It would be desirable if there existed a precise quantitative relationship between the shape of the optical transfer function and the power spectrum of the noise, so that one could state a priori when noise amplification would be a problem in the output. Such a relationship may be possible but has not been formulated. One obvious case, when the optical transfer function contains zero values, is usually handled manually by the image analyst prior to performing the inverse. If the optical transfer function contains no zeroes, yet the restored image is obviously disturbed by noise, then the analyst usually begins hand-tuning the inverse, that is, deliberately altering the numerical values of the inverse of the optical transfer function. The rule of thumb is to reduce any large inverse values at high spatial frequencies, because it is at these highest spatial frequencies that the noise usually is greatest and, thereby, the noise amplification by large inverse values most severe. Such alteration means that an exact inverse is no longer being performed; instead, an approximate inverse is being created, and the criterion of approximation is the preference of the individual in examining the output of the restoration.

Figure 2 illustrates a one-dimensional profile through an optical transfer function, the inverse of the OTF, and a possible profile for the inverse after hand tuning to eliminate the singularity and excessive high frequency amplification.

D. Wiener Filter Restoration

The alternative to treating the noise amplification problem with an inverse restoration is to engage in detailed and possibly tedious manipulation of the inverse of the optical transfer function. It would be preferable if it were possible to make sufficient a priori assumptions so that the control of noise amplification were automatic, and not left to the discretion of the analyst.

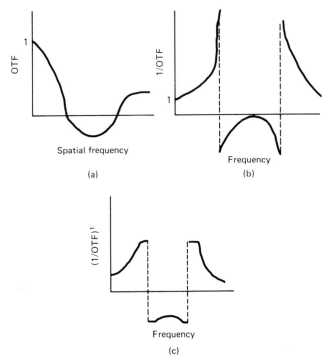

Fig. 2. (a) Optical transfer function of image degradation. (b) Transfer function of the inverse filter (note asymptotic singularities). (c) Transfer function of a hand-tuned inverse filter.

Such automatic control exists in the method of the Wiener filter image restoration.

The Wiener filter results from the following criterion. Let $\hat{f}(x,y)$ be an estimate of the original object distribution of radiant energy. The problem of noise amplification manifests itself in $\hat{f}(x,y)$. If we were able to compare $\hat{f}(x,y)$ and the actual $f(x,y)$, the noise amplification would cause the two to differ greatly. It is evident that we should do something to minimize this difference. If we had precise a priori knowledge about $f(x,y)$, then we might be able to conceive of ways to make this minimum most effective; however, such precise a priori knowledge is not at hand, which is one of the reasons we are seeking to deblur the image. Because of lack of precison in a priori knowledge, the best we can do is to adopt a stochastic viewpoint. That is, we can decide to minimize the difference between $f(x,y)$ and $\hat{f}(x,y)$ over some random ensemble of possible objects. This leads directly to the criterion

$$\text{minimize} \quad E\{[f(x,y) - \hat{f}(x,y)]^2\},$$

where E is the ensemble expectation operator. Because we wish to implement our restoration by a linear filter, we now seek a point spread function that, when convolved with the image, will achieve the minimum. Thus,

$$\hat{f}(x,y) = h_w(x,y) * g(x,y).$$

The problem becomes one of specifying h_w, either in the space or frequency domain. It is possible to show that the minimization is achieved by the transfer function

$$H_w(u,v) = \frac{H(\omega_1,\omega_2)^*}{|H(\omega_1,\omega_2)|^2 + \dfrac{S_n(\omega_1,\omega_2)}{S_f(\omega_1,\omega_2)}}, \tag{7}$$

where * superscript denotes complex conjugate, and S_n and S_f are the power spectra of noise and signal, respectively. See, for example, Andrews and Hunt [1] for the details in deriving Eq. (7).

The concept of power spectrum is valid for a stationary random process, which implies that the validity of Eq. (7) is predicated on the assumption that the object energy distribution can be modeled as a spatially stationary random process. Most images that we see in every-day life would seem to violate this assumption, and studies of images confirm that the autocorrelation statistics (and hence, power spectrum) of an image can vary widely in a single image. The validity of Eq. (7) would seem weak. It is a practical more than a theoretical issue, however.

Equation (7) exhibits the automatic control of noise amplification that we sought in our preceding discussion. For example, when any portion of $H(\omega_1,\omega_2)$ is equal to zero, the existence of the term involving S_n and S_f prevents the division by zero. Because the numerator contains $H(\omega_1,\omega_2)^*$, the net filter gain at any zero of $H(\omega_1,\omega_2)$ is identically zero. The filter also exhibits other useful behavior. For example, suppose that in a given region of the frequency spectrum the signal-to-noise ratio is quite high. In such a region the ratio of S_n to S_f will be small, and then

$$H_w(\omega_1,\omega_2) \rightarrow \frac{1}{H(\omega_1,\omega_2)} \quad \text{for} \quad S_n(\omega_1,\omega_2) \ll S_f(\omega_1,\omega_2). \tag{8}$$

Conversely, we see that for regions where the signal-to-noise ratio is small, then

$$H_w(\omega_1,\omega_2) \rightarrow 0 \quad \text{for} \quad S_n(\omega_1,\omega_2) \gg S_f(\omega_1,\omega_2), \tag{9}$$

and simultaneously, $S_n(\omega_1,\omega_2) \gg H_w(\omega_1,\omega_2)$. Thus, the Wiener filter avoids the excessive amplification of noise seen in the inverse filter.

The quantities responsible for the avoidance of noise amplification in the inverse filter are the power spectra of the object and noise. It is important to consider the realistic usage of the Wiener filter in the specification of these quantities, therefore.

The first point we take up is the matter of image stationarity. As previously mentioned, realistic images show spatial variations in the power spectrum computed in a local region. How critical are these variations? Part of the answer to this question lies in the way in which a model for the image power spectrum is derived. Extensive studies of image power spectra have been done in conjunction with development of methods for image data compression. These studies show that an excellent simple model for imagery statistics is that of a spatially isotropic first-order Markov process [16]. Therefore, the autocorrelation and power spectrum of the image can be derived directly, and it is only necessary to perform a statistical least-squares fit to determine the single correlation parameter that governs the autocorrelation function and the power spectrum. If the image possesses spatial variations in the power spectrum in local regions, this is reflected by variations between a global fit of the Markov parameter (i.e., a fit using data over the whole image) and a fit in a localized region. The global fit becomes an average Markov correlation parameter over the entire image region. Thus, the Wiener filter would be implemented with a Markov parameter that would not represent regions significantly different from the average. The practical effect is that in spatial regions in which the local correlation distance is much shorter than the average, the S_n/S_f term in the Wiener filter would be larger than necessary, thus limiting the overall filter gain in corresponding spatial frequencies. Fine or small structures in these spatial regions would not be as well restored. Conversely, in spatial regions where the local spatial correlation distance is much greater than the average, there would be too much gain in the filter and possible noise amplification.

In realistic terms, the variation in the correlation parameter for a typical image is probably 2:1. This does not usually translate to an appreciable change in the filter response. Indeed, studies of Wiener filters that have been constructed to vary spatially with the local correlation show little difference in the visual quality of images restored with and without localized processing. Because a filter with local variations is usually an order of magnitude more costly to compute, the usage of an average power spectrum for the entire image is justified in all but the few very pathological cases.

A second issue in Wiener filter implementation is the following. Note that in the denominator of Eq. (7) the power spectrum of the original object distribution S_f is modified by the optical transfer function. The solution to this dilemma is to recognize the basic relation, from linear systems theory, between the image and object power spectra, which is

$$S_g(\omega_1,\omega_2) = |H(\omega_1,\omega_2)|^2 S_f(\omega_1,\omega_2). \tag{10}$$

We are assuming that H is known; it obviously must be known to implement Eq. (7). Thus, if we utilize the image to estimate the average correlation and, hence, the power spectrum, then a Fourier transform of the correlation function can be divided by H to produce an estimate of S_f. If H contains a zero, we must deal with it in a manner analogous to hand-tuning the inverse filter, as discussed in the previous section.

Finally, suppose that a model of the power spectrum is not considered adequate, and a more direct measurement of it is desired. Such is possible, but we shall defer discussion of it until the later section on identification of blur in an image.

E. Power Spectral Equalization Restoration

The Wiener filter is derived from a criterion of minimizing the mean-square error between the original object distribution and its estimate but a different criterion can also be derived. Because we have no exact information about the original object distribution, it is possible to specify properties of either the original object or its image in terms of an average quantity. A very common average quantity in signal processing is the power spectrum, which describes the ensemble average distribution of signal power over spatial frequency. From Eq. (10) we know that the image-formation process modifies the object power spectrum to produce the power spectrum of the image, and to this power spectrum is added the noise power spectrum, which is not shown in Eq. (10). A meaningful criterion of restoration, therefore, is to require that the power spectrum of the estimated object be equal to the power spectrum of the original object. That is, we require that

$$S_{\hat{f}}(\omega_1,\omega_2) = S_f(\omega_1,\omega_2), \tag{11}$$

where the ^ symbol designates the estimate. From Eq. (10) we know that it is possible to write $S_{\hat{f}}$ in terms of S_g and the transfer function of the restoration filter:

$$S_{\hat{f}}(\omega_1,\omega_2) = |H_p(\omega_1,\omega_2)|^2 S_g(\omega_1,\omega_2)$$

$$= |H_p(\omega_1,\omega_2)|^2 [|H(\omega_1,\omega_2)|^2 S_f(\omega_1,\omega_2) + S_n(\omega_1,\omega_2)]. \tag{12}$$

By substituting from Eq. (12) into Eq. (11) and solving for H_p, which is the frequency response of the restoration filter that performs the power spectral equalization, we have

$$H_p(\omega_1,\omega_2) = \left\{ \frac{1}{|H(\omega_1,\omega_2)|^2 + \dfrac{[S_n(\omega_1,\omega_2)]}{[S_f(\omega_1,\omega_2)]}} \right\}^{1/2}. \tag{13}$$

We notice the following properties of this filter. First, it is not subject to ill-conditioned behavior in regions where $H(u,v) \to 0$, because of the denominator term S_n/S_f. Second, in a region of the spectrum where $H(\omega_1,\omega_2) \to 0$, the filter response does not approach zero. Instead it approaches the square root of the signal-to-noise spectrum. The net effect of this is seen in Figs. 3(a) and 3(b), which compare the filter response curves for a Wiener and a power spectral equalization filter. The important point is that because the Wiener filter is forced to zero at the singularity, the Wiener filter frequency response curve shows greater variation than the power spectral equalization filter. With wide variations in one domain being coupled to wide variations in the other domain, the point spread function of the Wiener filter will have more side lobes and fine structure than will the power spectral equalization filter. The usual result is a restored image with a more pleasing visual appearance, because the lobes and structure of the filter point spread function are usually seen as artifact patterns in the restored image. A final point is that because of the cutoff behavior at a singularity the power spectral equalization filter has higher gain at frequencies near the singularity, thus admitting more structure into the restored picture than the Wiener filter. Such structure is not, of course, optimal in the mean-square sense. However, experimental studies of Wiener and power spectral equalization filters show that the human viewer usually prefers images produced by power spectral equalization restorations. Because the human visual system is not necessarily a minimum mean-square error processor, it should not be surprising that the Wiener filter is not necessarily preferred. In general, the restoration for lower SNR with a power spectral equalization filter will appear slightly sharper than the corresponding Wiener filter. A slight increase in noise is associated with this increased sharpness, of course, but in general the human eye – brain combination is tolerant of some increased amounts of noise when information content appears to increase correspondingly.

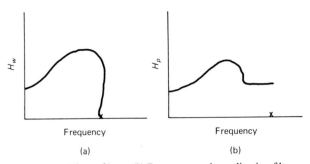

Fig. 3.   (a) Wiener filter.   (b) Power spectral equalization filter.

III. Indirect Techniques of Image Restoration

We called the previous section *direct* techniques of image restoration, because such techniques produce their results in a simple one-step fashion.
That is, a spatial filtering operation is executed with an appropriate spatial
filter, and the result is the desired restoration. Equivalently, the restoration
may be produced from the direct convolution of image and a suitable point
spread function. We wish now to discuss methods in which the results are
not so directly obtained and in which the final restoration is obtained only
after a number of steps. We term these *indirect* techniques, a more general
term within which are found nonlinear restoration techniques as well as
space-invariant restoration techniques.

The first technique in this class comes from the following observation.
The Wiener filter successfully controls the noise amplification that is associated with the ill-conditioned nature of the restoration problem. However,
the Wiener filter requires information that may be difficult to obtain, that is,
the power spectrum of object and noise. As we discussed previously, one
possibility is to use a suitable power spectral model. Is there any alternative
to the modeling process, however? The answer is to impose directly a
constraint on the restoration so as to enforce smoothness on the result. One
simple way to measure smoothness of a function is by its inner product,
that is,

$$t = \int_{-\infty}^{\infty} f(t) \cdot f(t) \; dt$$

is a measure of the smoothness of the function f, because the quantity t is less
for a function possessing small variation (say, a constant) than for a function
with many wiggles or kinks. Using the function α to denote a two-dimen-
sion inner product, that is,

$$\alpha(f_1, f_2) = \int_{-\infty}^{\infty} f_1(x,y)f_2(x,y) \; dx \; dy,$$

we formulate a restoration problem: Find the estimate $\hat{f}$ of the original object
distribution f such that $\alpha(f_1, f_2)$ is minimized.

The criterion formulated in the previous sentence is incomplete. An
obvious trivial solution is $f(x,y) = 0$ everywhere. It is necessary to add a side
constraint that embodies other information about the problem. The other
information available is the existence of noise. Suppose that an estimate
$\hat{f}$ were guessed, and the estimate were truly equal to the original object
distribution. Then by back-substituting this estimate into image formation
equations, we could *not* obtain the data recorded as the image, because the

actual recorded image was noisy. This suggests that the inner product of the residual between recorded data and a back-substituted solution should have a magnitude that is characteristic of the noise process. Thus,

$$r(x,y) = g(x,y) - h(x,y) * \hat{f}(x,y)$$

is the residual of the estimate $\hat{f}$, which must be nonzero because of noise. Our side constraint is

$$\alpha(r,r) = s,$$

where s is a function of known noise properties. In particular, for an image sampled as $N \times N$ pixels, it is direct to show [17] that

$$s = N^2\sigma_n^2 + \mu_n^2,$$

where σ_n and μ_n are the standard deviation and mean value, respectively, of the noise process.

The final formulation of our restoration problem is

$$\text{minimize} \quad \alpha(\hat{f},\hat{f})$$

$$\text{subject to the constraint} \quad \alpha(r,r) = N^2\sigma_n^2 + \mu_n^2. \tag{14}$$

As was the case for Wiener and power spectrum equalization filters, it is necessary to derive the process that satisfies the optimization problem stated in Eq. (14). It can be shown [11] that the restored image is generated by processing through a spatial filter that has the following transfer function:

$$H_c(\omega_1,\omega_2) = \frac{H(\omega_1,\omega_2)^*}{|H(\omega_1,\omega_2)|^2 + \delta}, \tag{15}$$

where δ is a parameter that must be set to satisfy the second part of Eq. (14). It is possible to show that δ can be found by successive iterations of Eq. (15), usually four to six separate spatial filterings with δ varied in each filter to satisfy the requirement on the residual equation [10].

This method of restoration is known as *constrained least-squares* restoration. The source of the term "least squares" is that the replacement of all quantities in Eq. (14) by their sampled versions for digital processing converts the inner product function into the square of the Euclidean norm of the resulting vectors, and we are minimizing the norm metric of the object estimate vector subject to a constraint on the residual metric.

In Eq. (15) we have a form that resembles the Wiener filter. Indeed, it is possible to show that the choice of proper weighting functions for the inner products in Eq. (14) leads to an identical form of the Wiener filter for Eq. (15) (see Hunt [9]). Thus, our previous comments on the Wiener filter, with respect to control of ill-conditioned behavior, are applicable.

It is possible to apply general weightings to the inner products in Eq. (14) to make them more or less sensitive to noise amplification. For example, to emphasize noise sensitivity, the derivative of the estimate $\hat{f}$ may be used, because the derivative will be influenced by the high-frequency discontinuities imposed by amplified noise. See [10] for a general discussion of constrained least-squares restorations and examples of images deblurred by same.

Constrained least squares is an example of image restoration in which the solution is obtained only after several iterations are carried out. Obviously, there is a larger computational requirement for employing this method. Two other methods of image restoration that also require greater computation are *maximum- a posteriori* restoration and *maximum-entropy* restoration. We shall briefly discuss these.

The Wiener filter introduced the concept of a statistical approach to restoration, with a criterion of mean-square error. Another statistical approach is to find the object distribution that is the most probable, given the recorded image. That is, in the nomenclature of probability, we wish to maximize $p(f(x,y)|g(x,y))$. From Bayes law we know that an equivalent statement of this is

$$\underset{f}{\text{maximize}} \quad \frac{p(g|f)p(f)}{p(g)}, \tag{16}$$

where the maximization is over f, because it is by appropriately choosing f that we accomplish the restoration. To be able to write equations that can be maximized requires that an appropriate statistical moel for f be assumed. The difficulty is that statistical image models are notoriously hard to formulate, because the amplitude statistics of image models is such a strong function of the actual object being imaged. Nonetheless, the following can be shown:

1. It is possible to formulate an appropriate model of an image as zero-mean Gaussian fluctuations of a stationary random process about a nonstationary mean [11].
2. The Gaussian model leads to a simple set of equations for a closed-form description of the maximization problem in Eq. (16).
3. The maximization problem in Eq. (16) can be solved by an iterative optimization. A number of forms of the iteration are possible. Because the Gaussian model concerns itself only with the statistics of f, the effect of any nonlinearity in destroying Gaussian behavior for g is of no concern. Thus, restoration problems can be solved in which the sensor nonlinearity s in Eq. (2) is truly nonlinear. The resulting iteration sequence can be expressed in the following form:

$$\hat{f}_{k+1} = \hat{f}_k - h * s_b\{\sigma_n^{-2}[g - s(h * \hat{f}_k)]\} - \sigma_f^{-2}(\hat{f}_k - \bar{f}), \tag{17}$$

where k is the iteration index, s_b is a function of derivatives of the sensor nonlinearity, σ_f^{-2} and σ_n^{-2} are reciprocals of the object variance and noise variance, respectively, and $\bar{f}$ is a nonstationary (spatially varying mean) about which the restoration takes place. (Experience shows that $\bar{f}$ can be a constant, but convergence to final solution requires more iterations.) We have eliminated the (x,y) arguments of the functions f, g, h in Eq. (17) for simplicity of notation. See [12] for complete details.

Equation (17) shows that a restored image can be calculated from a sequence of convolutions, the $*$ symbol in Eq. (17). So the implementation of Eq. (17) is straightforward, and any convolution–spatial filtering program can be adapted to evaluate Eq. (17).

Equation (17) is not unrelated to Wiener filtering. Indeed, it is possible to show that if the function s becomes linear, then the solution converges to a Wiener filter. Thus, there is no conceptual necessity to employ Eq. (17) except where the nonlinearity $s(\cdot)$ cannot be ignored. In such a case Eq. (17) gives a method for nonlinear Wiener filtering by iteratively converging to the solution. Even when $s(\cdot)$ is a linear function, it is sometimes useful to employ Eq. (17), for it allows the final restoration of the image to be evaluated as a sequence of restorations, because the choice of a preferred solution may be made before convergence is complete. Used in this way, restorations by the iterative sequence in Eq. (17) have been shown in human ranking and viewing experiments to be preferred to either Wiener filter or power spectrum equalization restorations [2].

Another method for image restoration that has no direct solution is the maximum entropy method of Frieden [4]. The method is derived on the assumption that there are a fixed number of quanta (e.g., photons of light or grains in a film) which must be distributed in the image in a way that is consistent with the image formation equation, Eq. (1). If Stirling's approximation is applied to the equations that describe the combinatorial form of the ways in which the quanta can be distributed, then an optimization problem results [1].

$$\begin{array}{ll} \underset{f,\,n}{\text{maximize}} & -\alpha(f,\ln(f)) - \alpha(n,\ln(n)) \\ \text{subject to} & g = h*f + n, \\ & \alpha(1,f) = p \end{array} \qquad (18)$$

where α is the inner product function introduced before, $\ln(\cdot)$ is the natural logarithm function, and p is the total number of quanta available. The maximum entropy description is obvious in that $\alpha(f,\ln(f))$ is the same as comentropy or thermodynamic entropy.

Equations (18) can be solved either as an optimization problem by direct computer maximizing according to the constraints or by using Lagrange multipliers to derive closed-form equations [1]. In either case, the solution is obtained as the result of a sequence of iterations. See the indicated references for further details.

Our final example of an indirect technique of restoration is for the restoration of images subject to spatially varying point spread functions. If we relax the assumption of spatial invariance in image formation as associated with Eq. (1), then the imaging equation becomes

$$g(x,y) = \int_{-\infty}^{\infty} h(x,x_1,y,y_1)f(x_1,y_1)\, dx_1\, dy_1 + n(x,y), \tag{19}$$

where $s(\cdot)$ being linear and equal to unity is also assumed. Equation (19) is no longer in the form of a convolution, and the solution of this image restoration problem by the application of Fourier computations is not possible.

In what situations do equations such as Eq. (19) arise? Actually, imaging by Eq. (19), which represents *space-variant image formation,* is much more common than the space-invariant form seen in Eq. (1). For example, Eq. (1) applies only to regions near the optical axis of a well-designed optical system. As the distance in the image from the optical axis increases, space-invariant image formation breaks down, and the image becomes subject to a number of space-variant aberrations, for example, the appearance of coma. Equation (19) also arises when there is motion between object and image plane during the image exposure time, and the motion is not uniform. That is, motion that possesses acceleration in any form, such as accelerated linear motion or rotation about a center, produces an image formation equation like Eq. (19).

Our ability to solve such problems as described by Eq. (19) is limited to cases in which the exact manner of change of the image formation in space is known can be described analytically. For example, image formation in the presence of coma is known to possess an analytic description. That is, the actual form of the point spread function h is known as a function of the coordinates x,y_1, y,y_1. As another example consider imaging with accelerated linear motion from left to right during the period of image exposure, using a semiconductor push-broom image scanner that scans left to right. Because the sensor is moving faster at the right-hand side of the image than at the left-hand side, the effect can be described in one of two equivalent ways: Objects at the right are more blurred than objects at the left, or a pixel at the right-hand side of the image is larger than at the left. In either case, if the acceleration is known, it is possible to write an equation that will describe the extent to which the imaging geometry differs within the image.

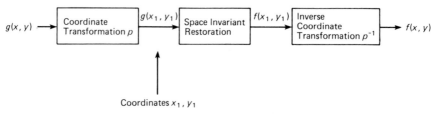

Coordinates x_1, y_1

Fig. 4.   Space-variant restoration.

The preceding example demonstrates a way of treating space-variant imaging.   If we know the manner in which the imaging system varies with space, then we may be able to find a spatial-coordinate transformation that converts the geometry of the spatially varying image system into a geometry with spatially invariant imaging.   In this new transformed geometry, the image can be restored by any of the techniques previously discussed.   Then another coordinate transformation is applied, which is the inverse of the first coordinate transform, in order to restore the geometry of the original object.

Figure 4 is a schematic diagram of this process of transformation and restoration.   For this process to be successful, the coordinate transformation must be invertible, that is, if $p(x,y,x_1,y_1)$ is the forward transformation, then $p^{-1}(x,y,x_1,y_1)$ must exist.

Space-variant restorations for which this method has been successfully applied represent some important real problems, such as optical coma and rotation about a fixed axis.   The difficulty of this method usually lies in obtaining an accurate description of the space-variant image formation process and then deriving the appropriate coordinate transformations [17].

IV. Identification of the Point Spread Function

In all the previous discussions it has been assumed that the point spread function is known.   Many situations arise, however, in which it is not known.   We suggest two methods of identifying the point spread function that have been successfully applied in a wide number of cases.

The first method is analysis of points and lines in the image.   In many cases an image will possess structures that can be identified as unresolved points or edges, that is, points or edges whose size is unresolved within the spatial extent of the point spread function of the image formation system.   Analysis of the images formed by such points and lines possesses direct information about the PSF.   For example, a point image, such as the image of a glint of sunlight off glass or metal, is the point spread function image by definition.   On the other hand, the image of an edge is the projection integral of the point spread function in the direction of the edge. In this latter case an additional assumption, such as rotational symmetry, is required to recover the PSF from the line spread image, which may be observed in the image itself. See [15], for example, for details or [1] for an example.

A second technique for identification of the point spread function is analysis of the power spectrum. As indicated before, there is a direct relation between the power spectra of image, object, and noise

$$S_g(\omega_1,\omega_2) = |H(\omega_1,\omega_2)|^2 S_f(\omega_1,\omega_2) + S_n(\omega_1,\omega_2). \tag{20}$$

If S_f and S_n are known, then at least the magnitude response of the optical transfer function can be computed. Often S_n can be inferred from a region of the image where the object structure is minimal, that is, the major contribution to the power spectrum in a region of constant intensity would be the noise. The object power spectrum will vary, of course, with the object. A useful course is to assume the first-order Markov image model discussed previously, making a fit to the correlation parameter, and evaluate S_f from the fit. The magnitude of H can then be estimated.

In many situations the image formation process will leave a characteristic signature in the power spectrum of the image. Consider images degraded by a focus blur. In this case a point is imaged as the cross section of the limiting pupil in the optical system. For example, a camera with a six-sided aperture stop images a point as a hexagon. The Fourier transform of this hexagon will also possess hexagonal symmetry, and the position of the zeroes in the hexagonal pattern is easy to observe in a display of the power spectrum as an image. It is possible to develop programs that will automatically analyze the zero patterns in a power spectrum and deduce the size and extent of a point spread function.

V. Assessment of Techniques

In these closing paragraphs we wish to give some general practical guidance to the individual who may be confronted with an image restoration problem.

The first practical comment is the following: If the image can be restored, then the chances are high that it can be restored with one of the simplest techniques, for example, inverse filter or Wiener filter. Thus, it is sound for the analyst to start out by trying an initial restoration using an inverse filter, possibly with hand-tuning of the inverse filter response if it becomes necessary to deal with noise amplification and ill-conditioned behavior in the neighborhood of a zero in the spectrum of the optical transfer function.

Usually this step of a first simple inverse restoration will furnish clues that will be useful in pursuing more complicated techniques of image restoration, for instance, clues as to how accurately the PSF is known, how well the SNR is known, or whether zeroes in the optical transfer function will be a problem.

If the simple inverse filter does not produce satisfactory results, the next step is to try a more sophisticated technique such as Wiener filtering or power spectral equalization. These techniques will require more extensive preparation before execution, for example, modeling and computation of the appropriate power spectral quantities. And if these techniques are not successful, the analyst can then decide the merits of further complexity, for example, an indirect technique such as discussed in Section III.

In the following, the author will make several assertions about the practical solution of image restoration problems, assertions that are offered with no proof but which have grown out of the author's experience in this area: If an image restoration problem can be solved, then about 75% of the time it can be treated with some of the simplest techniques, for example, inverse filter or Wiener filter. If the problem does not yield to these simpler techniques, there is a diminishing probability it will yeild to a more complex technique. Of the 25% of cases not solved by simpler techniques, perhaps only half of these will be amenable to solutions by a complex technique. There remains a core of problems that cannot be solved by any method of image restoration, problems associated with too much noise in the recorded image. A useful path of research would be the identification of images that cannot be restored by existing methods, thus making immediate the diagnosis of the prospects of success or failure *before* an investment in computing and manpower takes place.

VI. Bibliographical Notes

The utility of Fourier theory and the related concept of a linear system achieved great utility in classical analysis of linear circuits and filters. The applicability of this theory to imagery took place in parallel with applying it to linear circuits. Indeed, some early work in analyzing optical systems employed Fourier and linear systems theory prior to their appearance in circuit analysis. The books by Goodman, Gaskill, Papoulis, and Dainty and Shaw [3,5,7,15] contain references and bibliographic data that will make it possible for the interested reader to develop a historical and technical overview of this and closely related topics.

The discussion through most of this chapter focuses on the applicability of digital computers to the image restoration problem. However, the usage of optical methods to implement the relevant Fourier filters in image restoration is feasible for filters without singularities, such as the Wiener filter. The book by Goodman [7] contains simple examples and discussion of optical processors for imagery. It is important to note that optical methods usually require the preparation of an optical focal-plane mask, which encodes the filter information. The focal-plane mask is usually created by photographic technology, which leads to difficulties from the noise and nonlinear response of the photographic medium [3,5,7,15]. A continuing theme in optical processing of imagery is the search for spatial light modulators that have the information density of photographic film, without the nonlinear response, and which can be recycled, that is, written and read, erased, and reused [3]. The lack of recyclability in current modulators used for optical processing means that optical processors cannot be conveniently used for multiple pass or iterative restoration methods such as constrained least-squares [10].

It is a curious aspect of image restoration that although it was one of the first applications of digital computers to the processing of images [1,6], it has declined in relative importance to a number of other applications of digital image processing, such as data compression, feature extraction, and pattern recognition. The application of image restoration to images from NASA's remote space vehicles was the first success of digital image processing that was visible to the general public.

It is possible to obtain an incorrect impression about the extent to which image restoration can be carried out as an open-loop or closed-loop process. By open-loop we mean capable of being implemented without substantial human intervention; by closed-loop we mean by substantial human involvement. In reality, the restoration of images requires significant amounts of effort in human time and energy. A Wiener filter is easy to embody in a computer program, but to supply the proper operating parameters to the Wiener filter requires human ingenuity for power spectral modeling, estimation of SNRs, and estimation of the OTF of an optical system [1,9,10,16]. Increased complexity of an image restoration system further increases the investment of human effort required for successful operation. For example, an iterative method such as constrained least-squares or maximum entropy requires more sophistication on the part of the user in executing the computer programs that generate the solution [2,4,9,10,12,17]. The implementation of truly autonomous systems for image restoration will probably require integration of restoration filters with knowledge base or "expert" systems of artificial intelligence.

Acknowledgments

The author wishes to acknowledge the Fachgruppe Bildwissenschaft, Institute für Kommunikationstechnik, Eidgenösiche Technische Hochschule, Zürich, Switzerland for their support and extends special thanks to Professor Olaf Kübler for his friendship and support.

References

1. H. C. Andrews and B. R. Hunt, "Digital Image Restoration." Prentice-Hall, Englewood Cliffs, N.J., 1977.
2. T. M. Cannon, H. J. Trussell, and B. R. Hunt, A comparison of different image restoration methods, *Appl. Opt.* **17**, 1978, 2944–2951.
3. J. C. Dainty and R. Shaw, "Image Science." Academic Press, New York, 1976.
4. B. R. Frieden, Restoring with maximum likelihood and maximum entropy, *J. Opt. Soc. Am.* **62**, 1972, 511–518.
5. J. Gaskill, "Linear Systems, Fourier Transforms and Optics." Wiley, New York, 1978.
6. R. Gonzalez and P. Wintz, "Digital Image Processing." Addison-Wesley, New York, 1977.
7. J. W. Goodman, "An Introduction to Fourier Optics." McGraw-Hill, New York, 1968.
8. B. R. Hunt, Data structures and computational organization in digital image enhancement, *IEEE Proc.* **60**, 1972, 884–887.
9. B. R. Hunt, Deconvolution of linear systems by constrained regression and its relationship to the Wiener theory, *IEEE Trans. Autom. Control* **AC-17**, 1972, 703–705.
10. B. R. Hunt, The application of constrained least-squares estimation to image restoration by digital computer, *IEEE Trans. Comput.* **C-22**, 1973, 805–812.
11. B. R. Hunt, Nonstationary assumptions for Gaussian models of images, *IEEE Trans. Syst. Man Cybernet.* **SMC-6**, 1976, 876–882.
12. B. R. Hunt, Bayesian methods in nonlinear restoration of images by digital computer, *IEEE Trans. Comput.* **C-26**, 1977, 219–229.
13. B. R. Hunt, Report to the U.S. Congress Select Committee on Assassinations, Washington, D.C., 1978.
14. E. O'Neill, "Introduction to Statistical Optics." Addison-Wesley, New York, 1965.
15. A. Papoulis, "Systems and Transportation with Applications in Optics." McGraw-Hill, New York, 1968.
16. W. K. Pratt, "Digital Image Processing." Wiley, New York, 1978.
17. A. A. Sawchuck, Space-variant image motion degradation and restoration, *Proc. IEEE* **60**, 1972, 854–861.

3 Image Detection and Estimation

John W. Woods

Electrical, Computer, and Systems Engineering Department
Rensselaer Polytechnic Institute
Troy, New York

I. Introduction

This chapter treats the problems of detection and estimation on two-dimensional data using a stochastic or communication-theoretic approach. Specifically, we shall look at three idealized problems: the detection of known objects, the detection of random objects, and the estimation of the location of random curves. The word *object* here means a two-dimensional function that is concentrated in space, that is, it has significant energy over only a small part of the observation field. The first problem is to decide on the presence or absence of a known object. We then extend this problem to the case of random or unknown locations and present an approximate solution. The second problem concerns the detection of random objects. The objects are assumed to have significant random attributes, for example,

DIGITAL IMAGE PROCESSING TECHNIQUES

random amplitude, radius, velocity, shape, texture, and orientation. One can model the random object as an inhomogeneous random field with a known mean value for detection purposes.

Whereas these two idealized problems deal with the practical problem, loosely categorized as detecting blobs in images, a third idealized problem concerns estimation of the location of a random curve or boundary. The curve may close itself or may be open and extend over a large region. Methods based on sequential maximization of the a posteriori probability have been developed to deal with such problems. We assume that the curve is known to be present in the data, thus the only uncertainty is the precise location of the curve. This third problem thus differs from the first two problems, in which the presence of the object is uncertain. A real problem will most likely be a combination of these types.

Applications of the first, or known-object, problem are referred to as template matching [23] in the image processing literature, because the solution involves a two-dimensional version of the well-known matched filter [20,28]. Applications of template matching include machine recognition of typed or printed characters and automated visual inspection of IC chips [15,18]. Other applications are more closely related to the original problem formulation involving known signals in additive noise, for example, detecting objects in noisy IR images, low photon x-ray images, and radar images. We expect the technique to be more appropriate there. The random-object problem is often the better model when there are significant unknown aspects to the signal besides location, for example, locating vehicles in FLIR images [4] (where random orientation is a problem); and when detecting ocean eddies in satellite altimetry data [29], in which there is unknown amplitude, radius, and velocity. The third, or random curve problem, can be used to model the lung and cardiac outline in chest x rays, in which the noiseness of the image is inversely proportional to the radiation dose to the patient. Other potential applications include image segmentation; location of roads and linear geologic features from aerial reconnaissance data; and estimation of the precise location of oceanic fronts, such as the Gulf Stream, using satellite altimetry data.

This chapter contains one section on each of the preceding three idealized problems. Section II on the known-object problem treats the white noise case, the colored noise case with use of a noise-whitening filter, and the case of unknown location of an object using the approximate method of detection windows. Section III treats the detection of random objects with an estimator-correlator algorithm derived using a two-dimensional innovations sequence. First, the detector is derived; then, based on the use of AR signal models, the reduced update Kalman filter is suggested as an efficient signal estimator. Section IV discusses the estimation of the location of random curves using the maximum a posteriori probability approach of Cooper [4]. Two algorithms are presented, a simple ripple filter that achieves a local optimum of the likelihood function and a computationally complex sequential boundary finder that attempts to locate a global maximum of the likelihood function over a swath of data centered on the true boundary or curve. In each of these sections we shall attempt to describe the limitations of the techniques and present a realistic picture of how well they work.

II. Detecting Known Objects

Our approach to the solution of this problem will be communication-theoretic in nature. To simplify matters and at the same time present a result of wide applicability, we assume that the object is immersed in homogeneous (stationary) Gaussian noise in an additive manner. We initially assume that the noise is white and then relax this assumption through the use of a noise-whitening filter derived by means of spectral factorization of the noise covariance. The basic results are thus two-dimensional extensions of known matched filter results of the one-dimensional case [28]. This result is known in a two-dimensional context in the area of pattern recognition as template matching and has been used in a wide variety of image applications, many completely unrelated to the original problem formulation for which it was optimal. In some of these applications, such as character recognition, the background noise is mostly unwanted signals. In such cases the assumption of Gaussian background noise is clearly inappropriate, and we lose the optimality of the matched filter. On the other hand, if the background noise is the main problem, the known signal is additive to this noise, and the noise is nearly normally distributed; then the two-dimensional matched filter is a robust nearly optimal detector.

In applications the location is usually unknown, and part of "detection" is to estimate the unknown location. If the object were known to be present, then the peak of the matched filter output would constitute a maximum-likelihood estimate of the object's location [20,28]. However, if the presence of the object is not known, then we can obtain a generalized likelihood ratio test by first maximizing the likelihood over a two-dimensional shift parameter to estimate the location and then comparing the resulting generalized likelihood ratio to a threshold, thereby deciding the presence or absence of the signal. Though this is a common approach, one can encounter some problems, as discussed in Section 7 of [4]. See also Section III of this chapter.

If there is more than one object in a given data field, the previous approach can be modified by employing a detection window strategy. In this widely used but suboptimal approach, the known-signal problem is repetitively solved for each placement of a sliding or hopping detection window. We shall look at this practical approach in Section II.C.

A. White Noise Case

Assume an observation field $r(n_1, n_2)$ of size $N_1 \times N_2$ containing white Gaussian noise of mean zero and variance σ_w^2. Under Hypothesis 1 there is a known signal additively composed with the white noise. Under Hypothesis 0 there is white noise alone. In symbols we can write

$$r(n_1, n_2) = \begin{cases} w(n_1, n_2) & :H_0 \\ s(n_1, n_2) + w(n_1, n_2) & :H_1 \end{cases} \tag{1}$$

We assume a priori that Hypothesis 1 has probability P_1 and Hypothesis 0 has probability P_0. Then the well-known optimum Bayes statistic, which is associated with minimum probability of error, is

$$l = \sum_{n_1, n_2} s(n_1, n_2) r(n_1, n_2) - \frac{1}{2} \sum_{n_1, n_2} s^2(n_1, n_2). \tag{2}$$

We compare it to a threshold $l_0 \triangleq \sigma_w^2 \ln(P_0/P_1)$ to decide the presence or absence of the signal. This is no different from the one-dimensional case as presented in [20,28]. In Eq. (2) s is completely known, so it is convenient to redefine the likelihood statistic as

$$\tilde{l} = \Sigma s(n_1, n_2) r(n_1, n_2), \tag{3}$$

and to incorporate $\frac{1}{2}\Sigma s^2(n_1, n_2)$ into a modified threshold $\tilde{l}_0 = l_0 + \frac{1}{2}\Sigma s^2$. We note that in the a priori equally likely case, the threshold l_0 is half the energy in the signal.

The procedure just described can be extended to detect optimally any number M of known signals in white Gaussian noise. Because the signals will generally have different energies, it will be necessary to use Eq. (2) rather than Eq. (3) to correct properly for this variation in signal energy. The optimal decision is to choose the signal whose likelihood is largest (in the equal probability case). This is a better approach for character recognition.

The matched filter can also be derived by maximizing a filter's output SNR, as done in [23], for example. A shortcoming of this approach, however, is that it does not obtain the bias correcting energy term $\frac{1}{2}\Sigma s^2(n_1, n_2)$. In fact, it is stated in [23] that the matched filter thus obtained is more

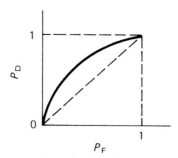

Fig. 1. Generic ROC.

sensitive to signal energy than to signal shape. The advice given there to correct the problem is to use a so-called derivative matched filter to emphasize the high-frequency or edge influence on the detection.

By varying the threshold $\tilde{l}_0$, we can get the continuum of optimum probability of detection P_D versus probability of false alarm P_F points following the Neyman–Pearson approach [20,28]. When $\tilde{l}$ is very small (i.e., near $-\infty$), the resulting $P_D \simeq 1$ and $P_F \simeq 1$. As the threshold is raised, P_D and P_F monotonically decrease until $P_D \simeq 0$ and $P_F \simeq 0$. Thus, we can in principle plot P_D versus P_F as a function of the parameter $\tilde{l}_0$ to summarize detector performance. This locus of probability pairs is known as the receiver operating characteristic (ROC) and is well known in one-dimensional detection theory. The line $P_D = P_F$ on an ROC plot can be obtained by flipping a biased coin and is thus indicative of no detection improvement. When the matched filter is appropriate to the problem at hand, the resulting ROC curve should always be above and to the left of the line $P_D = P_F$, as shown in Fig. 1.

Example 1 Here we present an example of the matched filter being used to detect the presence of a simulated ocean ring of known size, amplitude, and location. Ocean rings or eddies are areas of increased or decreased average height on the ocean surface. They can be nearly circular with heights of about one meter or less and radii of 50 to 100 kilometers. In this calculation the eddy is modeled as a known signal described by a raised and truncated J_0 Bessel function [29]. We assume that the noise, composed of a random surface height profile, is spatially white. This assumption will be modified in the next section, in which we treat colored noise. The ROC for a 50-km eddy is shown below as Fig. 2. The signal amplitude is 0.35 cm, and the noise power is 0.0751 cm^2. The sampling is 21 km along track and 45 km across track.

The implementation of the matched filter is very simple. It may be

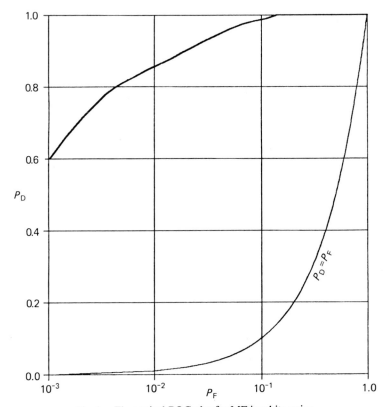

Fig. 2. Theoretical ROC plot for MF in white noise.

thought of as the dot product of two $N_1 \times N_2$ image matrices:

$$\mathbf{S} = \{s(n_1, n_2)\},$$

$$\mathbf{R} = \{r(n_1, n_2)\},$$

$$\mathbf{S} \cdot \mathbf{R} = \sum_{n_1, n_2} s(n_1, n_2) \cdot r(n_1, n_2).$$

In practice the noise does not have to be white over the full band; its power spectral density must only be approximately flat over the signal band. If this is not the case, a noise-whitening filter will be necessary, as discussed in the next section.

B. Colored Noise Case

The preciding two-dimensional matched filtering approach can be extended to the case of colored Gaussian noise by the use of a two-dimensional whitening filter [8]. Let the noise power spectral density satisfy $\epsilon \leq S_n(\omega_1,\omega_2) \leq M$ for some $\epsilon > 0$ and $M < \infty$, with the noise correlation function $R_n(n_1,n_2) = \text{IFT}[S_n(\omega_1,\omega_2)$ satisfying $\Sigma|R_n(n_1,n_2)| < \infty$. Then the power density function $S_n(\omega_1,\omega_2)$ can be factored into the product

$$S_n(\omega_1,\omega_2) = \sigma^2|B_{\oplus+}(\omega_1,\omega_2)|^2, \tag{4}$$

where $B_{\oplus+}$ is a stable two-dimenstional filter that is causal in the nonsymmetric half-plane (NSHP) sense and has a stable and causal inverse [9]. Thus $B_{\oplus+}^{-1}$ can be used as a causal and causally invertible whitening filter to transform the given problem into an equivalent white noise problem,

$$\tilde{r}(n_1,n_2) = \begin{cases} \tilde{s}(n_1,n_2) + w(n_1,n_2) & :H_1 \\ w(n_1,n_2) & :H_0 \end{cases}, \tag{5}$$

where $\tilde{s}(n_1,n_2)$ is the signal $s(n_1,n_2)$ after transformation by the noise-whitening filter $B_{\oplus+}^{-1}(w_1,w_2)$. Also, the variance of the white noise $w(n_1,n_2)$ is σ^2, the parameter defined by the spectral factorization. This transformation is shown diagramatically in Fig. 3.

It has been known for some time that the whitening filter $B_{\oplus+}^{-1}$ will almost always be infinite order (i.e., nonrational) even if the noise power spectral density is rational; thus, some approximation stage is necessary between a spectral modeling stage and a practical whitening filter. One possibility is the window approximation method suggested in [9], or alternatively, one can use any general two-dimensional filter design algorithm to design a rational filter to approximate the ideal spectral factor $B_{\oplus+}^{-1}(w_1,w_2)$ in the magnitude and phase sense, as was done in [33].

In [7] it is shown that the least-squares inverse polynomials obtained in [3] with respect to a correlation matrix constructed from $R_n(n_1,n_2)$ converge to the previous spectral factor as the NSHP order tends to (∞,∞). In fact, this least-squares inverse procedure is equivalent to solving the linear least mean square error predictor problem with NSHP support for the correlation function $R_n(n_1,n_2)$. This linear prediction whitening filter is not guaranteed to have a stable inverse [30] and will not have the correlation matching property. In practice, however, the linear prediction whitening filter will tend to have a stable inverse as its order increases in both dimensions, although this is not guaranteed. Also, one may need to add a little white noise to the original data as a regularization, especially if the filter order is too low. Intuitively, one needs the predictor order large enough to decorrelate the data well, thus producing a fairly white residual. Then the filter

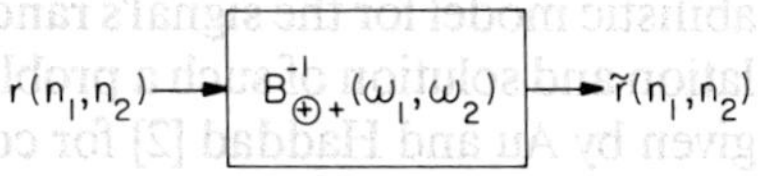

$$r(n_1,n_2) \longrightarrow \boxed{B_{\oplus+}^{-1}(\omega_1,\omega_2)} \longrightarrow \tilde{r}(n_1,n_2)$$

Fig. 3. Two-dimensional noise-whitening filter.

should have a stable inverse; and even more importantly, it will be a good inverse filter to whiten the colored noise.

One problem with this approach is that the ideal whitening filter will spread out the signal. However, the approximate support (say, 95% of the energy) has been found to be only modestly increased ($\approx 20\%$) in several practical problems. Truncation of the whitened signal can then be safely performed over this modestly increased support.

Example 2 This example concerns the use of the matched filter to detect an eddy in colored noise generated using an isotropic signal model and an approximate low-pass noise model [29]. The sample spacing is 21 km along track and 45 km across track in this simulation of satellite altimetry data. A 3×3 order NSHP whitening filter was derived by the method of linear prediction and applied both to the data and to the raised J_0 signal. Then the matched filtering was performed on the noise-whitened data. Figure 4 shows an empirically determined ROC for this case. This ROC, although determined by simulation, is comparable to that of Example 1, which was determined analytically. Evidently, the coloring of the noise did not change the performance significantly. Examination of the signal and noise spectra has revealed the reason for this unusual occurrence. Because the overall noise power is not changed, the colored noise has more power at low frequencies and less power at high frequencies than the white noise. The overall effect on signal detectability has approximately balanced out in this case.

C. Random Location

We now assume that the signal is known up to its location or shift, which is unknown. The optimal solution in this case is not the matched filter. The optimal detector of one signal with random unknown parameters involves a composite likelihood ratio [20,28] that averages over the a priori distribution of the random location. In fact, this is a nonlinear average of the outputs of an infinite bank of matched filters, each matched to the signal at a possible shift. The detector output will not give the found object's location in the possibly large data field.

An additional problem in practice is that not only is the possible signal's location unknown, but also there is a random number of such signals present. In this case, an optimal detector needs to have specified the probabilistic model for the signal's random occurrence. A near-optimal formulation and solution of such a problem in the one-dimensional case has been given by Au and Haddad [2] for continuous time and an exponential signal

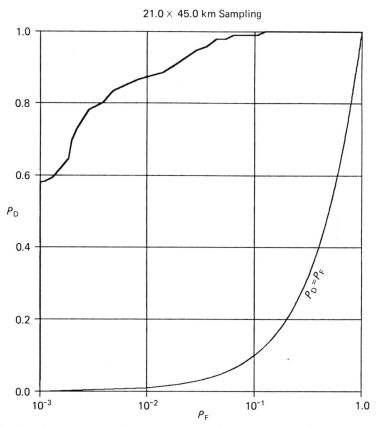

Fig. 4. Experimental ROC plot for MF in colored noise. Detectorsize is three tracks of three samples; 96 detectors contain rings, 15960 do not.

interarrival density. Their approach is computationally intensive and has not yet been extended to the two-dimensional case.

The usual practical method to deal with the random location problem is the suboptimal one of using a detection window and sliding or hopping it along the data field. For each window location, the known-signal hypothesis problem of the previous section is solved. The detector thus produces a binary-valued field of decisions indicating a decision on the presence or absence of the signal at that location. One can then utilize a false alarm rate per unit area to describe the detector performance at a given level of P_D. For this detection window approach to succeed, the shift of the window must be small with respect to structure in the signal; otherwise, there will be significant detection loss due to the residual signal mismatch. Also, the size of the detection window should be large enough to include the approximate signal support after noise whitening. Because the detector is basically working with the signal's energy, we want at least 90% of the energy in the detection window. These two considerations set a computational requirement.

An indication of the nonoptimality of the window technique is that there will often be many false alarms clustered about a true detection. One can deal with this artifact by processing the clusters to retain the maximum likelihood in the cluster and reject the other, presumably false, detections. If strong signals are widely space, this approach will be sufficient, but clearly we shall miss some closely spaced targets. One possible solution for this problem is to subtract off signals based on prior detections, as is done, for example, in [2].

Example 3   We consider the same raised J_0 signal model as before, but its location is now randomly distributed in space. We use a 3×3 sliding window detector that tests for signal presence in its center element. Because the pixel spacing is 21 km along track and 45 km across track, there is considerable uncertainty in the location of the 50-km radius ocean ring. We ignore false alarms due to neighboring rings on the assumption that the resulting cluster of detections will not be a problem. Effectively, we have the case of Example 2 with the addition of random signal location. The matched filter performance thus degrades as shown in Fig. 5 as compared with Fig. 4.

D. Additional Random Parameters

Often in image processing, the object has not only unknown location but also unknown orientation, amplitude, size, or any combination of these. The best approach is based on a composite likelihood ratio computed with each detection window, $\mathcal{L} = \int p(\Theta) \exp l(\Theta) \, d\Theta$. Here $l(\Theta)$ is the likelihood given by Eq. (2), preceded by a noise-whitening filter if necessary, based on the signal parameter vector Θ, that is $s(n_1,n_2)$ is replaced by $s(n_1,n_2; \Theta)$. Practically, the above can be approximated by K terms, $\mathcal{L} \simeq \Sigma_{k=1}^{K} P(\Theta_k) \exp l(\Theta_k)$, where $P(\Theta_k) \triangleq \int_{\Theta_k} p(\Theta) \, d\Theta$. This is just a pointwise nonlinear combination of K two-dimensional matched filter outputs. The choice of K depends on how much $l(\Theta)$ varies with changes in Θ. Our experience has been that good performance can be obtained in this way for reasonable values of K, such as $8 \leq K \leq 16$. However, this depends on many things including the resolution of the data. In general, higher data resolution will require a larger value of K. Note that K can grow very quickly, because four orientations, two sizes, and two amplitudes already give us 16 values for Θ. If we also must locate the object within the detection window, we shall have

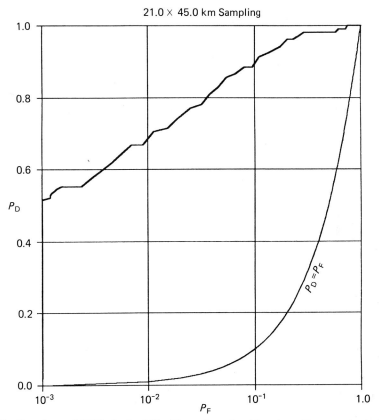

Fig. 5. Experimental ROC plot for MF with random signal location. Detector size is three tracks of three samples; 105 detectors contain rings, 15945 do not.

$K = 64$ with just four candidate locations. One should note, however, that the matched filters corresponding to different signal amplitudes are simply related, so we only need to compute one directly. Still, if the required K is too large or the randomness in the signal is not simply parametrized, one has recourse to an estimator – correlator approach (see III.A.) that can be viewed as a generalization of the simple two-dimensional matched filter.

III. Detecting Random Objects

This section treats the problem of detecting a random signal. We assume that this randomness goes beyond the unknown parameter vector problem of the previous section, so that no simple extension of the matched filter method will serve for this purpose. Remarkably, what does happen is that the matched filter generalizes into an estimator – correlator structure for the case of white Gaussian observation noise. In this structure, it is as though the known signal has been replaced by its minimum mean square error (MMSE) estimate under Hypothesis 1. This method reduces the problem to one of developing approximate MMSE estimators to use in the estimator – correlator detector and was motivated by the continuous time, one-dimensional results of Kailath in [14]. The approach is easily extended to the colored noise case through use of a noise-whitening filter, as previously described.

We first derive the estimator – correlator detector using the two-dimensional innovations approach [26]. Then we look at the two-dimensional Kalman recursive estimator. This is followed by a discussion of the problems of a moving window detector implementation. We then present some examples of estimator – correlator detector performance. Finally, there is a short presentation on the use of nonlinear estimators.

A. Estimator – Correlator Detector

We initially consider the following problem: On Hypothesis 1, a random signal is present in homogeneous white Gaussian noise. Under Hypothesis 0, only the noise is present. We assume that the signal is also Gaussian and independent of the noise. The signal mean value is a specified nonzero function $m_s(n_1, n_2)$. We thus have the following detection problem:

$$r(n_1, n_2) = \begin{cases} s(n_1, n_2) + w(n_1, n_2) & :H_1 \\ w(n_1, n_2) & :H_0 \end{cases}.$$

We assume that r is observed over the rectangle $[0, N_1 - 1] \times [0, N_2 - 1]$, although the shape of the region is not restrictive. We also assume sequential, causal processing in a line-by-line manner; thus, we are lead to compute the two-dimensional innovations of $r(n_1, n_2)$ under each hypothesis,

$$v_i(n_1, n_2) \triangleq r(n_1, n_2) - E[r(n_1, n_2) | r \text{ on } \mathcal{P}_{\oplus+}(n_1, n_2), H_i],$$

where $i = 0,1$. Here $\mathcal{P}_{\oplus+}(n_1, n_2)$ denotes the past at (n_1, n_2) and consists of the points $(k_1, k_2) \in \{k_1 < n_1, k_2 = n\} \cup \{k_1, k_2 < n_2\}$. The innovation $v_1(n_1, n_2)$ can be calculated from a causal predictor. Because the noise is white, the innovation v_0 is just r itself. The calculation of the innovation sequences is thus seen to be causal and causally invertible, so we can use these sequences equivalently for detection purposes. The likelihood ratio is then seen to be equivalent to

$$\exp\left[-\frac{1}{2}\sum \frac{v_1^2(n_1, n_2)}{\sigma_1^2}\right] \Big/ \exp\left[-\frac{1}{2}\sum \frac{v_0^2(n_1, n_2)}{\sigma_0^2}\right],$$

where $\sigma_i^2 \triangleq E[v_i^2 | H_i]$ and $E[v_i] = 0$. It should be noted that near the boundaries, or in the case of an inhomogeneous signal model, σ_1^2 will not be constant but will depend on position. However, away from the boundaries and when the signal model is homogeneous, it has been experimentally observed that σ_1^2 will be approximately constant at its steady-state value.

On defining $\hat{r}_p \triangleq E[r(n_2,n_2) | r$ on $\mathcal{P}_{\oplus+}(n_1,n_2), H_1]$ and noting $\hat{s}_p = \hat{r}_p$ because of the white noise assumption, we have

$$v_i(n_1,n_2) = \begin{cases} r(n_1,n_2) - \hat{s}_p(n_1,n_2) & i = 1 \\ r(n_1,n_2) & i = 0 \end{cases},$$

where $\hat{s}_p(n_1,n_2)$ is the (1,0) step MMSE prediction estimate, of the signal under Hypothesis 1. For this reason it is sometimes called a pseudoestimate. If we ignore boundary effects and assume that σ_1^2 is constant, we can write the likelihood ratio as equivalent to

$$\tilde{l} = \frac{1}{2}\sum \left\{ \frac{r^2(n_1,n_2)}{\sigma_0^2} - \frac{[r(n_1,n_2) - \hat{s}_p(n_1,n_2)]^2}{\sigma_1^2} \right\}$$

$$= \frac{1}{\sigma_1^2}\left[\sum \hat{s}_p(n_1,n_2)r(n_1,n_2) - \frac{1}{2}\hat{s}_p^2(n_1,n_2) \right] + \frac{1}{2}\sum \left(\frac{\sigma_1^2 - \sigma_0^2}{\sigma_1^2\sigma_0^2} \right)r(n_1,n_2). \quad (6)$$

If this discrete-space problem arose from sampling a continuous-space field consisting of a smooth signal in spatially white noise, then the sampling must be done with an integral averager. If the area of this averager is A, then asymptotically the discrete-space signal variance will be proportional to A^2 and the noise variance to A. Now the innovation $v_1 = r - \hat{r}_p = r - \hat{s}_p = (s - \hat{s}_p) + w$, and noting that $s - \hat{s}_p$ and w are independent, we obtain $\sigma_1^2 = E[(s - \hat{s}_p)^2] + \sigma_w^2$. Because $E[(s - \hat{s}_p)^2] \le E[s^2]$, if the sample spacing is dense enough, we obtain $\sigma_1^2 \approx \sigma_0^2 = \sigma_w^2$, the variance of the white noise. In this case,

$$\tilde{l} \approx \frac{1}{\sigma_w^2}\left[\sum \hat{s}_p(n_1,n_2)r(n_1,n_2) - \frac{1}{2}\sum \hat{s}_p^2(n_1,n_2) \right]. \quad (7)$$

Comparing Eqs. (7) and (2), we see that the unknown signal is replaced by an estimate, hence the name estimator–correlator for the statistic. This is exactly analogous to the one-dimensional case treated in [24]. To calculate this estimator–correlator likelihood, we need an MMSE (1,0) step predictor $\hat{s}_p(n_1,n_2)$. This can be obtained by the two-dimensional Kalman filter [34] if the signal model is of the NSHP type and AR. If the signal is ARMA, then the method of [31] can be used. If the region $[0,N_1 - 1] \times [0,N_2 - 1]$ is very large, the innovations sequences can be approximated by the outputs of two whitening filters G_0 and G_1 derived using spectral factorization,

$$G_0(\omega_1,\omega_2) = 1 \quad (8a)$$

and

$$G_1(\omega_1,\omega_2) = B_{\oplus+}^{-1}(\omega_1,\omega_2), \tag{8b}$$

where $S_r(\omega_1,\omega_2) = \sigma^2|B_{\oplus+}(\omega_1,\omega_2)|^2$. Because the noise is white, the spectrum of r under Hypothesis 0 is just σ_w^2, so we choose $B_{\oplus+}^{-1} = 1$ and $\sigma_0 = \sigma_w$. Under Hypothesis 1, the spectrum is $S_r = S_s + S_w$, so we choose $G_1 = B_{\oplus+}^{-1}$ and $\sigma_1 = \sigma$.

A third possibility is to use the causal Wiener filter approach of [8] to get the (1,0) step MMSE predictor. However, this (as well as the second approach) would yield some optimality near the boundary and would be restricted to signal models that are covariance homogeneous. Also, one could use the method of [26].

i. Colored Noise Case

In the case of colored noise, we can preceed the estimator–correlator detector by a noise-whitening filter. The overall system is shown in Fig. 6. Effectively, the signal spectrum is altered by the transmission through the noise-whitening filter. The signal model for the estimator should be designed to match this noise-whitened signal spectrum.

ii. Noncausal Representation

The preceding causal estimator has the advantage that the likelihood can be calculated recursively as the data are being received, in a line-by-line manner. If this is not an issue, the unrealizable estimator can be employed in a noncausal representation for the likelihood ratio. If we order the observed data onto the vector $\mathbf{r}$ and write the corresponding signal and noise vectors $\mathbf{s}$ and $\mathbf{w}$, respectively, then the likelihood ratio is well known to be equivalent to

$$\exp - (1/2)\,\mathbf{r}^T(\mathbf{R}_s + \sigma_w^2\mathbf{I})^{-1}\mathbf{r}/\exp - (1/2\sigma_w^2)\mathbf{r}^T\mathbf{r},$$

which in turn is equivalent to $\tilde{l} = \mathbf{r}^T[(\mathbf{R}_s + \sigma_w^2\mathbf{I})^{-1} - \sigma_w^{-2}\mathbf{I}]\mathbf{r}$, which after some matrix algebra becomes $\tilde{l} = \mathbf{r}^T[\mathbf{R}_s(\mathbf{R}_s + \sigma_w^2\mathbf{I})^{-1}]\mathbf{r}$, where we recognize the

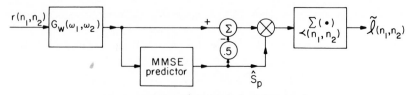

Fig. 6. Estimator–Correlator for colored noise.

well-known best linear estimate, $E[s\,|\,\mathbf{r}] = \hat{\mathbf{s}}_u$. Thus we have $\tilde{l} = \mathbf{r}^T\hat{\mathbf{s}}_u$, which can be put into two-dimensional notation as

$$\tilde{l} = \sum \hat{\mathbf{s}}_u(n_1,n_2)\mathbf{r}(n_1,n_2). \tag{9}$$

B. Kalman Recursive Estimator

This section will review some of the important features of the two-dimensional reduced update Kalman filter (RUKF) derived in [34]. The notation is that used in [34] and [9].

The signal model is Markov and is given as an NSHP recursion,

$$s(n_1,n_2) = \sum_{\mathcal{R}_{\oplus+}} c_{k_1,k_2}\, s(n_1 - k_1,n_2 - k_2) + v(n_1,n_2), \tag{10}$$

where $v(n_1,n_2)$ is a white Gaussian noise field. Also, $\mathcal{R}_{\oplus+} = \{k_1 \geq 0, k_2 \geq 0\} \cup \{k_1 < 0, l_1 > 0\}$ and $c_{00} = 0$. The model can be thought of as a two-dimensional recursive filtering of the input field v. We assume this filter is $(M_1 \times M_2)$th order. We assume that the signal mean value is zero. The case of nonzero mean will be treated later.

The observation equation is

$$r(n_1,n_2) = s(n_1,n_2) + w(n_1,n_2), \tag{11}$$

where w is a white Gaussian source. The model and observation equations can be written in vector form as

$$\mathbf{s}(n_1,n_2) = \mathbf{C}\mathbf{s}(n_1 - 1, n_2) + \mathbf{v}(n_1,n_2) \tag{12}$$

and

$$r(n_1,n_2) = \mathbf{h}^T\mathbf{s}(n_1,n_2) + w(n_1,n_2), \tag{13}$$

with $\mathbf{h}^T = (1,0,\ldots,0)$ and

$$\mathbf{s}(n_1,n_2) \triangleq [s(n_1,n_2), s(n_1 - 1,n_2), \ldots, s(0,n_2);$$
$$s(N_1 - 1,n_2 - 1), \ldots, s(0,n_2 - 1); \ldots;$$
$$s(N_1 - 1,n_2 - M_2), \ldots, s(n_1 - M_1 + 1,n_2 - M_2)]^T. \tag{14}$$

We solve these equations in the rectangular region with known deterministic boundary values on the top and sides. For convenience these boundary values are taken to be zero. If the known boundary values were nonzero, the filtering equations could be straightforwardly modified to account for this deterministic input. Then through the assumed line-by-line scanning we convert this two-dimensional problem into an equivalent one-dimensional problem with global state vector $\mathbf{s}$. When we speak of causal filtering, we

refer to this ordering. The Kalman filtering equations with the above interpretation of the **s** vector can immediately be written. The difficulty with
these equations is the amount of computations and memory requirements
associated with them. By limiting the update procedure only to those
elements near the present point, the wasteful computations can be avoided,
thereby increasing the efficiency of the algorithm [34]. A reduced update
constraint is then applied to update only the local state vector,

$$\mathbf{s}_1(n_1,n_2) \triangleq [s(n_1,n_2), s(n_1-1,n_2), \dots, s(n_1-M_1+1,n_2);$$

$$s(n_1+M_1+1,n_2-1), \dots, s(n_1-M_1+1,n_2-1); \dots;$$

$$s(n_1+M_1+1,n_2-M_2), \dots, s(n_1-M_1+1,n_2-M_2)]^{\mathrm{T}}. \quad (15)$$

Also, we define $\mathbf{s}_2(n_1,n_2)$ to be the remainder of $s(n_1,n_2)$. The resulting
assignment of the points is as shown in Fig. 7.

Given this constraint, the gains are then recursively pointwise optimized
in [34] to produce the RUKF equations corresponding to the scalar signal
and observation models given by Eqs. (10) and (11). The resulting recursive
filter equations, when put into scalar form, are

$$\hat{s}_b^{(n_1,n_2)}(n_1,n_2) = \sum_{\mathcal{R}_{\oplus+}} c_{k_1 k_2} \hat{s}_a^{(n_1-1,n_2)}(n_1-k_1,n_2-k_2) \quad (16)$$

$$\hat{s}_a^{(n_1,n_2)}(i_1,i_2) = \hat{s}_b^{(n_1,n_2)}(i_1,i_2) + k^{(n_1,n_2)}(n_1-i_1,n_2-i_2)$$

$$\cdot [r(n_1,n_2) - \hat{s}_b^{(n_1,n_2)}(n_1,n_2)]; \qquad (i_1,i_2)\epsilon\mathcal{R}_{\oplus+}^{(n_1,n_2)}. \quad (17)$$

The error covariance and gain equations are

$$R_b^{(n_1,n_2)}(n_1,n_2;k_1,k_2) = \sum_{\mathcal{R}_{\oplus+}} c_{l_1,l_2} R_a^{(n_1,n_2)}(n_1-l_1,n_2-l_2;k_1,k_2);$$

$$(k_1,k_2)\epsilon\mathcal{S}_{\oplus+}^{(n_1,n_2)}, \quad (18)$$

$$R_b^{(n_1,n_2)}(n_1,n_2;n_1,n_2) = \sum_{\mathcal{R}_{\oplus+}} c_{l_1,l_2} R_b^{(n_1,n_2)}(n_1,n_2;n_1-l_1,n_2-l_2) + \sigma_v^2, \quad (19)$$

$$k^{(n_1,n_2)}(i_1,i_2) = R^{(n_1,n_2)}(n_1,n_2;n_1-i_1,n_2-i_2)/$$

$$(R_b^{(n_1,n_2)}(n_1,n_2;n_1,n_2) + \sigma_w^2); \qquad (i_1,i_2)\epsilon\mathcal{R}_{\oplus+}, \quad (20)$$

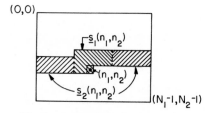

Fig. 7. Assignment of **s** to $\mathbf{s}_1$ and $\mathbf{s}_2$.

$$R_a^{(n_1,n_2)}(i_1,i_2;k_1,k_2) = R_b^{(n_1,n_2)}(i_1,i_2;k_1,k_2)$$
$$-k^{(n_1,n_2)}(n_1-i_1,n_2-i_2)R^{(n_1,n_2)}(n_1,n_2;k_1,k_2);$$
$$(i_1,i_2)\epsilon\mathcal{R}_{\oplus+}^{(n_1,n_2)}, \quad (k_1,k_2)\epsilon\mathcal{S}_{\oplus+}^{(n_1,n_2)}. \quad (21)$$

The superscript in these equations represents the step in the processing or filtering, and the argument represents the position in the data field. The approximate reduced update equations are obtained by replacing the region $\mathcal{S}$ in Eqs. (18) and (21) by a much smaller $\mathcal{T}$ of fixed size independent of the image size $M_1 \times N_2$. This is shown in Fig. 8. This approximation assumes negligible source correlation at large distances and thus requires a stable dynamical model. The approximation works well in practice for the many models tested. Region $\mathcal{T}$ is typically chosen as a few columns wider then the updated region.

The RUKF can provide a fixed-delay smoothed estimate that has been experimentally found to approximate $\hat{s}_u$ for use in the noncausal likelihood, Eq. (9), or we can use the (1,0) step predicted estimate $\hat{s}_p$ in the causal likelihood, Eq. (6), without any approximation. We have not observed a distinct advantage for either formulation in practice.

Because of the approximations in the derivation of the approximate RUKF, that is, replacing $\mathcal{S}$ by $\mathcal{T}$ in the nonlinear error covariance equations, instabilities can develop in the solution procedure. This has been a problem when the initial error covariances on the boundary were far from the steady-state values. This effect can be ameliorated by combinations of the following: increasing the size of the covariance update region $\mathcal{T}$, using a window function to taper the update, and windowing the initial error covariances. See [32] for more information on boundary conditions.

i. Signal Mean

A spatially varying signal mean can be incorporated in the above zero mean estimate in a straightforward manner. One simply subtracts the known mean prior to processing with the RUKF and then adds it back at the output as shown in Fig. 9. Thus, in the limiting case of a purely determinis-

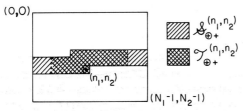

Fig. 8. $\mathcal{T}$ and $\mathcal{S}$ regions.

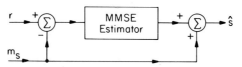

Fig. 9.   Linear estimation with nonzero mean.

tic signal, the estimator–correlator degenerates to the simple matched filter. The spatial variation in the signal mean is just a consequence of the structure of the deterministic part of the signal. If this deterministic or known part were a constant, we should have a constant mean. Otherwise, we have a space-variant mean signal.

If we model the random component of the signal as homogeneous and Gaussian, then we obtain a *covariance homogeneous Gaussian* field. In this case the two-dimensional RUKF in Fig. 9 is processing zero mean, homogeneous data. Thus, one would expect an approximate steady state to develop away from the boundaries. Then the RUKF estimator can be replaced by a constant coefficient filter when the observations are not near the boundaries. Covariance stationary Gaussian processes have been considered in [16], in which spectral distance measures have been obtained to bound the probability of error. These measures can be extended to multidimensions using the multidimensional version of the Toeplitz distribution theorem. It should be pointed out that these bounds are useful in the case of the estimator–correlator because of the lack of computationally attractive closed form expressions for the exact probability of error for this random signal detector even in one dimension. The only alternative is to compute an estimated ROC by simulation, which may have to be quite extensive to estimate low error probabilities reliably.

ii. Moving-Window Case

As mentioned in Section II.C., the signal often has an unknown location as part of the hypothesis structure of the problem. In other words, one wants to know not only that there is a signal in the observation field but also its approximate location. As in the known-signal case, the standard approach is to divide the observation field into detection windows and then perform the binary signal–no signal hypothesis problem at each detection window location separately. We can also think of this detection window as sliding or hopping through the data. The signal hypothesis at any fixed detection window location is that a signal's location corresponds to the center of the window. The amount of fine structure in the signal then sets an upper limit on how much we can allow the detection window to move between detection locations. If the signal locations are continuously distributed, it may also be necessary to search for the fine part of the signal delay within the detection window.

The following simplification occurs for the estimator – correlator with a signal model that is covariance homogeneous and Gaussian. The entire observation field can initially be processed by the RUKF, designed on the signal present hypothesis, without the need to estimate each window separately. Then one calculates a *detector bias field* to add to this estimate while doing the correlation in each detection window [29]. This detector bias field is the space-variant mean centered on the detection window plus the negative of the RUKF output due to the space-variant mean. The correctness of this procedure is understood by noting the easily perceived equivalence between Figs. 9 and 10. Note that in all cases the RUKF is designed for the zero mean signal $(s - m_s)$, which in this case is assumed homogeneous. If the detection window support approximates the support of average signal objects, then the covariance homogeneous assumption, that is, the assumption that the Gaussian random part $(s - m_s)$ is homogeneous, may be a good approximation. In any event, it greatly simplifies the estimator filter.

Example 4 We continue the examples of the simulation of detectability of ocean eddies by allowing the raised J_0 model to have varying amplitude, radius, and velocity. The effect of the varying velocity is to distort the elliptical contours due to the time required for the satellite to scan the data field. Figure 11 shows image data for four frames of the ocean ring signal simulation at 45 km track separation. Column a is the noise-free signal, column b is the signal and ocean noise, column c is the noise-whitened surface height data, column d shows the estimator outputs, and column e shows the estimator – correlator outputs prior to thresholding.

We computed the ROCs for the simulated detection performance for two detection strategies at a closer 35-km track separation. This closer track separation requires a longer ellapsed time for the satellite ground track to sweep out a given area; however, the closer track spacing makes the problem *more two-dimensional*. We present a comparison of the matched filter and the estimator – correlator for two detection window strategies.

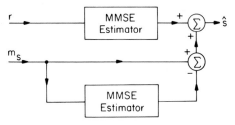

Fig. 10. Equivalent linear estimator.

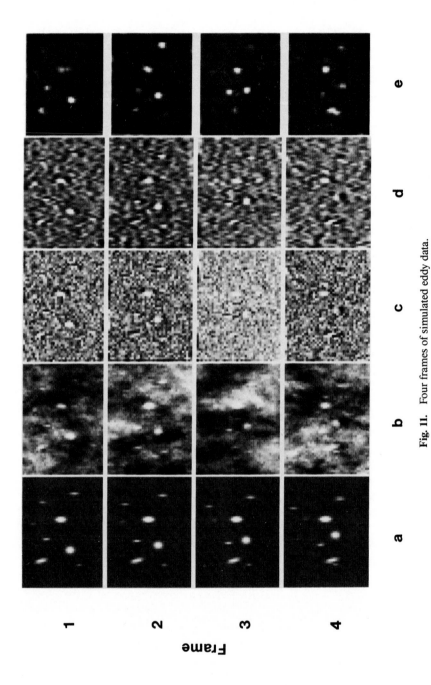

Fig. 11. Four frames of simulated eddy data.

In the first strategy, the 5×5 detection windows attempt to detect the signal in their 3×3 center pixels and are then hopped through the data, advancing by three samples per hop. Figure 12(a) shows the experimental ROC curve for the matched filter. Figure 12(b) shows the corresponding curve for the estimator–correlator. Here the design SNR has been varied to optimize performance on the data, which consisted of 810 windows containing signals in their 3×3 center pixels and 20,595 windows not containing any signal at all. The detection performance of the estimator–correlator is much better than that of the matched filter, because the signal centers are uniformly distributed over the 3×3 central pixels, whereas the matched filter corresponds to a signal at the center pixel. Similarly, good detection results can be obtained by combining the outputs of nine matched filters, each one matched to a signal centered at one of the nine pixels in the hopping detection window.

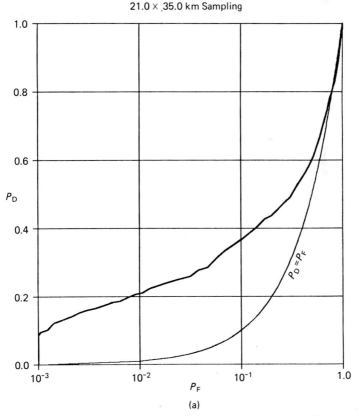

(a)

Fig. 12. Hopping window detector ROC curves. Detector has five trades of five samples; 810 detectors have rings, 20595 do not. (a) *ME* (b) *EC.*

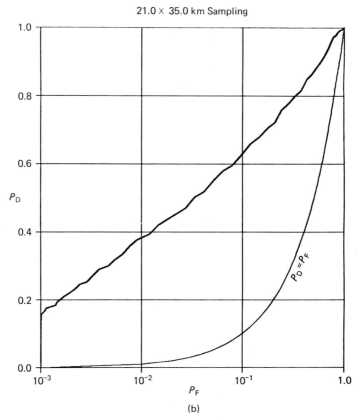

21.0 × 35.0 km Sampling

(b)

Fig. 12. *(Continued)*

The second detection strategy used a sliding 5 × 5 detection window. Here the detection was attempted for only the center pixel, and the window was, therefore, slid along rather than hopped. Figure 13(a) shows the matched filter ROC as experimentally determined from simulation. The matched filter performance greatly improved for the sliding versus the hopping detector. This improvement can be understood from an analysis of the bandwidth of the matched filter output and, hence, its Nyquist sampling rate.

Figure 13(b) shows the estimator–correlator result for the same data set. Evidently the estimator–correlator offers only a small improvement here with respect to the matched filter. The main reason is that the uncertainty in pixel delay is here restricted to one pixel size, thus effectively reducing the random component in the signal. In the hopping detector, the signal had a 9 pixel uncertainty that led to a rather large random component, which in turn led to the estimator – correlator's advantage. For a signal with more random attributes, such as random orientation and eccentricity, the random component of the signal would be large even for the sliding window detector. In such a case, we should expect the estimator – correlator to significantly outperform the matched filter. In any event, even with the isotropic eddy signals, the estimator – correlator has the advantage of providing a signal pseudoestimate, fewer required matched filter computations and the slightly better performance.

C. Nonlinear Estimators

Within the estimator – correlator framework, and motivated by some results in one-dimensional continuous time theory [14], one is lead to use

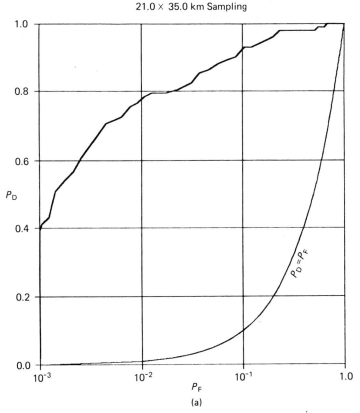

Fig. 13. Sliding window detector ROC curves. Detector has five tracks of five samples; 102 detectors have rings, 20595 do not. (a) *ME* (b) *EC.*

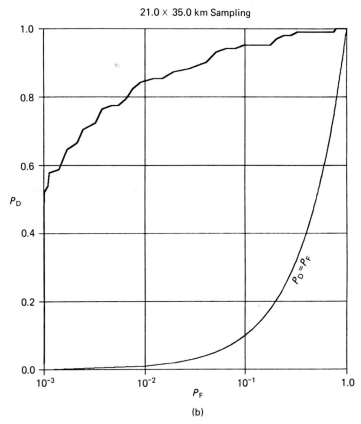

21.0 × 35.0 km Sampling

(b)

Fig. 13. *(Continued)*

nonlinear estimators in place of the linear fixed-lag estimator used in the last section. The most obvious candidate is the extended Kalman filter [1]; however, this approach would be complicated and probably would not offer much improvement due to its necessary approximations. A more promising approach in the case of a random parametrized signal would be the use of a maximum likelihood (ML) signal estimate. This type of estimator – correlator is then equivalent to the generalized likelihood ratio test [20,28]. If, instead, we use a maximum a posteriori (MAP) estimate, we should expect improved signal estimates and, hence, improved detection performance through use of the known a priori distribution of the random parameter. A comparison of these estimator – correlator detectors and the composite hypothesis test of Section II.D is contained in [25].

IV. Estimating Random Curves

This section deals with the estimation of the position of random curves in noisy data. It shares with the previous two problems a stochastic formulation, and it shares with the preceding problem a random signal. But there are two important differences. (1) Here we assume that we know that the random curve is present in the data, that is, we are not trying to decide on its presence or absence. We merely want to estimate its position, and this is done using the MAP approach of Cooper *et al.* [4] and Elliott *et al.* [10]. (2) The model for the randomness in the signal (the random curve) is substantially different from that of the previous section. We here assume that the signal is a known (constant) function with a random border. This random border or curve defines the random signal, which is observed in white Gaussian noise. If the curve closes on itself, it defines a random object that is constant inside a finite support region of random shape. A Markov process model is used to model the random boundary. Two estimators are developed to generate the boundary estimate. These are the *ripple filter* and the *sequential boundary finder,* to be discussed.

This MAP approach was originally motivated by the heuristic tree searching algorithm of Martelli [19]. As such, it lies between the rather classical statistical approaches that we have seen in our first two problems and the heuristic image analysis approach to the problem. In Cooper and Elliott's approach we have the simple, constant object characteristic of image analysis combined with a stochastic model for the boundary. Rather than incorporating heuristics into his algorithm, he searches for a boundary that will maximize the a posteriori probability calculated by means of a stochastic model. Both techniques have the advantage over the simple image analysis gradient techniques that an entire contour is calculated without gaps. Applications of either Martelli's or Cooper and Elliott's technique include locating outlines of objects in aerial reconnaissance images, biomedical images, radar images, and satellite radar-altimeter images; surveying of the open ocean; and so on.

A. Problem Statement

The noise-free signal is assumed to be constant to one side of the curve. If the curve closes on itself, we say that the signal is constant inside the curve. If the curve is open and extends across the whole image, we say that this random edge separates two regions of constant value. In [10] it is assumed that the signal is parametrized by Δ as follows

$$s(n_1,n_2) \triangleq \begin{cases} +\Delta/2 & \text{for} \quad (n_1,n_2) \quad \text{inside the curve} \\ -\Delta/2 & \text{for} \quad (n_1,n_2) \quad \text{outside the curve.} \end{cases} \tag{22}$$

The random boundary of the signal is specified in terms of a clockwise scanning of the pixel edge elements labeled $t_1, \ldots, t_M$ for a boundary of length M. Each t_i is four-valued and is defined as

$$t_i \triangleq \begin{cases} 0 & \text{up} \\ 1 & \text{right} \\ 2 & \text{down} \\ 3 & \text{left.} \end{cases} \tag{23}$$

A Markov model is postulated for t_i,

$$P_B(\mathbf{x}_i|\mathbf{x}_{i-1}), \tag{24}$$

where the K-dimensional state $\mathbf{x}_i$ is given as

$$\mathbf{x}_i \triangleq (t_i, t_{i-1}, \ldots, t_{i-K+1})^{\mathrm{T}}. \tag{25}$$

The transition probability can then be parametrized as a decreasing function of curvature. Two parameters and $K = 7$ were used in [10] to model Martelli's curvature function. Alternatively, a second-order difference equation can generate, through a quantization approximation, random contours of nearly elliptical shape [10], thereby incorporating more global information into the boundary model.

The probability of a starting state $\mathbf{x}_1$ is specified as $P_s(\mathbf{x}_1)$ along with the probability $P_L(M)$ of the boundary having length M. This random signal $s(n_1,n_2)$ is then additively immersed in white Gaussian noise $w(m,n)$ of mean 0 and variance σ^2 to produce the observed data sequence $r(n_1,n_2) = s(n_1,n_2) + w(n_1,n_2)$. Then the overall joint probability can be written and logarithmically transformed to yield the likelihood

$$\ln \mathcal{L} = \ln \mathcal{L}_B + \ln \mathcal{L}_{P|B}, \tag{26}$$

where $\mathcal{L}_B$ is the likelihood of the boundary edge sequence and $\mathcal{L}_{P|B}$ is the likelihood of the picture data r given the boundary edge sequence.

Upon removing irrelevant terms and simplifying, we obtain the following likelihood to maximize,

$$\mathcal{L} = \ln P_s(\mathbf{x}_1) + \sum_{i=2}^{M} P_B(\mathbf{x}_i|\mathbf{x}_{i-1}) + \ln P_L(M)$$

$$+ \sum_{\substack{(n_1,n_2) \in \\ \text{object}}} \frac{\Delta}{2\sigma^2} r(n_1,n_2) - \sum_{\substack{(n_1,n_2) \in \\ \text{background}}} \frac{\Delta}{2\sigma^2} r(n_1,n_2) \tag{27}$$

The estimate of the location of the random curve is then determined by maximizing Eq. (27) over the observed data $r(n_1, n_2)$ on $[0, N_1 - 1] \times [0, N_2 - 1]$.

These MAP boundary estimation algorithms assume simple objects with no occlusions. Martelli presents some examples of using heuristics to deal with the problems of touching or occluding objects [19]. The MAP approach could be extended by including signal modeling of the occlusion. Also, if the algorithms were run on modest-sized windows and the curves pieced together as in [5], then probably most windows would not contain occlusions, and the MAP estimate would need to be modified only in those few windows with occlusions.

B. Solution Techniques

Two solution techniques were presented in Cooper and Elliott *et al.* [4]: the *ripple filter,* which proceeds repetitively along the contour investigating pixel-by-pixel perturbations, and the *sequential boundary finder,* which proceeds causally along the boundary, calculating the likelihood over a swath approximately 4 pixels wide and searching the tentative paths using the A^* algorithm [19]; see also [22].

i. Ripple Filter

It is initially assumed that a complete though rough boundary estimate has been provided through course but reliable pixels, that is, by replacing 2×2 or 4×4 pixels by their averages or perhaps by human interaction with a light pen at a video terminal. The ripple algorithm then proceeds around its initial boundary estimate, interchanging each hypothesized boundary pixel one by one from object to background, or vice versa, to increase the likelihood $\tilde{l}$. The tours around the boundary are repeated until no further boundary changes take place. A stable equilibrium will occur, although it may be only at a local maximum of the likelihood. This procedure works well when the signal-to-noise ratio $\Delta/\sigma \geq 3$. The approach has been generalized into a parallel multiple-window boundary finder, described in [5], in which pixels of varying coarseness have been used to permit more reliable estimation at lower signal-to-noise ratios.

ii. Sequential Boundary Finder

This approach formulates an approximation to the a posteriori probability and the corresponding likelihood function by sequentially maximizing a likelihood computed over a data swath of width $\simeq 4$ pixels. Both the boundary and the object data are nearly optimally considered within this swath. A sequential tree searching routine, the A^* or branch-and-bound algorithm of the artificial intelligence literature [19,22], is used to obtain a near global maximum of the approximate likelihood function computed over the swath. As described in [10], the algorithm recursively calculates the likelihood for possible extension branches to the decision tree at depth i as

$$\tilde{l}(\mathbf{x}_i + 1) = \tilde{l}(\mathbf{x}_i) + \ln P_B(\mathbf{x}_{i+1} | \mathbf{x}_i) + D(\mathbf{x}_{i+1}), \qquad (28)$$

where P_B is the same state transition probability defined earlier and D is a change function used to describe the change in data likelihood in the swath caused by adding node $\mathbf{x}_{i+1}$ to the path. A method is also described for recursively calculating the function D for a total of 15 possible reasonable extensions of the present path from the center of a 4×4 data block. It is assumed that the starting state is specified on input. Also, the effect of the overall path length that is, $P(M)$, is considered in the sequential maximization [10], even though this is not apparent in Eq. (28), by including an entropy-derived factor to avoid favoring short paths. See Section 5.2 of [4] for a discussion of this factor.

The modified A^* algorithm maintains a list of paths with their associated likelihoods. The path with the greatest likelihood at any point in the search is the one to be extended next. The states are viewed as nodes in a graph with branches valued by the incremental likelihood. No heuristics are included in the cost functions. The modified A^* algorithm finds a path from the initial node to a goal node that posesses a suitably large likelihood. This tree search algorithm is not guaranteed to find an optimal boundary but has been found to work well in practice.

Example 5 We present some examples from [4] of the boundary estimation algorithms just discussed. The true boundary of a perturbed ellipse is shown as an overlay in Fig. 14 at an SNR of $\Delta/\sigma = 2$. The noisy data and

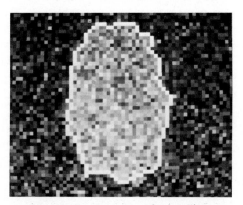

Fig. 14. True boundary of noisy ellipse.

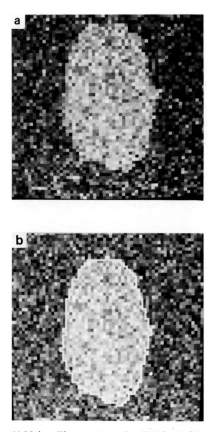

Fig. 15.   (a) Noisy ellipse at $\Delta/\sigma = 2$.   (b) Ripple filter estimate.

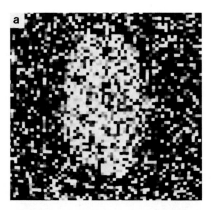

Fig. 16.   (a) Noisy ellipse at $\Delta/\sigma = 1$.   (b) Ripple filter estimate.

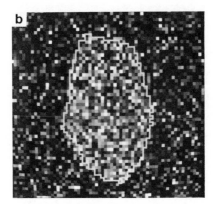

Fig. 16. *(Continued)*

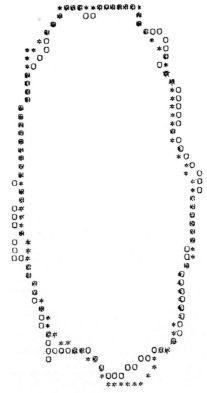

Fig. 17. Sequential boundary finder estimate at $\Delta/\sigma = 1$.

ripple filter estimate are shown in Fig. 15(a) and (b), respectively. The estimate is quite good, but it does deviate from the true boundary of Fig. 14. The results of Fig. 16 are at $\Delta/\sigma = 1$. Here the ripple estimate has degraded markedly with respect to the true boundary. Figure 17 shows a printer plot of the sequential boundary finder estimate at $\Delta/\sigma = 1$. The asterisks are the boundary estimates and the zeros are the true boundary. It can be seen that the boundary estimate of Fig. 17 is better than that of Fig. 16(b). No quantitative error measures were available.

V. Conclusions

In this chapter we have investigated three main topics relating to the two-dimensional image detection and estimation problems. We have looked at the detection of known objects, the detection of random objects, and the estimation of the location of random curves.

The software necessary to detect known signals consists mainly of a program to design the noise-whitening filter and a two-dimensional linear convolution routine. Any of the two-dimensional filter design procedures may be used to design the noise-whitening filter, the simplest being FIR windowing procedures. A FORTRAN program for the least squares linear predictive whitening filter used in Sections II and III is available from this author.

The random signal detector using the estimator–correlator method needs the above software plus estimator design and implementation programs. These are available in FORTRAN for the reduced update Kalman filter from this author. However, many other two-dimensional estimators have been reported in the literature, for instance, [26] and [8], and the software is often available from the authors.

The curve location estimators presented in Section IV were developed by D. B. Cooper of Brown University and his co-workers. Though neither the ripple filter nor the sequential boundary finder software is presently available for distribution, interested potential users should contact Professor Cooper about the availability of a dynamic programming variant using multiple windows [5] for increased parallelism.

VI. Bibliographical Notes

The analog two-dimensional matched filter was used for pattern recognition of handprinted characters by Highleyman [13] in 1961. Spatial matched filters were used in a detection-in-noise context by Vander Lugt [27] in 1964 and by Kozma and Kelly [17] in 1965. Both these papers considered mainly the optical implementation problem but did include some examples of the detection of known patterns in noise. Optical spatial filtering was also considered by Cutrona et al. [6] in 1960, but the detection problem was not directly addressed.

The detection of two-dimensional random signals was considered by Pratt [23] using a so-called stochastic matched filter designed to maximize the SNR at the filter output under H_1. This approach would be expected to be somewhat suboptimal to the two-dimensional estimator – correlator derived here.

The image estimation problem has been studied by Helstrom [12] who derived the unrealizable two-dimensional Wiener filter and by Ekstrom [8] who derived the NSHP causal Wiener filter using the spectral factorization technique of Ekstrom and Woods [9]. Two-dimensional recursive filters were studied by Habibi [11], Strintzis [26], and Woods and Radewan [34] who developed the reduced update Kalman filter used in Section III.

Early work on curve and line detection was done by Montanari [21] using a dynamic programming algorithm. Later work by Martelli [19] incorporated heuristic search methods and employed the A^* algorithm. This branch-and-bound algorithm was presented by Nilsson in [22]. Cooper and Elliott et al. [4] modified the method of Martelli by incorporating a simple model and dynamic Markov curve model. They then formulated the MAP estimate for the resulting curved edge appearing in noise. This approach, in turn, has been extended and modified in [5].

Acknowledgments

The author would like to thank David B. Cooper of Brown University for providing the examples of random boundary estimation. He would also like to thank his two students, Thomas M. Watson, III and Michael F. Scully for their help in developing the simulation results of image detection.

References

1. B. D. O. Anderson and J. B. Moore, "Optimal Filtering." Prentice-Hall, New Jersey, 1979.
2. S. Au and A. H. Haddad, Suboptimal sequential estimation-detection scheme for poisson driven linear systems, Inf. Sci. 16, 1978, 95 – 113.
3. H. Chang and J. K. Aggarwal, Design of two-dimensional semicausal recursive filters, IEEE Trans. Circuits Syst. CAS-25, 1978, 1051 – 1054.
4. D. B. Cooper, H. Elliott, F. Cohen, L. Reiss, and P. Symosek, Stochastic boundary estimation and object recognition, Comput. Graph. Image Process. 12, 1980, 326 – 356.

5. D. B. Cooper, F. Sung, and P. S. Schenker, "Toward a Theory of Multiple-Window Algorithms for Fast Adaptive Boundary Finding in Computer Vision," Brown University, Technical Report #ENG PRMI 80-3., July 1980.

6. L. J. Cutrona *et al.*, Optical data processing and filtering systems, *IRE Trans. Inf. Theory* **IT-6**, 1960, 386–400.

7. P. Delsarte, Y. V. Genin, and Y. G. Kamp, Half-Plane Toeplitz Systems, *IEEE Trans. Inf. Theory* **IT-26**, 1980, 465–474.

8. M. P. Ekstrom, Realizable Wiener filtering in two-dimensions, *IEEE Trans. Acoust. Speech, Signal Process.* **ASSP-30**, 1982, 31–40.

9. M. P. Ekstrom and J. W. Woods, Two-dimensional spectral factorization with applications in recursive digital filtering, *IEEE Trans. Acoust. Speech Signal Process.* **ASSP-24**, 1976, 115–128.

10. H. Elliott, D. B. Cooper, F. S. Cohen, and P. F. Symosek, Implementation, interpretation, and analysis of a suboptimal boundary finding algorithm, *IEEE Trans. Pattern Anal. Mach. Intelligence* **PAMI-4**, 1982, 167–182.

11. A. Habibi, Two-dimensional Bayesian estimate of images, *Proc. IEEE* **60**, 1972, 878–883.

12. C. W. Helstrom, Image restoration by the method of least squares, *Opt. Soc. Am. J.* **57**, 1967, 297–303.

13. W. H. Highleyman, An analog method for character recognition, *IRE Trans. Electron. Comput.* **EC-10**, 1961, 502–512.

14. T. Kailath, A general likelihood-ratio formula for random signals in Gaussian noise, *IEEE Trans. Inf. Theory* **IT-15**, 1969, 350–361.

15. S. Kashioka, M. Ejiri, and Y. Sakamoto, A transistor wire bonding system utilizing multiple local pattern machine, *IEEE Trans. Syst. Man Cybernet.* **SMC 6**, 1976 P. 562–570.

16. D. Kazakos and P. P. Kazakos, Spectral distance measures between Gaussian processes, *IEEE Trans. Autom. Control* **AC-25**, 1980, 950–959.

17. A. Kozma and D. L. Kelly, Spatial filtering for detection of signals submerged in noise, *Appl. Opt.* **14**, 1965, 387–392.

18. R. P. Kruger and W. B. Thompson, A technical and economic assessment of computer vision for industrial inspection and robotic assembly, *Proc. IEEE* **69**, 1981, 1524–1538.

19. A. Martelli, An application of Heuristic search methods to edge and contour detection, *Commun. Assoc. Comput. Mach.* **19**, 1976, 73–83.

20. J. L. Melsa, D. L. Cohn, "Decision and Estimation Theory." McGraw-Hill, New York, 1978.

21. U. Montanari, On the optimal detection of curves in noisy pictures, *Commun. ACM,* **14**, 1971, 335–345.

22. N. Nilsson, "Problem-Solving Methods in Artificial Intelligence," pp. 54–70. McGraw-Hill, New York, 1971.

23. W. K. Pratt, "Digital Image Processing." Wiley, New York, 1978.

24. L. L. Scharf and L. W. Nolte, Likelihood ratios for sequential hypothesis testing on Markov sequences, *IEEE Trans. Inf. Theory* **IT-23**, 1977, 101–109.

25. M. F. Scully and J. W. Woods, "Evaluation of Methods Enhanced Automatic Detection of Ocean Eddy Signals," RPI Image Processing Laboratory Technical Report No. IPL-TR-83-039, April 1983.

26. M. G. Strintzis, Dynamic representation and recursive estimation of cyclic and two-dimensional processes, *IEEE Trans. Autom. Control* **AC-23**, 1978, 801–809.

27. A. Vander Lugt, Signal detection by complex spatial filtering, *IEEE Trans. Inf. Theory* **IT-10**, 139–145.

28. H. L. VanTrees, "Detection, Estimation and Modulation Theory" (Part I). Wiley, New York, 1968.

29. T. M. Watson and J. W. Woods, "Automatic Detection of Ocean Ring Signals," RPI Image Processing Laboratory Technical Report No. IPL-TR-81-018, November 1981.

30. J. W. Woods, Stability of DPCM codes for television, *IEEE Trans. Commun.* **COM-23,** 1975, 845–846.
31. J. W. Woods and S. Dravida, Two-dimensional recursive estimation for ARMA signal models, *Proc. ICASSP 1982,* Paris, France, pp. 1150–1153.
32. J. W. Woods and V. K. Ingle, Kalman filtering in two-dimensions: Further results, *IEEE Trans. Acoust. Speech, Signal Process.* **ASSP-29,** 1981, 188–197.
33. J. W. Woods, I. Paul, and N. Sangal, 2-D recursive filter design with magnitude and phase, *Proc. 1981 ICASSP,* Atlanta, GA, pp. 700–703.
34. J. W. Woods and C. H. Radewan, Kalman filtering in two-dimensions, *IEEE Trans. Inf. Theory* **IT-23,** 1977, 473–482.

4 Image Reconstruction from Projections

A. C. Kak

School of Electrical Engineering
Purdue University
West Lafayette, Indiana

I. Introduction

There are many applications for reconstructing images fom their projections. The application that has revolutionized diagnostic medicine is that of computed tomography in which one reconstructs cross sections of the human body from projection data taken with x rays. The spatial resolution and tissue discrimination capability of this technique far surpass what could be achieved with conventional x-ray imaging. Image reconstruction from projections is also used in nuclear medicine to map the distribution of the concentration of gamma-ray emitting radionuclides in a given cross section

DIGITAL IMAGE PROCESSING TECHNIQUES

of the human body. Another medical application area is computed tomographic imaging with ultrasound [41]. Due to the radiation hazards associated with x rays, ultrasonic imaging is attracting considerable attention. Although for imaging, ultrasound is mostly used in the pulse–echo mode like radar, a number of investigations have recently focussed on using it for computed tomography because of the quantitative nature of this latter approach. Computed tomography with ultrasound can only be achieved for soft tissue structures, because the distortion caused by beam refraction in the presence of bone and air is too severe. Thus, this imaging approach seems limited to the important case of breast cancer detection. Nonmedical applications in which images can be reconstructed from projections include radioastronomy, optical interferometry, electron microscopy, and geophysical exploration.

Mathematically the problem can be stated as follows. Let $g(x,y)$ represent a two-dimensional function. A line running through $g(x,y)$ is called a *ray* (Fig. 1). The integral of $g(x,y)$ along a ray is called a *ray integral,* and a set of ray integrals forms a *projection.* Ray integrals and projections can be mathematically defined as follows. The equation of line AB in Fig. 1 is given by

$$x \cos \theta + y \sin \theta = t_1, \tag{1}$$

where t_1 is the perpendicular distance of the line from the origin. The integral of the function $g(x,y)$ along this line may be expressed as

$$P_\theta(t_1) = \int_{\text{ray AB}} g(x,y)\, ds = \int_{-\infty}^{\infty} \int_{-\infty}^{\infty} g(x,y)\delta(x \cos \theta + y \sin \theta - t_1)\, dx\, dy. \tag{2}$$

The function $P_\theta(t)$ as a function of t (for a given value of θ) is the parallel projection of $g(x,y)$ for angle θ The two-dimensional function $P_\theta(t)$ is also called the *Radon transform* of $g(x,y)$. A projection taken along a set of parallel rays is called a *parallel projection,* two examples of which are shown in Fig. 2. One can also generate projections by integrating a function along a set of lines emanating from a point source as shown in Fig. 3. Such projections are called *fan-beam projections.*

We shall be concerned with reconstructing an image $g(x,y)$ from its projections taken at various angles. We have emphasized the computational aspect of reconstruction and have presented in Section II each of the major algorithms as an annotated sequence of computation steps. The underlying theory has been presented, although a bit tersely, in Section III. A somewhat more detailed exposition of the theory can be found in the second edition of Rosenfeld and Kak [62]. In this chapter we have not dealt with the artifacts produced in reconstructed images by polychromaticity effects of x rays, refraction and multipath effects with ultrasound, and so on. A survey of these appears in [19,43]. Also, the algorithms presented in this

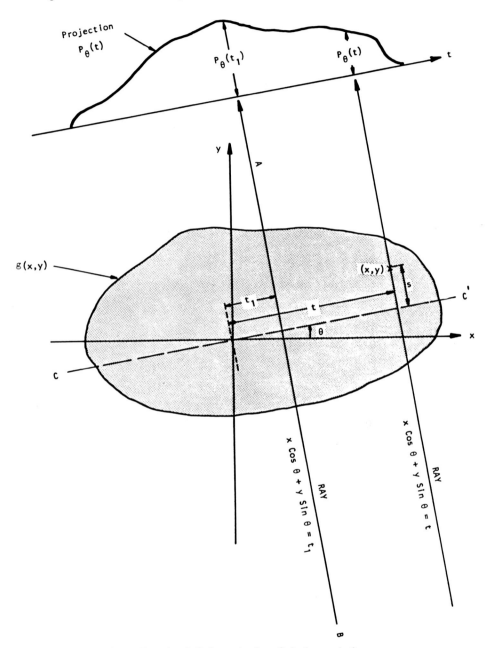

Fig. 1. Function $P_\theta(t)$ the projection of $g(x,y)$ at angle θ.

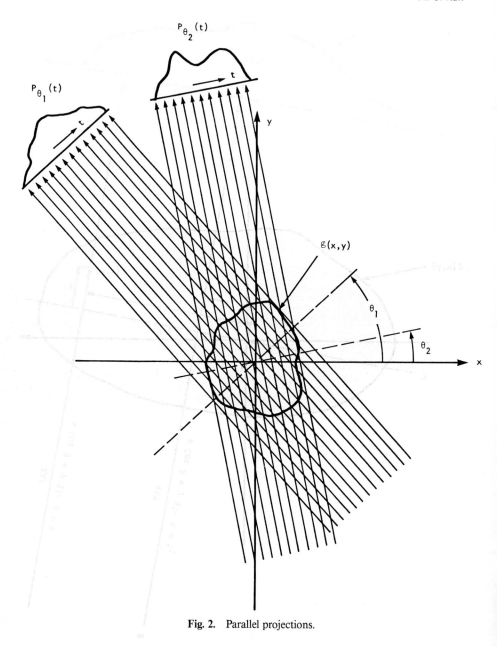

Fig. 2. Parallel projections.

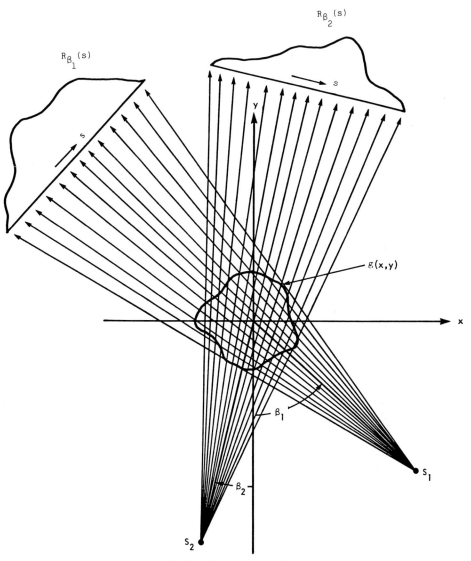

Fig. 3. Fan-beam projections.

chapter only apply when the radiation used for imaging travels in straight lines, or approximately so. Moderate amounts of ray bending can be taken into account by combining the reconstruction algorithms presented here with digital ray tracing and ray linking techniques discussed in [1,26]. If the size of the inhomogeneities within the object is comparable to the wavelength of the radiation used, one must then use the techniques of diffraction tomography, which has been an area of considerable activity during the last few years [22,23,55].

Most of the computer simulation results in this chapter will be shown for the image in Fig. 4(a). This is the well-known Shepp and Logan [67] "head phantom," so called because of its use in testing the accuracy of reconstruction algorithms for their ability to reconstruct cross sections of the human head with x-ray tomography. (The human head is believed to place the greatest demands on the numerical accuracy and the freedom from artifacts of a reconstruction method.) The image in Fig. 4(a) is composed of 10 ellipses, as illustrated in Fig. 4(b). The parameters of these ellipses are given in Table I.

A major advantage of using an image like Fig. 4(a) for computer simulation is that one can write analytical expressions for the projections. On account of the linearity of Radon transform, a projection of an image composed of a number of ellipses is simply the sum of the projections for each of the ellipses. We shall now present expressions for the projections of a single ellipse. Let $g(x,y)$ be as shown in Fig. 5(a), that is,

$$g(x,y) = \rho \quad \text{for } x^2/A^2 + y^2/B^2 \leq 1 \quad \text{inside the ellipse}$$
$$= 0 \quad \text{otherwise} \quad \text{outside the ellipse.} \quad (3)$$

TABLE I.

Component Ellipses of the Shepp and Logan Head Phantom

Ellipse	Coordinates of the center	Major A axis	Minor B axis	Rotation α^a angle	Gray ρ level
a	(0,0)	0.92	0.69	0	2
b	(0, − 0.0184)	0.874	0.6624	0	− 0.98
c	(0.22, 0)	0.31	0.11	72°	− 0.02
d	(− 0.22, 0)	0.41	0.16	108°	− 0.02
e	(0, 0.35)	0.25	0.21	0	0.01
f	(0, 0.1)	0.046	0.046	0	0.01
g	(0, − 0.1)	0.046	0.046	0	0.01
h	(− 0.08, − 0.605)	0.046	0.046	0	0.01
i	(0, − 0.605)	0.023	0.023	0	0.01
j	(0.06, − 0.605)	0.046	0.023	90°	0.01

^{a} The rotation angle α, shown in Fig. 5(b), is measured counterclockwise from the x axis.

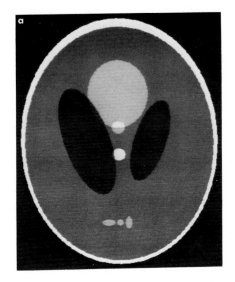

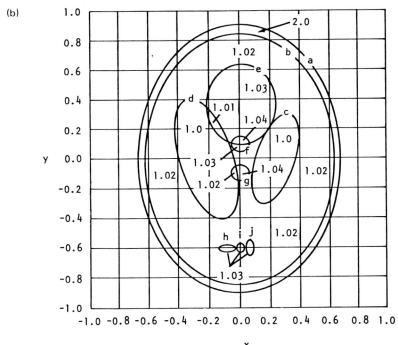

Fig. 4. (a) The Shepp and Logan head-phantom image; computer simulation results shown in this chapter were generated on this image. (b) The Shepp and Logan phantom, a superposition of 10 ellipses. On a scale of 0 to 2, the various gray levels in the phantom are also shown.

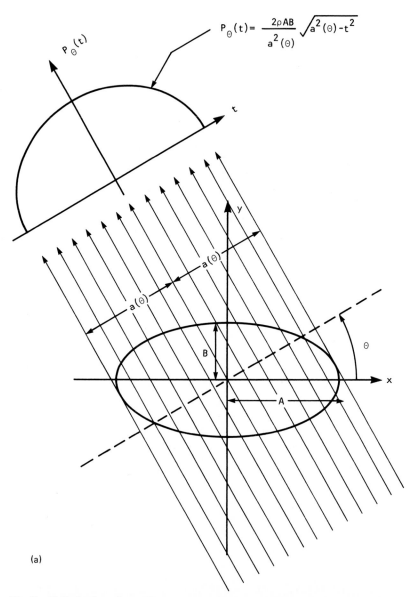

Fig. 5. (a) Projection of an ellipse; the gray level in the interior of the ellipse is ρ and zero outside. (b) An ellipse with its center located at (x_1, y_1) and its major axis rotated by α.

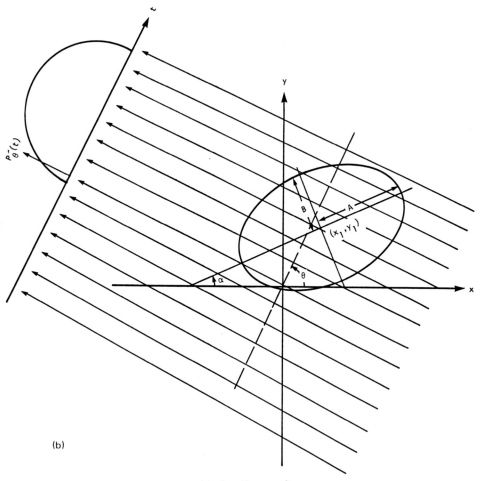

(b)

Fig. 5.   *(Continued)*

One can show that the projections of such a function are given by

$$P_\theta(t) = [2\rho AB/a^2(\theta)] \sqrt{a^2(\theta) - t^2} \qquad \text{for} \quad |t| \le a(\theta)$$
$$= 0 \qquad\qquad\qquad\qquad \text{for} \quad |t| > a(\theta), \qquad (4)$$

where $a^2(\theta) = A^2 \cos^2 \theta + B^2 \sin^2 \theta$. The quantity $a(\theta)$ is equal to the projection half-width as shown in Fig. 5(a).

If the center of the ellipse is moved to (x_1, y_1) and its orientation is rotated by an angle α as shown in Fig. 5(b), the resulting projections, denoted by

$P'_\theta(t)$, are related to the original $P_\theta(t)$ by

$$P'_\theta(t) = P_{\theta-\alpha}[t - s \cos(\gamma - \theta)], \tag{5}$$

where $s = (x_1^2 + y_1^2)^{1/2}$ and $\gamma = \tan^{-1}(y_1/x_1)$.

II. Computational Procedures for Image Reconstruction

This section is primarily intended for those readers who are interested in the implementation of the reconstruction algorithms but not in their underlying theory. The relevant theory will be presented in Sections III and IV. In this section each algorithm is presented as an annotated sequence of computational steps, followed by computer simulation results to give the reader some idea of the reconstruction accuracy.

A. Filtered-Backprojection Algorithm for Parallel Data

The reconstruction procedure that is most commonly used on CT scanners is based on the filtered-backprojection algorithm. This algorithm has been shown to be extremely accurate and amenable to fast implementation. It is presented here for the case of parallel projection data. Fan projections will be considered in the next two subsections.

This algorithm is based on the relationships

$$g(x,y) = \int_0^\pi Q_\theta(x \cos \theta + y \sin \theta) \, d\theta, \tag{6}$$

where $Q_\theta(t)$, called filtered projections, are related to the projections $P_\theta(t)$ by

$$Q_\theta(t) = \int_{-\infty}^\infty P_\theta(\alpha) h_1(t - \alpha) \, d\alpha, \tag{7}$$

where the filter impulse response $h_1(t)$ is the inverse Fourier transform of the $|w|$ function in the frequency domain over the bandwidth of the system†,

$$h_1(t) = \int_{-W}^W |w| \exp(j2\pi wt) \, dw. \tag{8}$$

---

† We have used the symbol W to represent the frequency. If t is measured in cm, the dimensions of W are cycles/cm.

Here W is the frequency beyond which the spectral energy in any projection can be assumed to be zero. These equations suggest the following steps for a digital implementation of the algorithm.

Step 1 (Filtering)   Let us say that each projection is sampled with a sampling interval of τ cm. In order that the sampled projections do not suffer from aliasing distortion, this implies that $W = 1/2\tau$. Substituting this value of W in Eq. (8), we get for the impulse response

$$h_1(t) = \frac{1}{2\tau^2}\left(\frac{\sin(2\pi t/2\tau)}{2\pi t/2\tau}\right) - \frac{1}{4\tau^2}\left(\frac{\sin(\pi t/2\tau)}{\pi t/2\tau}\right)^2. \tag{9}$$

Since the data are measured with a sampling interval of τ and are presumably bandlimited, for digital processing the impulse response need only be known with the same bandwidth and, hence, the same sampling interval. We obtain from Eq. (9)

$$h_1(n\tau) = \begin{cases} 1/4\tau^2, & n = 0, \\ 0, & n \text{ is even,} \\ -(1/n^2\pi^2\tau^2), & n \text{ is odd,} \end{cases} \tag{10}$$

where n takes both negative and positive integer values. At the sampling points $n\tau$ the filtered projection values may be obtained from Eq. (7):

$$Q_\theta(n\tau) = \tau \sum_{m=-\infty}^{\infty} P_\theta(m\tau)h_1[(n-m)\tau]. \tag{11}$$

For bandlimited functions satisfying $W = 1/2\tau$, the summation in Eq. (11) results in an exact evaluation of $Q_\theta(t)$ at $t = n\tau$. In practice each projection is of finite extent. Suppose that each $P_\theta(m\tau)$ is zero outside the index $m = 0,...,N - 1$. We can now write the following equivalent forms of Eq. (11),

$$Q_\theta(n\tau) = \tau \sum_{m=0}^{N-1} P_\theta(m\tau)h_1[(n-m)\tau] \tag{12a}$$

or

$$Q_\theta(n\tau) = \tau \sum_{m=-(N-1)}^{N-1} P_\theta[(n-m)\tau]h_1(m\tau), \tag{12b}$$

for $n = 0,1,2,...,N - 1$. Both these equations imply that to determine $Q_\theta(n\tau)$ over the N sampling points of $P_\theta(t)$, the length of the sequence $h_1(n\tau)$ used should be from $n = -(N - 1)$ to $n = (N - 1)$.

Although the discrete convolution in Eq. (12) may be implemented directly on a general purpose computer, it is much faster to carry it out in the frequency domain using FFT algorithms. [With specially designed hardware, direct implementations of Eq. (12) can be made as fast or faster than the frequency domain implementation.] For the frequency domain implementation, one has to keep in mind the fact that one can now only perform periodic (circular) convolutions, whereas the convolution required in Eq. (12) is aperiodic. To eliminate the interperiod interference artifacts inherent to the periodic convolution, we pad both the projection data and the impulse response function with a sufficient number of zeroes. It can be shown that if we pad $P_\theta(m\tau)$ with zeroes so that it is $(2N - 1)$ elements long, we avoid interperiod interference over the N samples of $Q_\theta(n\tau)$. Of course, if one wants to use the popular base-2 FFT algorithm, the sequences $P_\theta(m\tau)$ and $h_1(m\tau)$ have to be zero padded so that each is $(2N - 1)_2$ elements long, where $(2N - 1)_2$ is the smallest integer that is a power of 2 and greater than $2N - 1$. The frequency domain implementation may be expressed as

$$Q_\theta(n\tau) = \tau \times \text{IFFT}\{\text{FFT}[P_\theta(n\tau) \text{ with ZP}] \times \text{FFT}[h_1(n\tau) \text{ with ZP}]\}, \quad (13)$$

where FFT and IFFT denote, respectively, the fast Fourier transform and the inverse fast Fourier transform; and ZP stands for zero padding. One usually obtains superior reconstructions when some smoothing is also incorporated in the filtering process. For example, in Eq. (13) smoothing may be incorporated by multiplying the product of the two FFTs by a Hamming window [33]. When such a window is used, the frequency domain implementation of Eq. (12) can be written as

$$Q_\theta(n\tau) = \tau \times \text{IFFT}\{\text{FFT}[P_\theta(n\tau) \text{ with ZP}] \times \text{FFT}[h_1(n\tau) \text{ with ZP}]$$
$$\times \text{ smoothing window}\}. \quad (14)$$

Step 2 (Backprojection) The second step deals with reconstructing the image from the filtered projections using a digital approximation to the integral in Eq. (6). When the number of projections M_{proj} is large and uniformly distributed over 180°, Eq. (6) may be approximated by

$$\hat{g}(x,y) = \frac{\pi}{M_{\text{proj}}} \sum_{i=1}^{M_{\text{proj}}} Q_{\theta_i}(x \cos \theta_i + y \sin \theta_i). \quad (15)$$

This equation calls for each filtered projection Q_{θ_i} to be *backprojected*. This can be explained as follows. To every point (x,y) in the image frame there corresponds a value of $t = (x \cos \theta + y \sin \theta)$ for a given value of θ. The contribution that Q_{θ_i} makes to the reconstruction at (x,y) is its value for the corresponding value of t, as illustrated in Fig. 6. It is easily shown that for the indicated angle θ_i, the value of $t = (x \cos \theta_i + y \sin \theta_i)$ is the same for all (x,y) on line LM. Therefore, the filtered projections Q_{θ_i} will make the same contribution to the reconstruction at all these points. This may be visualized as the smearing of Q_{θ_i} over the image frame. The sum (multiplied by π/k) of all such smearings results in the reconstructed image.

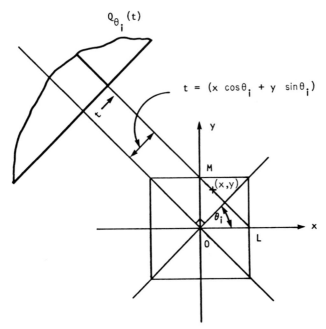

Fig. 6. The filtered projection $Q_{\theta_i}(t)$ in reconstruction. It makes the same contribution to all points (pixels) on line LM in the xy plane.

In Eq. (15), $\hat{g}(x,y)$ is the reconstructed approximation to the original function $g(x,y)$. In backprojecting $Q_{\theta_i}(t)$ to a point (x,y), we need to know it for $t = x\cos\theta_i + y\sin\theta_i$. This value of t may not correspond to one of the discrete values for which $Q_{\theta_i}(t)$ is known, in which case interpolation is called for and often linear interpolation is adequate. To eliminate the computations required for interpolation, preinterpolation of the functions $Q_\theta(t)$ is also used [46, 65]. In this technique, prior to backprojection, the function $Q_\theta(t)$ can be preinterpolated onto 10 to 100 times the number of points at which it was originally known. From this dense set of points one retains the nearest neighbor to obtain the value of Q_{θ_i} at $x\cos\theta_i + y\sin\theta_i$. In an FFT-based procedure the preinterpolation can be done rapidly by zero padding in the transform domain.

With preinterpolation and with appropriate programming, backprojection for parallel data can be implemented with *virtually no multiplications*.

Using the projection filtering implementation in Eq. (13), we show in Fig. 7(b) the reconstructed values on the line $y = -0.605$ for the image in Fig. 4(a). Figure 7a shows the complete reconstruction. The number of rays in each projection was 127 and the number of projections 100. To make convolutions aperiodic, the projection data were padded with zeroes to make each projection 256 elements long.

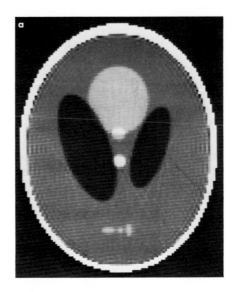

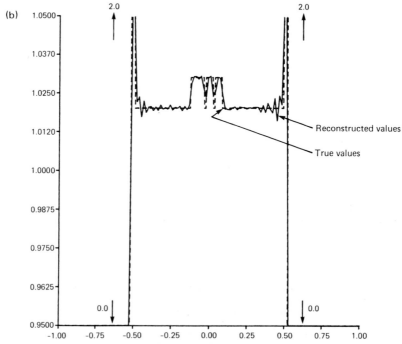

Fig. 7. (a) A parallel-beam reconstruction of the image of Fig. 4(a) with 100 projections over 180° and 127 rays in each projection. The display matrix size is 128 × 128. (b) A numerical comparison of the true and reconstructed values on the $y = -0.605$ line.

B. Filtered-Backprojection Algorithm for Fan Data Generated by Equiangular Rays

Almost all fast CT scanners use fan beams for tomographic reconstructions. In this and the next subsection we shall discuss only algorithms that *directly* reconstruct an image from fan-beam projections. It is also possible to rearrange the fan projection data into parallel projections and then use the algorithm described in the preceding subsection. This rearrangement can be done "on the fly" provided the angular interval between fan projections and the sampling interval in each projection satisfy certain conditions, which are discussed in Section II.D.

There are two types of fan-beam projections: those that take the projection data with an equiangular set of rays and those that utilize a set of rays that result in equispaced detector locations on a straight line. In this subsection we shall present a digital implementation for the former case. When the rays are equiangular, the detectors for measuring projection data are equispaced along the arc $D_1 D_2$ shown in Fig. 8(a). The radius of this arc is $2D$ where D is the source to center distance.

To present the formulas that form the basis of the digital implementation, we first represent the image in polar coordinates and denote it by $g(r,\phi)$. [This is done only for convenience in representing the formulas. In the computer implementation of the formulas, the image is always reconstructed on a rectangular matrix.] The projection data will now be denoted by $R_\beta(\gamma)$ where the angle γ gives the angular location of a ray in the projection taken at angle β, as in Fig. 8(b). To facilitate our presentation we also need to define two new parameters L and γ', shown in Fig. 8(c). In a given projection at angle β, L is the distance from the source S to the pixel at (r,ϕ).

$$L(\beta,r,\phi) = \sqrt{D^2 + r^2 + 2Dr\sin(\beta - \phi)}. \qquad (16)$$

The variable γ' will give us the angular location of the ray that passes through the pixel (r,ϕ) in the given projection at angle β,

$$\gamma'(\beta,r,\phi) = \tan^{-1}(\overline{FC}/\overline{SF}) = \tan^{-1}[r\cos(\beta - \phi)/\{D + r\sin(\beta - \phi)\}]. \qquad (17)$$

The image $g(r,\phi)$ and the fan-beam projections $R_\beta(\gamma)$ can be shown to be related by

$$g(r,\phi) = \int_0^{2\pi} \frac{1}{L^2} \int_{-\gamma_m}^{\gamma_m} D\cos\gamma R_\beta(\gamma) \frac{1}{2}\left(\frac{\gamma' - \gamma}{\sin(\gamma' - \gamma)}\right)^2 h_1(\gamma' - \gamma)\, d\gamma\, d\beta, \qquad (18)$$

where $-\gamma_m$ and γ_m are angles for the extreme rays in each projection and where the function $h_1(\gamma)$ is the same as in Eq. (9) (with argument t replaced by γ). A derivation of this equation is presented in Section III.C.

The relationship in Eq. (18) can be translated into the following weighted filtered-backprojection algorithm for computer implementation.

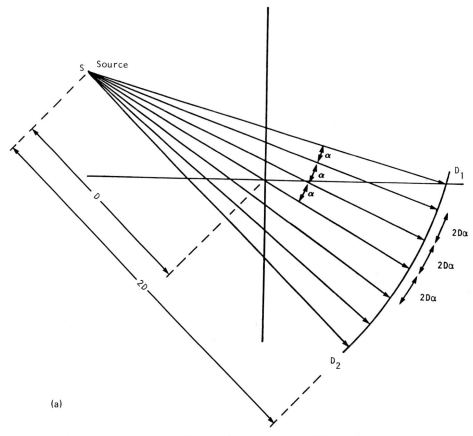

(a)

Fig. 8. (a) A fan-beam projection taken with equiangular rays. (b) The various parameters used in the derivation of the fan-beam reconstruction algorithm for the case of equiangular rays. (c) Illustration that L is the distance of the pixel at location (x,y) from the source S and γ' is the angle that the source-to-pixel line subtends with the central ray.

Step 1 (Modify Each Projection) Assume that each fan-beam projection $R_\beta(\gamma)$ is sampled with an angular sampling interval of α radians and that the sampled data $R_\beta(n\alpha)$ are free of aliasing errors. The first step consists of obtaining for each projection $R_\beta(n\alpha)$ a corresponding modified projection $R'_\beta(n\alpha)$, as follows:

$$R'_\beta(n\alpha) = R_\beta(n\alpha)D\cos n\alpha. \tag{19}$$

Note that $n = 0$ corresponds to the center ray in each projection.

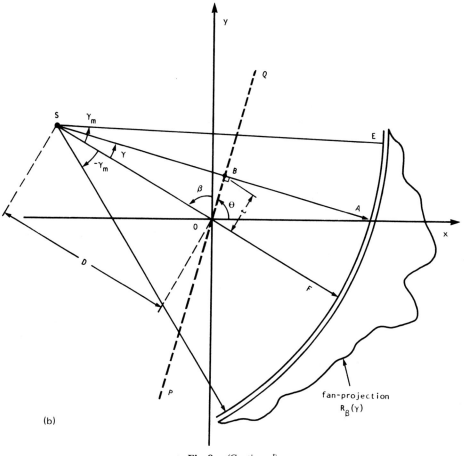

Fig. 8.   *(Continued)*

Step 2 (Filtering)   In this step we filter each modified projection by convolving it with an impulse response $h_2(\gamma)$ defined by

$$h_2(\gamma) = \tfrac{1}{2}\,(\gamma/\sin\gamma)^2 h_1(\gamma). \tag{20}$$

In the continuous domain this operation of filtering is described by

$$Q_\beta(\gamma) = \int_{-\gamma_m}^{\gamma_m} R'_\beta(\sigma)h_2(\gamma - \sigma)\,d\sigma, \tag{21}$$

where $Q_\beta(\gamma)$ is the filtered projection corresponding to $R'_\beta(\gamma)$. Its digital implementation is given by (exact for the bandlimited case and when α

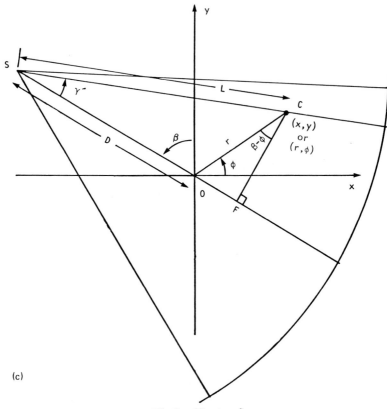

Fig. 8. *(Continued)*

satisfies the Nyquist condition)

$$Q_\beta(n\alpha) = \alpha \sum_m R'_\beta (m\alpha) h_2[(n - m)\alpha]. \tag{22}$$

The discrete impulse response $h_2(n\alpha)$ is given by the samples of Eq. (20),

$$h_2(n\alpha) = \begin{cases} 1/8\alpha^2, & n = 0, \\ 0, & n \text{ is even}, \\ -[1/2\pi^2(\sin n\alpha)^2], & n \text{ is odd}, \end{cases} \tag{23}$$

where Eq. (9) was used for $h_1(\,\cdot\,)$. The convolution in Eq. (22) possesses a frequency domain implementation similar to Eq. (13). The comment we made there regarding the desirability of incorporating some smoothing also applies here.

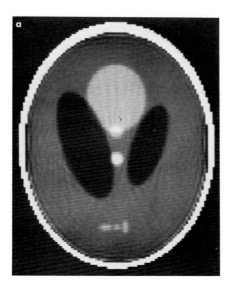

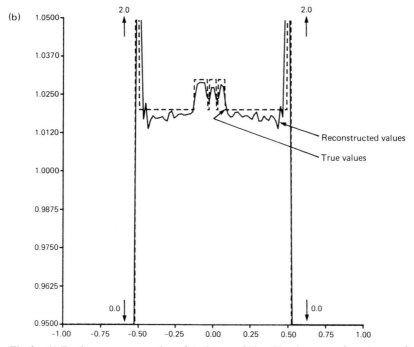

Fig. 9. (a) Fan-beam reconstruction of the image of Fig. 4(a) using an equiangular set of rays. The fan angle is 45°, the number of projections is 200 over 360°, and the number of rays in each projection is 100. (b) A numerical comparison of the true and reconstructed values on the $y = -0.605$ line.

Step 3 (Fan Backprojection) This last step consists of implementing the remaining integration in Eq. (18). In terms of filtered projections this step may be expressed as

$$g(r,\phi) = \int_0^{2\pi} \frac{1}{L^2(\beta,r,\phi)} Q_\beta(\gamma') \, d\beta. \tag{24}$$

For digital implementation, when the number of projections M_{proj} is large and uniformly distributed over 360°, Eq. (24) may be expressed as

$$\hat{g}(x,y) = \frac{2\pi}{M_{proj}} \sum_{i=1}^{M_{proj}} \frac{1}{L^2(\beta_i,x,y)} Q_{\beta_i}(\gamma'), \tag{25}$$

where $x = r \cos \phi$ and $y = r \sin \phi$. The function $\hat{g}(x,y)$ is the reconstructed approximation to the image expressed as a function of Cartesian coordinates. To find the contribution of the filtered projection Q_{β_i} to the reconstruction at a pixel at (x,y), we must first find L and γ' using Eqs. (16) and (17). [This is in contrast to the case of parallel projections in which to compute such a contribution we only had to find $x \cos \theta_i + y \sin \theta_i$.] Computation of such contributions at *all* pixels from *one* Q_{β_i} is called a *fan backprojection*. The sum of all such fan backprojections multiplied by $2\pi/M_{proj}$ gives $\hat{g}(x,y)$. For a given pixel, the computed value of γ' may not correspond to one of $n\alpha$ at which Q_{β_i} is known from Eq. (22), in which case interpolation becomes necessary. In most cases linear interpolation is adequate.

Using this implementation, we show in Fig. 9(a) a reconstruction of the image in Fig. 4(a) from 200 fan projections over 360° with 127 rays in each projection. The fan angle was 45°, and each projection was padded with zeroes to make it 256 elements long. Figure 9(b) shows the reconstructed values on the line $y = -0.605$.

C. Filtered-Backprojection Algorithm for Fan Data Generated by Equally Spaced Collinear Detectors

Figure 10(a) shows the geometry for data collection. Although the projections are measured on a line such as $D_1 D_2$, for theoretical purposes it is more efficient to assume the existence of an imaginary detector line $D_1' D_2'$ passing through the origin. We associate the ray integral along SB with point A on $D_1' D_2'$ as opposed to point B on $D_1 D_2$.

Each projection is now denoted by $R_\beta(s)$, where the distance s along the imaginary detector line $D_1' D_2'$ identifies the location of a ray in the fan projection at angle β (Fig. 10(b)). Again, let D be the source-to-center distance and $g(r,\phi)$ the image we seek to reconstruct in polar coordinates.

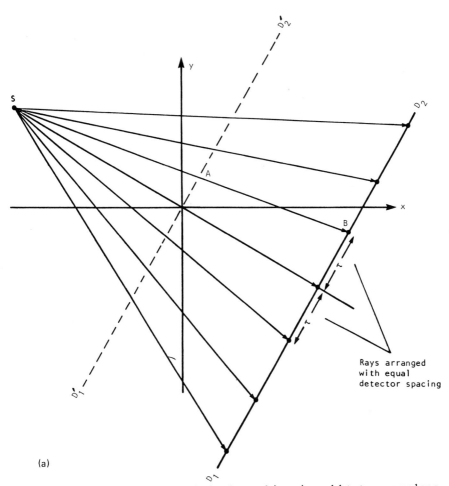

(a)

Fig. 10. (a) The rays in a fan-beam projection that result in equispaced detectors arranged on a straight line. (b) The various parameters used in the fan-beam reconstruction algorithm for the case of equispaced collinear detectors. (c) For a pixel at polar coordinates (r,ϕ), the variable U is the ratio of the distance SP, which is the projection of the source-to-pixel line on the central ray, to the source-to-center distance.

For ease of presentation we introduce two more parameters U and s'; see Fig. 10(c). The parameter U is for each pixel at (x,y), or equivalently at (r,ϕ), the ratio of $\overline{SP}$ to the source-to-origin distance, where $\overline{SP}$ is the projection of the source-to-pixel distance on the central ray.

$$U(r,\phi,\beta) = [(\overline{SO} + \overline{OP})/D] = [D + r \sin(\beta - \phi)]/D. \qquad (26)$$

The other parameter s' is the value of s that corresponds to the ray passing

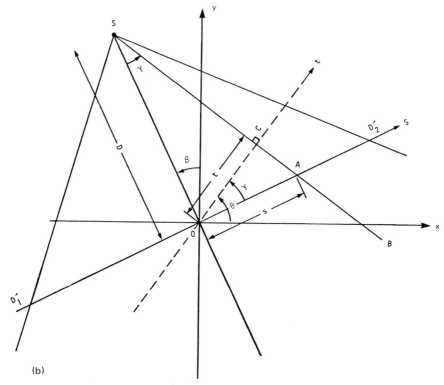

Fig. 10. *(Continued)*

through the pixel (r,ϕ). In Fig. 10(c) s' is equal to the distance $\overline{OF}$.

$$s'(r,\phi,\beta) = \overline{OF} = D[r\cos(\beta - \phi)/(D + r\sin(\beta - \phi))]. \tag{27}$$

With U and s' defined as in Eqs. (26) and (27), the image $g(r,\phi)$ and its projections $R_\beta(\gamma)$ can be shown to be related by

$$g(r,\phi) = \int_0^{2\pi} \frac{1}{U^2} \int_{-s_m}^{s_m} R_\beta(s) \frac{D}{(D^2 + s^2)^{1/2}} \tfrac{1}{2} h_1(s' - s)\, ds\, d\beta, \tag{28}$$

where the function $h_1(\cdot)$ is the same as in Eq. (9). A derivation of this equation, which is similar to that presented in Section III.C for Eq. (18), can be found in Chapter 8 of [62].

For the purpose of computer implementation, Eq. (28) can be interpreted as a weighted filtered-backprojection algorithm. To show this we rewrite Eq. (28) as

$$g(r,\phi) = \int_0^{2\pi} (1/U^2) Q_\beta(s')\, d\beta, \tag{29}$$

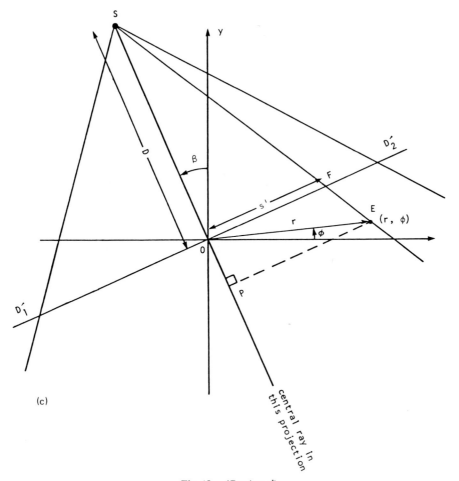

Fig. 10. *(Continued)*

where

$$Q_\beta(s) = R'_\beta(s) * h_3(s), \tag{30}$$

with

$$R'_\beta(s) = R_\beta(s)\, D/(D^2 + s^2)^{1/2} \quad \text{and} \quad h_3(s) = \tfrac{1}{2} h_1(s). \tag{31}$$

Computer implementation of Eqs. (29) through (31) can be carried out as follows.

Step 1 (Modify Each Projection) Assume that each projection $R_\beta(s)$ is sampled with a sampling interval of a. The known data then are $R_\beta(na)$

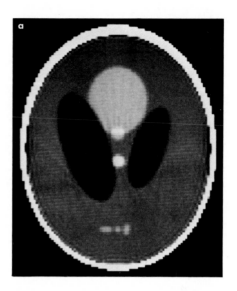

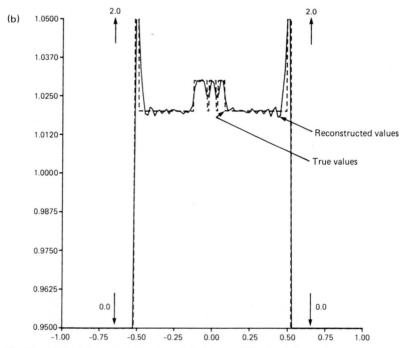

Fig. 11. (a) Fan-beam reconstruction of the image of Fig. 4(a) using an equispaced set of rays. The fan angle is 45°, the number of projections is 200 over 360°, and the number of rays in each projection 100. (b) A numerical comparison of the true and reconstructed values on the $y = -0.605$ line.

where n takes integer values with $n = 0$ corresponding to the central ray passing through the origin; β_i are the angles for which fan projections are known. The first step is to generate for each fan projection $R_{\beta_i}(na)$ the corresponding modified projection $R'_{\beta_i}(na)$ given by

$$R'_{\beta_i}(na) = R_{\beta_i}(na) \, D/(D^2 + n^2a^2)^{1/2}. \qquad (32)$$

Step 2 (Filtering) Convolve each modified projection $R'_{\beta_i}(na)$ with $h_3(na)$ to generate the corresponding filtered projection,

$$Q_{\beta_i}(na) = R_{\beta_i}(na) * h_3(na), \qquad (33)$$

where the sequence $h_3(na)$ is given by

$$h_3(na) = \tfrac{1}{2} h_1(na). \qquad (34)$$

Substituting in this the sampled values of $h_1(\cdot)$ given in Eq. (10), we obtain for the impulse response of the convolving filter,

$$h_3(na) = \begin{cases} 1/8a^2, & n = 0 \\ 0, & n \text{ even.} \\ -(1/2n^2\pi^2a^2), & n \text{ odd} \end{cases} \qquad (35)$$

The convolution in Eq. (33) possesses a frequency domain implementation similar to Eq. (13). The comment we made there regarding the desirability of incorporating some smoothing also applies in Eq. 33.

Step 3 (Fan Backprojection) Perform a *weighted* backprojection of each filtered projection along the corresponding fan. The sum of all the backprojections is the reconstructed image

$$\hat{g}(x,y) = \frac{2\pi}{M_{\text{proj}}} \sum_{i=1}^{M} \frac{1}{U^2(x,y,\beta_i)} Q_{\beta_i}(s'), \qquad (36)$$

where s' identifies the ray that passes through (x,y) in the fan for the source located at angle β_i. If for a given pixel, the computed value of s' does not correspond to one of na at which Q_{β_i} is known, interpolation is called for, the linear type being adequate in most cases.

Figure 11(a) shows a reconstruction obtained by using the steps outlined here. The number of projections is 100, the number of rays in each projection 127, and the fan angle is 45°. The original image is shown in Fig. 4(a). Figure 11(b) shows the reconstructed values on the line $y = -0.605$.

D. A Resorting Algorithm for Fan Data

We shall now describe a procedure that rapidly resorts equiangular fan-beam projection data into equivalent parallel-beam projection data. After such resorting one can use the filtered-backprojection algorithm for the parallel-projection data to reconstruct the image. This fast resorting algorithm does place constraints on the angles at which the fan-beam projections must be taken and, also, on the angles at which projection data must be sampled within each fan-beam projection.

Referring to Fig. 8(b), we find the relationships between the independent variables of the fan-beam and parallel projections to be

$$t = D \sin \gamma \quad \text{and} \quad \theta = \beta + \gamma. \tag{37}$$

If, as before, $R_\beta(\gamma)$ denotes a fan-beam projection taken at angle β, and $P_\theta(t)$ denotes a parallel projection taken at angle θ, using Eq. (37) we can then write

$$R_\beta(\gamma) = P_{\beta+\gamma} D \sin \gamma. \tag{38}$$

Let $\Delta\beta$ denote the angular increment between successive fan-beam projections, and let $\Delta\gamma$ denote the angular interval used for sampling each fan-beam projection. We will assume that the condition

$$\Delta\beta = \Delta\gamma = \alpha \tag{39}$$

is satisfied, which implies that β and γ in Eq. (38) are equal to $m\alpha$ and $n\alpha$, respectively, for some integer values of the indices m and n. We can, therefore, write Eq. (38) as

$$R_{m\alpha}(n\alpha) = P_{(m+n)\alpha} D \sin n\alpha. \tag{40}$$

This equation, which serves as the basis of the fast resorting algorithm, expresses the fact that the nth ray in the mth radial projection is the same as the nth ray in the $(m + n)$th parallel projection. Of course, because of the sin $n\alpha$ factor on the right-hand side in Eq. (40), the parallel projections obtained are not uniformly sampled. This can usually be rectified by interpolation.

E. Algebraic Reconstruction Algorithm

In Fig. 12 we have superimposed a square grid on the image $g(x,y)$, and we shall assume that in each cell $g(x,y)$ is constant. Let x_m denote this constant value in the mth cell, and let N be the total number of cells. For algebraic techniques, a ray is defined somewhat differently. A ray is now a "fat" line running through the xy plane. To illustrate this we have shaded the ith ray in Fig. 12, in which each ray is of width τ. In most cases the ray width is approximately equal to the image cell width. A ray integral will now be called a *ray sum*.

Like the image, the projections will also be given a one-index representation. Let p_m be the ray sum measured with the mth ray as shown in Fig. 12.

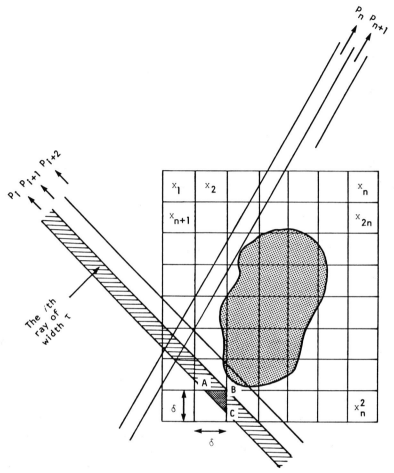

Fig. 12. A square grid, superimposed over the unknown image in algebraic reconstruction methods. Image values are assumed to be constant within each cell of the grid. For the second cell from the left in the bottom-most row, the value of the weighting factor is given by $w_{ij} =$ (area of ABC)$/\delta^2$.

The relationship between x_i and p_i may be expressed as

$$\sum_{j=1}^{N} w_{ij} x_j = p_i, \qquad i = 1, 2, ..., M, \tag{41}$$

where M is the total number of rays (in all the projections), and w_{ij} is the weighting factor that represents the contribution of the jth cell to the ith ray integral. The factor w_{ij} is equal to the fractional area of the jth image cell intercepted by the ith ray, as shown for one of the cells in Fig. 12. Note that most w_{ij} are zero, because only a small number of cells contribute to any given ray sum.

If M and N were small, one could use conventional matrix theory methods to invert the system of equations in Eq. (41). However, in practice N may be as large as 65,000 (for 256 × 256 images), and in most cases for images of this size, M will also have the same magnitude. For these values of M and N, the size of the matrix $[w_{ij}]$ in Eq. (41) is 65,000 × 65,000, which precludes any possibility of direct matrix inversion. Of course, when noise is present in the measurement data and when $M < N$, even for small N it is not possible to use direct matrix inversion, and some least-squares methods may have to be used. When both M and N are large, such methods are also computationally impractical.

For large values of M and N, there exist very attractive iterative methods for solving Eq. (41) that are based on the *method of projections* first proposed by Kaczmarz [42]. To explain the computational steps involved in these methods, we first write Eq. (41) in an expanded form,

$$w_{11}x_1 + w_{12}x_2 + w_{13}x_3 + \cdots + w_{1N}x_N = p_1$$
$$w_{21}x_1 + w_{22}x_2 + w_{23}x_3 + \cdots + w_{2N}x_N = p_2$$
$$\cdot$$
$$\cdot \qquad\qquad (42)$$
$$\cdot$$
$$w_{M1}x_1 + w_{M2}x_2 + w_{M3}x_3 + \cdots + w_{MN}x_N = p_M.$$

A grid representation with N cells gives an image N degrees of freedom. Therefore, an image as represented by $(x_1, x_2, ..., x_N)$ may be considered to be a single point in an N dimensional space. In this space each of the above equations represents a hyperplane. When a unique solution to these equation exists, the intersection of all these hyperplanes is a single point giving that solution. This concept is further illustrated in Fig. 13, in which for the purpose of display we have considered the case of only two variables x_1 and x_2 satisfying

$$w_{11}x_1 + w_{12}x_2 = p_1,$$
$$w_{21}x_1 + w_{22}x_2 = p_2. \qquad\qquad (43)$$

The computational procedure for locating the solution in Fig. 13 consists of starting with a guess, projecting this initial guess on the first line, reprojecting the resulting point on the second line, and then projecting back onto the first line, and so on, as shown in Fig. 13. If a unique solution exists, the iterations will always converge to that point. (Section IV contains a more detailed discussion on the effects of observation noise, the angle between the

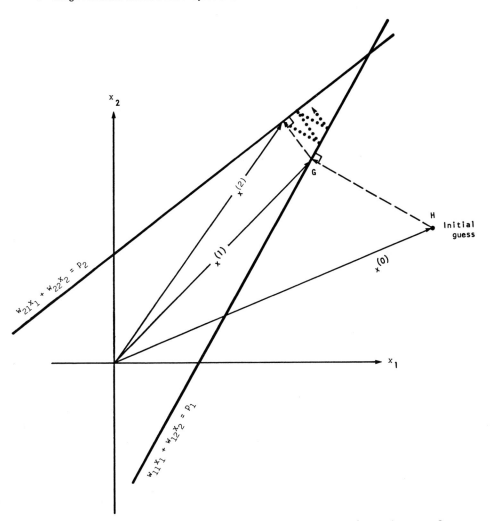

Fig. 13. Kaczmarz method of solving algebraic equations for the case of two unknowns. One starts with an initial guess and then projects onto the line corresponding to the first equation. The resulting point is now projected onto the line representing the second equation. If there are only two equations, this process is continued back and forth, as illustrated by the dots in the figure, until convergence is achieved.

hyperplanes, and so forth, on the convergence of the procedure.) A computer implementation of this procedure follows.

Step 1 One makes an initial guess at the solution. This guess, denoted by $x_1^{(0)}, x_2^{(0)}, \ldots, x_N^{(0)}$, is represented vectorially by $\mathbf{x}^{(0)}$ in the N-dimensional space. In most cases one assigns a value of zero to all the x_j. This initial guess is projected on the hyperplane represented by the first of the equations in Eq. (42) giving $\mathbf{x}^{(0)}$, as illustrated in Fig. 13 for the two-dimensional case. Then $\mathbf{x}^{(0)}$ is projected on the hyperplane represented by the second equation in Eq. (42) to yield $\mathbf{x}^{(2)}$, and so on. When $\mathbf{x}^{(j-1)}$ is projected on the hyperplane represented by the jth equation to yield $\mathbf{x}^{(j)}$, the process can be mathematically described by

$$\mathbf{x}^{(j)} = \mathbf{x}^{(j-1)} - \frac{(\mathbf{x}^{(j-1)} \cdot \mathbf{w}_j - p_j)}{\mathbf{w}_j \cdot \mathbf{w}_j} \mathbf{w}_j, \tag{44}$$

where $\mathbf{w}_j = (w_{j1}, w_{j2}, ..., w_{jN})$.

The difficulty with using Eq. (44) can be in the calculation, storage, and fast retrieval of the weight coefficients w_{ij}. Consider the case in which we wish to reconstruct an image on a 128×128 grid from 150 projections with 150 rays in each projection. The total number of weights w_{ij} needed in this case is $128 \times 128 \times 150 \times 150$ ($\approx 2.7 \times 10^8$), which can pose problems in fast storage and retrieval in applications in which reconstruction speed is important. This problem could be somewhat eased by making approximations, such as considering w_{ij} to be only a function of the perpendicular distance between the center of ith ray and the center of the jth cell. This perpendicular distance could then be computed at run time.

To get around the implementation difficulties caused by the weight coefficients, a myriad of other algebraic approaches have been suggested, many of which are approximations to Eq. (44). To discuss these more implementable approximations we shall first recast Eq. (44),

$$x_m^{(j)} = x_m^{(j-1)} + \left[\left(p_j - q_j\right) / \sum_{k=1}^{N} w_{jk}^2\right] w_{jm}, \tag{45}$$

where

$$q_j = \mathbf{x}^{(j-1)} \cdot \mathbf{w}_j = \sum_{k=1}^{N} x_k^{(j-1)} w_{jk}. \tag{46}$$

These equations say that when we project the $(j$-1)th solution onto the jth hyperplane [jth equation in Eq. (42)], the gray level of the mth element, whose current value is $x_m^{(j-1)}$, is obtained by correcting this current value by $\Delta x_i^{(j)}$, where

$$\Delta x_m^{(j)} = x_m^{(j)} - x_m^{(j-1)} = \frac{p_j - q_j}{\sum_{k=1}^{N} w_{jk}^2} w_{jm}. \tag{47}$$

Note that whereas p_j is the measured ray sum along the jth ray, q_j can be considered to be the computed ray sum for the same ray based on the $(j$-1)th solution for the image gray levels. The correction Δx_m to the mth cell is obtained by first calculating the difference between the measured ray sum and the computed ray sum, normalizing this difference by $\Sigma_{k=1}^{N} w_{jk}^2$, and then assigning this value to all the image cells in the jth ray, each assignment being weighted by the corresponding w_{jm}. This is illustrated in Fig. 14.

In one approximation to Eq. (47), which leads to one type of the so-called algebraic reconstruction technique, (ART) algorithms, the w_{jk} are simply replaced by 1's and 0's, depending on whether the center of the kth image cell is within the jth ray. This makes the implementation easier, because such a decision can easily be made at computer run time. Clearly, in this case the denominator in Eq. (47) is given by $\Sigma_{k=1}^{N} w_{jk}^2 = N_j$, which is the number of image cells whose centers are within the jth ray. The correction to the mth image cell from the jth equation in Eq. (47) may now be written as

$$\Delta x_m^{(j)} = (p_j - q_j)/N_j \qquad (48)$$

for all the cells whose centers are within the jth ray. We are essentially smearing back the difference $(p_j - q_j)/N_j$ over these image cells. (This is analogous to the concept of backprojection in the filtered-projection algorithms.) In Eq. (48), the q_j are calculated using the expression in Eq. (46), except that one now uses the binary approximation for the w_{jk}.

The approximation in Eq. (48), although easy to implement, may lead to artifacts in the reconstructed images if N_j is not a good approximation to the denominator. Superior reconstructions may be obtained if Eq. (48) is replaced by

$$\Delta x_m^{(j)} = p_j/L_j - q_j/N_j, \qquad (49)$$

where L_j is the length (normalized by δ, see Fig. 14) of the jth ray through the reconstruction region.

Step 2   The process of taking projections on different hyperplanes is continued until we get $\mathbf{x}^{(M)}$, which is obtained by taking the projection on the last equation in Eq. (42). This completes one cycle of this iterative procedure.

In steps 1 and 2, as each new equation in Eq. (42) is taken up, the values of all the affected pixels are updated. In a related approach, which may lead to fewer artifacts, one again computes the changes $\Delta x_m^{(j)}$ caused by the jth equation in Eq. (42). However, the value of the mth cell is not changed at this time. Before making any changes, one goes through all the equations, and then only at the end of each iteration are the cell values changed, the change for each cell being the average value of all the computed changes for that cell.

Step 3   One now iterates by projecting $\mathbf{x}^{(M)}$ on the first hyperplane again. [For example, for the two-dimensional case one reprojects $\mathbf{x}^{(2)}$ on the first

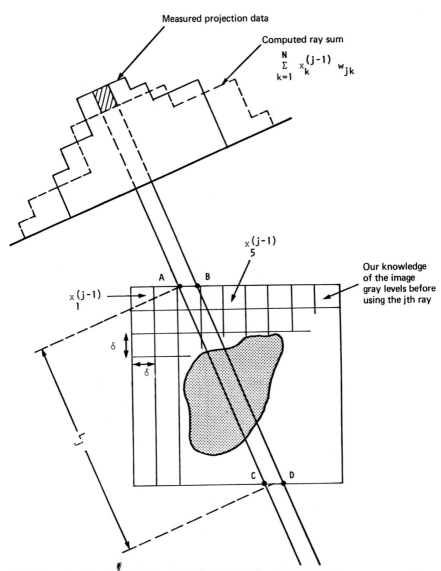

Fig. 14. Another way of looking at the Kaczmarz method. After using the $(j\text{-}1)$th ray, we first compute the ray sum for all the cells (including fractions thereof) in the jth ray. The computed ray sum is subtracted from the measured projection data for that ray. After normalization, this difference is assigned to all the cells in the jth ray.

hyperplane (in this case a line) to get $\mathbf{x}^{(3)}$ (Fig. 13).] This process continues until all the M hyperplanes have again been cycled through, resulting in $\mathbf{x}^{(2M)}$. The second iteration is started by projecting $\mathbf{x}^{(2M)}$ onto the first hyperplane, and so on. Iterations are stopped when the computed changes of the image cell values are negligible fractions of their current values.

The computational steps given above constitute the standard implementations of the ART algorithms. Although these algorithms enjoy a rapid convergence in the root-mean-squared error criterion, the reconstructed images, however, exhibit a very noisy salt-and-pepper characteristic. Smoother images are obtained by the simultaneous iterative reconstruction technique (SIRT) algorithm, in which the corrections Δx_m in Eqs. (48) or (49) are computed due to all the equations in Eq. (42) *before altering any pixel values.* Only then are the pixel values updated by their average corrections. This constitutes one iteration. In the second iteration, one goes back to the first equation in Eq. (42), and the process is repeated, and so on. In contrast with ART reconstructions, SIRT methods yield smoother images, although at the expense of slower convergence. [With ART also one can obtain smoother images by using relaxation, which consists of retaining only a fraction of the computed correction term in each iteration. This, however, slows down the convergence.]

In ART algorithms, the salt-and-pepper noise is caused by the inconsistencies introduced in the set of equations by the approximations commonly used for w_{jk}. This results in the computed ray sums in Eq. (46) being a poor approximation to the corresponding line integrals. The effect caused by such inconsistencies is exacerbated by the fact that as each equation corresponding to a ray in a projection is taken up, it changes some of the pixels just altered by the preceding equation in the same projection. The SIRT algorithm also suffers from these inconsistencies in the forward process [appearing in the computation of q_i in Eq. (46)]; but by eliminating the continual pixel update as each new equation is taken up, one obtains smoother reconstructions.

We have recently developed a variation on these algebraic approaches, which we call the simultaneous algebraic reconstruction technique (SART), that seems to combine the best of ART and SIRT. In only one iteration we can obtain reconstructions of good quality and numerical accuracy. In SART, discussed in detail in [2], we have improved the quality of the forward process by writing the equations in Eq. (41) as

$$p_j = \sum_{m=1}^{P_j} g(s_{jm})\, \Delta x, \tag{50}$$

where the s_{jm} are an equispaced set of points along the jth ray in the xy plane as shown in Fig. 15, and P_j is the total number of such points along the ray.

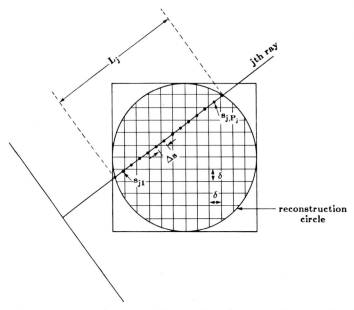

Fig. 15. A ray sum expressed as a sum of image values along a set of equispaced points on a straight line through the reconstruction circle. For the implementation of SART, the object is assumed to be contained within the circle shown. The physical length L_θ of a ray is given by the intercept of the staight line and the circle. Only the pixels contained within the circle are used for formulating the algebraic equations.

[Note that $g(x,y)$ denotes the image values as a function of the Cartesian coordinates in the xy plane.] The value $g(s_{jm})$ is now represented in terms of the four nearest x_n by bilinear interpolation. In general, we may write

$$g(s_{jm}) = \sum_{i=1}^{N} d_{ijm}x_i. \tag{51}$$

In bilinear interpolation, for each s_{jm} only the four nearest d_{ijm} are nonzero, and their sum equals unity. Combining the above two equations, we obtain

$$p_j = \sum_{i=1}^{N} w_{ij}x_i, \tag{52}$$

where

$$w_{ij} = \sum_{m=1}^{P_j} d_{ijm} \, \Delta s. \tag{53}$$

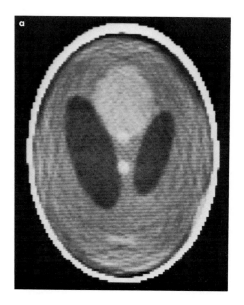

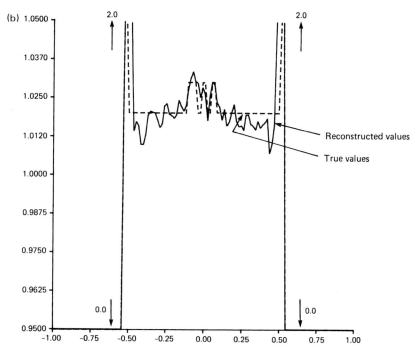

Fig. 16.   (a) A single-iteration algebraic reconstruction of the image of Fig. 4(a) with 127 projections and 100 rays in each projection.   The reconstruction is on a 128 × 128 matrix.   (b) A numerical comparison of the true and reconstructed values on line $y = -0.605$.

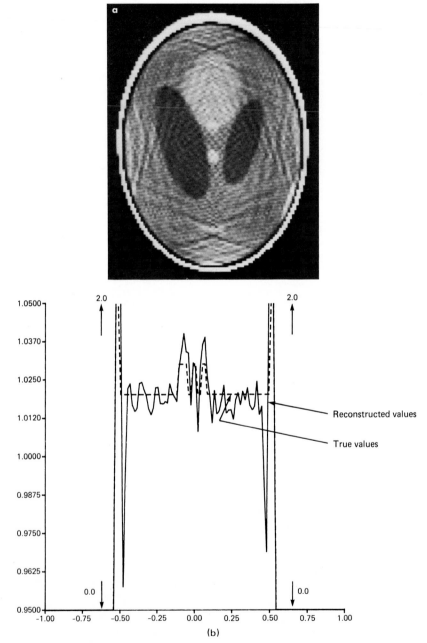

Fig. 17. (a) A two-iteration version of the reconstruction shown in Fig. 16(a). (b) A numerical comparison of the true and reconstructed values on line $y = -0.605$.

For the overall accuracy of this procedure, it is important that the weights corresponding to the pixels at the beginning and the end of each ray be such that $\Sigma_{i=1}^{N} w_{ij}$ be equal to the actual physical length L_j. As far as the choice of Δs (and, therefore, P_j) is concerned, we have found that setting it equal to half the spacing δ of the sampling lattice provides a good tradeoff between the accuracy of representation and the computational cost.

In addition to the above procedure for computing w_{ij}, we deviate from the traditional ART and SIRT algorithms in two other ways. All the pixel corrections Δx_m are computed for *one* complete projection, and only then are the pixels updated. Also, the back-distribution of the pixel corrections is longitudinally weighted with a Hamming window, the maximum of the window being at the center of the reconstruction circle.

Using the SART implementation, we have shown in Fig. 16(a) a *single* iteration algebraic 128 × 128 reconstruction of the test image of Fig. 4(a). The data used for the reconstruction consisted of 100 parallel projections over 180° with 127 rays in each projection. Figure 16(b) shows the reconstructed values on the line $y = -0.605$. Figure 17 shows the corresponding results for two iterations. Note that the improvement over a single iteration is marginal.

III. The Theory of Filtered-Backprojection Algorithms

In this section we shall present the theory underlying the filtered-backprojection algorithms for which the computational procedures were presented in the preceding section. Basic to the filtered-backprojection algorithms is the Fourier slice theorem.

A. The Fourier Slice Theorem

The Fourier slice theorem relates the one-dimensional Fourier transform of a projection of a function $g(x,y)$ to its two-dimensional Fourier transform. Let $G(u,v)$ be the Fourier transform of the image (x,y), which implies that

$$G(u,v) = \int_{-\infty}^{\infty} \int_{-\infty}^{\infty} g(x,y) \exp[-j2\pi(ux + vy)] \, dx \, dy \tag{54}$$

and

$$g(x,y) = \int_{-\infty}^{\infty} G(u,v) \exp[j2\pi(ux + vy)] \, du \, dv. \tag{55}$$

Also let $S_\theta(w)$ be the Fourier transform of the projection $P_\theta(t)$, that is

$$S_\theta(w) = \int_{-\infty}^{\infty} P_\theta(t) \exp(-j2\pi wt)\, dt. \qquad (56)$$

Let us first consider the values of $G(u,v)$ on the line $v = 0$ in the uv plane. From Eq. (54),

$$
\begin{aligned}
G(u,0) &= \int_{-\infty}^{\infty} \int_{-\infty}^{\infty} g(x,y) \exp(-j2\pi ux)\, dx\, dy \\
&= \int_{-\infty}^{\infty} \left[\int_{-\infty}^{\infty} g(x,y) dy \right] \exp(-j2\pi ux)\, dx \\
&= \int P_0(t) \exp(-j2\pi ut)\, dt = S_0(w),
\end{aligned}
\qquad (57)
$$

because $\int g(x,y)\, dy$ gives the projection of the image for $\theta = 0$. Note that for this projection x and t are the same.

The preceding result indicates that the values of the Fourier transform $G(u,v)$ on the line defined by $v = 0$ can be obtained by Fourier transforming the vertical (along y) projection of the image. This result can be generalized to show that if $G(w,\theta)$ denotes the values of $G(u,v)$ along a line at an angle θ with the u axis, as shown in Fig. 18, and if $S_\theta(w)$ is the Fourier transform of the projected $P_\theta(t)$, then

$$G(w,\theta) = S_\theta(w). \qquad (58)$$

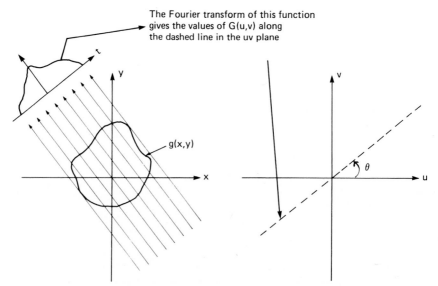

The Fourier transform of this function gives the values of G(u,v) along the dashed line in the uv plane

Fig. 18. Illustration of the Fourier slice theorem.

This can be proved as follows.   Let $g(t,s)$ be the function $g(x,y)$ in the rotated coordinates system (t,s) in Fig. 18.   The coordinates (t,s) are related to the (x,y) coordinates by

$$\begin{bmatrix} t \\ s \end{bmatrix} = \begin{bmatrix} \cos\theta \ \sin\theta \\ -\sin\theta \ \cos\theta \end{bmatrix} \begin{bmatrix} x \\ y \end{bmatrix}. \tag{59}$$

Clearly,

$$P_\theta(t) = \int_{-\infty}^{\infty} g(t,s) \, ds. \tag{60}$$

Therefore,

$$S_\theta(w) = \int_{\infty}^{\infty} P_\theta(t) \exp(-j2\pi wt) \, dt$$

$$= \int_{-\infty}^{\infty} \int_{-\infty}^{\infty} g(t,s) \, ds \, \exp(-j2\pi wt) \, dt. \tag{61}$$

Transforming the right-hand side of the above equation into xy coordinates, we obtain

$$S_\theta(w) = \int_{-\infty}^{\infty} \int_{-\infty}^{\infty} g(x,y) \exp[-j2\pi w(x\cos\theta + y\sin\theta)] \, dx \, dy$$

$$= G(u,v) \qquad \text{for} \quad \begin{cases} u = w\cos\theta \\ v = w\sin\theta \end{cases}$$

$$= G(w,\theta), \tag{62}$$

which prove Eq. (58).   This result is also known as the *projection slice theorem.*

B. Derivation of Filtered-Backprojection Equations for Parallel Data

If, as before, (w,θ) are the polar coordinates in the uv plane, the integral in Eq. (55) can be expressed as

$$g(x,y) = \int_0^{2\pi} \int_0^{\infty} G(w,\theta) \exp[j2\pi w(x\cos\theta + y\sin\theta)] \, w \, dw \, d\theta$$

$$= \int_0^{\pi} \int_0^{\infty} G(w,\theta) \exp[j2\pi w(x\cos\theta + y\sin\theta)] w \, dw \, d\theta$$

$$+ \int_0^{\pi} \int_0^{\infty} G(w,\theta + 180°) \exp\{j2\pi w[x\cos(\theta + 180°)$$

$$+ y\sin(\theta + 180°)]\}w \, dw \, d\theta.$$

Using the property $G(w,\theta + 180°) = G(-w,\theta)$, we can write the preceding expression for $g(x,y)$ as

$$g(x,y) = \int_0^\pi \left[\int_{-\infty}^\infty G(w,\theta)|w|\exp(j2\pi wt)\,dw \right] d\theta$$

$$= \int_0^\pi \left[\int_{-\infty}^\infty S_\theta(w)|w|\exp(j2\pi wt)\,dw \right] d\theta, \tag{63}$$

where, as noted before, $t = x \cos \theta + y \sin \theta$ and where we have used Eq. (62). The integral in Eq. (63) may be expressed as

$$g(x,y) = \int_0^\pi Q_\theta(x \cos \theta + y \sin \theta)\,d\theta, \tag{64}$$

where

$$Q_\theta(t) = \int_{-\infty}^\infty S_\theta(w)|w|\exp(j2\pi wt)\,dw. \tag{65}$$

These two equations form the basis of the algorithm discussed in Section II.A.

C. Derivation of Filtered-Backprojection Equations for Equiangular Fan Data

In this section, we shall present a derivation for Eq. (18) of Section II.B. Consider the ray SA in Fig. 8(b). If the projection data were generated along a set of parallel rays, then ray SA would belong to a parallel projection $P_\theta(t)$ for θ and t given by

$$\theta = \beta + \gamma \quad \text{and} \quad t = D \sin \gamma, \tag{66}$$

where D is the distance of the source S from the origin O. The relationships in Eq. (66) are derived by noting that all the rays in the parallel projection at angle θ are perpendicular to the line PQ and that along such a line the distance OB is equal to the value of t. We know from the preceding subsection that from the parallel projections $P_\theta(t)$ we may reconstruct $g(x,y)$ by

$$g(x,y) = \int_0^\pi \int_{-t_m}^{t_m} P_\theta(t)h_1(x \cos \theta + y \sin \theta - t)\,dt\,d\theta, \tag{67}$$

where t_m is the value of t for which $P_\theta(t) = 0$ with $|t| > t_m$ in all projections. This equation requires that the parallel projections only be collected over 180°. However, if one would like to use the projections generated over 360°, this equation may be rewritten as

$$g(x,y) = \tfrac{1}{2} \int_0^{2\pi} \int_{-t_m}^{t_m} P_\theta(t) h_1(x \cos \theta + y \sin \theta - t) \, dt \, d\theta. \tag{68}$$

Derivation of the algorithm becomes easier when the point (x,y), marked C in Fig. 8(c), is expressed in polar coordinates (r,ϕ), that is, $x = r \cos \phi$ and $y = r \sin \phi$. The expression in Eq. (68) can now be written as

$$g(r,\phi) = \tfrac{1}{2} \int_0^{2\pi} \int_{-t_m}^{t_m} P_\theta(t) h_1[r \cos(\theta - \phi) - t] \, dt \, d\theta. \tag{69}$$

Using the relationships in Eq. (66), one can express the double integration in terms of γ and β as

$$g(r,\phi) = \tfrac{1}{2} \int_{-\gamma}^{2\pi-\gamma} \int_{-\sin^{-1}(t_m/D)}^{\sin^{-1}(t_m/D)} P_{\beta+\gamma}(D \sin \gamma) h_1[r \cos(\beta + \gamma - \phi)$$

$$- D \sin \gamma] D \cos \gamma \, d\gamma \, d\beta, \tag{70}$$

where we have used $dt \, d\theta = D \cos \gamma \, d\gamma \, d\beta$. A few observations about this expression are in order. The limits $-\gamma$ to $2\pi - \gamma$ for β cover the entire range of 360°. Because all the functions of β are periodic (with period 2π), these limits may be replaced by 0 and 2π, respectively. $\sin^{-1}(t_m/D)$ is equal to the value of γ for the extreme ray SE in Fig. 8(b). Therefore, the upper and lower limits for γ may be written as γ_m and $-\gamma_m$, respectively. The expression $P_{\beta+\gamma}(D \sin \gamma)$ corresponds to the ray integral along SA in the parallel-projection data $P_\theta(t)$. The identity of this ray integral in the fan-projection data is simply $R_\beta(\gamma)$. Introducing these changes into Eq. (70), we obtain

$$g(r,\phi) = \tfrac{1}{2} \int_0^{2\pi} \int_{-\gamma_m}^{\gamma_m} R_\beta(\gamma) h_1(r \cos(\beta + \gamma - \phi) - D \sin \gamma) D \cos \gamma \, d\gamma \, d\beta. \tag{71}$$

To express the reconstruction formula given by Eq. (71) in a filtered-back-projection form we first examine the argument of the function h_1. The argument can be rewritten as

$$r \cos(\beta + \gamma - \phi) - D \sin \gamma = r \cos(\beta - \phi) \cos \gamma - [r \sin(\beta - \phi) a D] \sin \gamma. \tag{72}$$

Let L be the distance from the source S to a point (x,y) [or (r,ϕ) in polar coordinates], such as C in Fig. 8(c). Clearly, L is a function of three variables r, ϕ, and β. Also, let γ' be the angle of the ray that passes through this point (r,ϕ). One can now show that

$$L \cos \gamma' = D + r \sin(\beta - \phi),$$

and $\hspace{9cm}$ (73)

$$L \sin \gamma' = r \cos(\beta - \phi).$$

Note that the pixel location (r,ϕ) and the projection angle β completely determine both L and γ':

$$L(r,\phi,\beta) = \sqrt{[D + r \sin(\beta - \phi)]^2 + [r \cos(\beta - \phi)]^2}$$

and $\hspace{9cm}$ (74)

$$\gamma' = \tan^{-1} \frac{r \cos(\beta - \phi)}{D + r \sin(\beta - \phi)}.$$

Using Eq. (73) in Eq. (72), we obtain for the argument of h

$$r \cos(\beta + \gamma - \phi) - D \sin \gamma = L \sin(\gamma' - \gamma), \tag{75}$$

and substituting this in Eq. (71), we obtain

$$g(r,\phi) = \tfrac{1}{2} \int_0^{2\pi} \int_{-\gamma_m}^{\gamma_m} R_\beta(\gamma) h_1[L \sin(\gamma' - \gamma)] D \cos \gamma \, d\gamma \, d\beta. \tag{76}$$

We shall now express the function $h_1[L \sin(\gamma' - \gamma)]$ in terms of h_1. Note that $h_1(t)$ is the inverse Fourier transform of $|w|$ in the frequency domain:

$$h_1(t) = \int_{-\infty}^{\infty} |w| \exp(j2\pi wt) \, dw.$$

Therefore,

$$h_1(L \sin \gamma) = \int_{-\infty}^{\infty} |w| \exp(j2\pi wL \sin \gamma) \, dw. \tag{77}$$

Using the transformation

$$w' = wL \sin \gamma / \gamma, \tag{78}$$

we can write

$$h_1(L \sin \gamma) = \left(\frac{\gamma}{L \sin \gamma} \right)^2 \int_{-\infty}^{\infty} |w'| \exp(j2\pi w'\gamma) \, dw'$$

$$= \left(\frac{\gamma}{L \sin \gamma} \right)^2 h_1(\gamma). \tag{79}$$

Therefore, Eq. (76) may be written as

$$g(r,\phi) = \int_0^{2\pi} \frac{1}{L^2} \int_{-\gamma_m}^{\gamma_m} R_\beta(\gamma) h_2(\gamma' - \gamma) D \cos \gamma \, d\gamma \, d\beta, \tag{80}$$

where

$$h_2(\gamma) = \tfrac{1}{2}\,(\gamma/\sin\,\gamma)^2 h_1(\gamma). \tag{81}$$

Equation (80) formed the basis of the reconstruction algorithm discussed in Section II.B.

IV. The Theory of Algebraic Algorithms

To see how Eq. (44) comes about, we write the first equation of Eq. (42) or Eq. (41) as

$$\mathbf{w}_1 \cdot \mathbf{x} = p_1. \tag{82}$$

The hyperplane represented by this equation is perpendicular to the vector $\mathbf{w}_1$. This is illustrated in Fig. 19, in which the vector $\mathbf{OD}$ represents $\mathbf{w}_1$. This equation says simply that the projection of a vector $\mathbf{OC}$ (for any point C on the hyperplane) on the vector $\mathbf{w}_1$ is of constant length. The unit vector $\mathbf{OU}$ along $\mathbf{w}_1$ is given by

$$\mathbf{OU} = \mathbf{w}_1/\sqrt{\mathbf{w}_1 \cdot \mathbf{w}_1}, \tag{83}$$

and the perpendicular distance of the hyperplane from the origin, which is equal to the length of $\overline{\mathbf{OA}}$ in Fig. 19, is given by $\mathbf{OC} \cdot \mathbf{OU}$:

$$|\mathbf{OA}| = \mathbf{OU} \cdot \mathbf{OC} = (1/\sqrt{\mathbf{w}_1 \cdot \mathbf{w}_1})(\sqrt{\mathbf{w}_1 \cdot \mathbf{w}_1}\;\mathbf{OU} \cdot \mathbf{OC})$$
$$= (1/\sqrt{\mathbf{w}_1 \cdot \mathbf{w}_1})(\mathbf{w}_1 \cdot \mathbf{x}) = p_1/\sqrt{\mathbf{w}_1 \cdot \mathbf{w}_1}. \tag{84}$$

To obtain $\mathbf{x}^{(1)}$ we have to subtract from $\mathbf{x}^{(0)}$ the vector $\mathbf{HG}$:

$$\mathbf{x}^{(1)} = \mathbf{x}^{(0)} - \mathbf{HG}, \tag{85}$$

where the length of the vector $\mathbf{HG}$ is given by

$$|\mathbf{HG}| = |\mathbf{OF}| - |\mathbf{OA}|$$
$$= \mathbf{x}^{(1)} \cdot \mathbf{OU} - \mathbf{OA}. \tag{86}$$

Substituting Eqs. (78) and (79) in this equation, we obtain

$$|\mathbf{HG}| = (\mathbf{x}^{(0)} \cdot \mathbf{w}_1 - p_1)/\sqrt{\mathbf{w}_1 \cdot \mathbf{w}_1}. \tag{87}$$

Because the direction of $\mathbf{HG}$ is the same as that of the unit vector $\mathbf{OU}$, we can write

$$\mathbf{HG} = |\mathbf{HG}|\,\mathbf{OU} = \frac{\mathbf{x}^{(0)} \cdot \mathbf{w}_1 - p_1}{\mathbf{w}_1 \cdot \mathbf{w}_1}\,\mathbf{w}_1. \tag{88}$$

Substituting Eq. (88) in Eq. (85), we obtain Eq. (44).

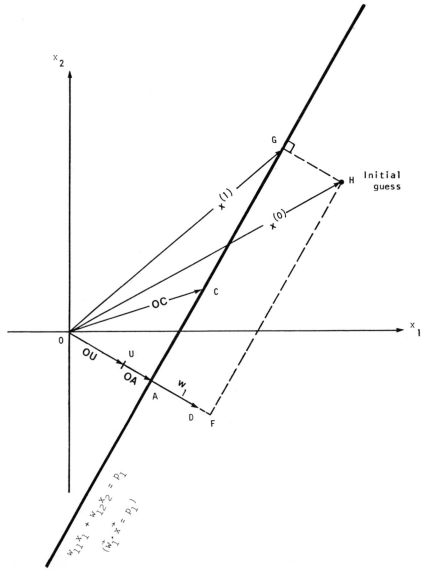

Fig. 19. The hyperplane $\mathbf{w}_1 \cdot \mathbf{x} = p_1$ (represented by a line in this two-dimensional figure), perpendicular to the vector $\mathbf{w}_1$.

The computational procedure in Section II.D consisted of starting with an initial guess for the solution, taking successive projections on the hyperplanes represented by the equations in Eq. (42), and eventually obtaining $\mathbf{x}^{(M)}$. In the next iteration, $\mathbf{x}^{(M)}$ was projected onto the hyperplane represented by the first equation in Eq. (42), and successively onto the rest of the hyperplanes in Eq. (42), to yield $\mathbf{x}^{(2M)}$, and so on. Tanabe [72] has shown that if a unique solution $\mathbf{x}_s$ to the system of equations in Eq. (42) exists, then

$$\lim_{k \to \infty} \mathbf{x}^{(kM)} = \mathbf{x}_s. \tag{89}$$

A few comments about the convergence of the algorithm are in order here. If in Fig. 13 the two hyperplanes are perpendicular to each other, the reader can show that given (for an initial guess) any point in the (x_1, x_2) plane, it is possible to arrive at the correct solution in only two steps like Eq. (44). On the other hand, if the two hyperplanes have only a very small angle between them, the value of k in Eq. (89) may acquire a large value (depending on the initial guess) before the correct solution is reached. Clearly, the angles between the hyperplanes influence the rate of convergence to the solution. If the M hyperplanes in Eq. (42) could be made orthogonal with respect to one another, the correct solution would be obtained with only one pass through the M equations (assuming a unique solution exists), and the value of k in Eq. (89) would have to equal 1 for us to get the correct solution. Although such orthogonalization is theoretically possible using, for example, the Gram–Schmidt procedure, in practice it is not computationally feasible. Full orthogonalization will also tend to enhance the effects of the ever-present measurement noise in the final solution. Ramakrishnan *et al.* [61] have suggested a pairwise orthogonalization scheme that is computationally easier to implement and at the same time considerably increases the speed of convergence.

A common situation in image reconstruction is that of an overdetermined system in the presence of measurement noise. That is, we may have $M > N$ in Eq. (42) and $p_1, p_2, ..., p_m$ corrupted by noise. No unique solution exists in this case. In Fig. 20 we show a two-variable system represented by three noisy hyperplanes. The dashed line represents the course of the solution as we successively implement Eq. (44). The solution does not converge to a unique point but will oscillate in the neighborhood of the region of the intersections of the hyperplanes.

When $M < N$, a unique solution of the set of linear equations in Eq. (42) does not exist; and, in fact, an infinite number of solutions are possible. For example, suppose only the first of two equations in Eq. (43) is given to use for the two unknowns x_1 and x_2; then the solution can be anywhere on the line corresponding to this equation. Given the initial guess $\mathbf{x}^{(0)}$ (see Fig. 13), the

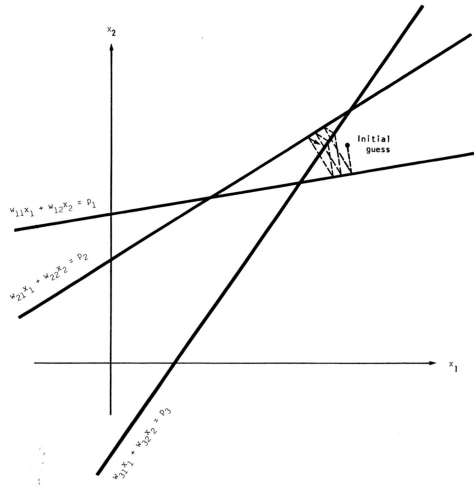

Fig. 20. The case in which the number of equations is greater than the number of unknowns. the lines do not intersect at a single point, because the observations p_1, p_2, p_3 are assumed to be corrupted by noise. No unique solution exists in this case, and the final solution will oscillate in the neighborhood of intersections of the three lines.

best one can probably do is to draw a projection from $\mathbf{x}^{(0)}$ on the line and call the resulting $\mathbf{x}^{(1)}$ the solution. Note that the solution obtained in this way corresponds to the point on the given line that is closest to the initial guess. This result has been rigorously proved by Tanabe [72], who has shown that when $M < N$, the iterative approach described above converges to a solution, call it $\mathbf{x}'_s$, such that $|\mathbf{x}^{(0)} - \mathbf{x}'_s|$ is minimized.

Besides its computational efficiency, another attractive feature of the iterative approach presented here is that it is now possible to incorporate into the solution some types of a priori information about the image one is reconstructing. For example, if it is known a priori that the image $g(x,y)$ is nonnegative, then in each of the solutions $x^{(k)}$, successively obtained by using Eq. (44), one may set the negative components equal to zero. One may similarly incorporate the information that $g(x,y)$ is zero outside a certain area, if this is known.

V. Aliasing Artifacts

Figure 21 shows 16 reconstructions of an ellipse with various values of K, the number of projections, and N, the number of rays in each projection. The projections for the ellipse were generated by using Eq. (4). The gray level inside the ellipse, ρ in Eq. (2), was 1 and the background 0. To bring out all the artifacts, the reconstructed images were windowed between 0.1 and -0.1. (In other words, all the gray levels above 0.1 were set at white and all below -0.1 at black.) The images in Fig. 21 are displayed on a 128×128 matrix. Figure 22 is a graphical depiction of the reconstructed numerical values on the middle horizontal lines for two of the images in Fig. 21. From Figs. 21 and 22 the following artifacts are evident: Gibbs phenomenon, streaks and Moiré patterns. In this section we shall discuss these artifacts and their sources.

The streaks evident in Fig. 21 for the cases when N is small and K is large are caused by aliasing errors in the projection data. Note that a fundamental problem with tomographic images in general is that the objects (in this case an ellipse) and, therefore, their projections are not bandlimited. In other words, the bandwidth of the projection data exceeds the highest frequency that can be recorded at a given sampling rate. To illustrate how aliasing errors enter the projection data, we assume that the Fourier transform $S_\theta(f)$ of a projection $P_\theta(t)$ looks as shown in Fig. 23(a). The bandwidth of this function is B as shown there. We choose a sampling interval τ for sampling the projection and associate with this interval a measurement bandwidth W, which is equal to $1/2\tau$. We assume that $W < B$. On the other hand, the Fourier transform of the *samples* of the projection data is given by Fig. 23(b). We see that the information within the measurement band is contaminated by the tails (shaded areas) of the higher and lower replications of the original Fourier transform. This contaminating information constitutes the aliasing errors in the sampled projection data. These contaminating frequencies constitute the aliased spectrum.

Backprojection is a linear process, so the final image can be thought of as made up of two functions. One is the image made from the bandlimited projections, degraded primarily by the finite number of projections. The

Number of projections (K)

64   128   256   512

Samples per projection (N)

64

128

256

512

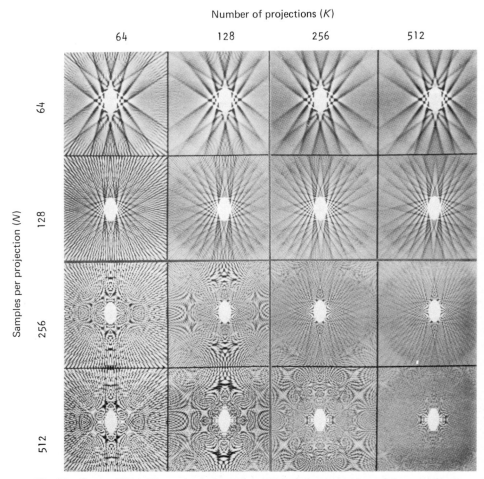

Fig. 21. Sixteen 128 × 128 reconstructions of an ellipse for various values of the number of projections and the number of rays in each projection. The reconstructions were windowed for the purpose of display to bring out the aliasing streaks and Moiré artifacts.

second is the image made from the aliased portion of the spectrum in each projection.

The aliased portion of the reconstruction can be isolated by subtracting the transforms of the sampled projections from the corresponding theoretical transforms of the original projections. Then, if this result is filtered as before, the final reconstructed image will be that of the aliased spectrum. We performed a computer simulation study along these lines for an elliptical

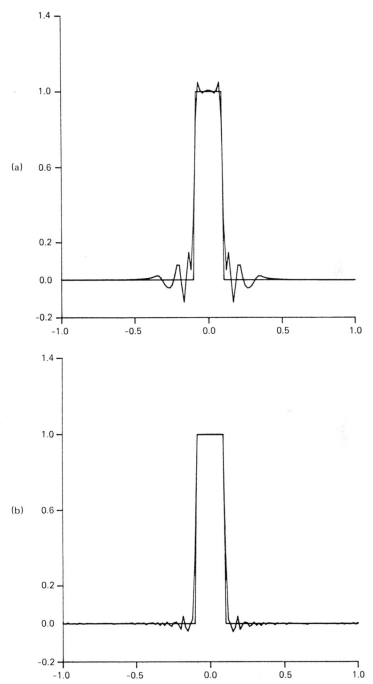

Fig. 22. A numerical comparison of the true and reconstructed values on the middle horizontal lines in two of the reconstructions in Fig. 19. The jagged lines are the reconstructed values, whereas the straight lines are the true values. (a) 512 projections with 64 rays in each. (b) 512 projections with 512 rays in each.

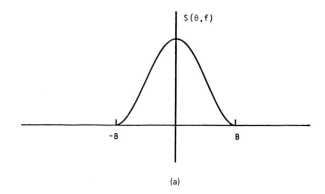

(a)

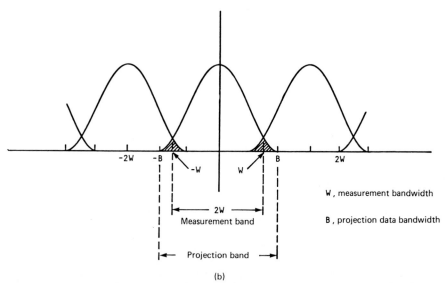

(b)

Fig. 23. Depiction of aliasing distortion. (a) $S(\theta,f)$ is the Fourier transform of the true projection at angle θ. (b) Some of the replications of $S(\theta,f)$. Sum of these replications constitute the Fourier transform of the sampled projection data.

object. To present the result of this study we first show in Fig. 24(a) the reconstruction of an ellipse for $N = 64$. (The number of projections was 512 and will remain the same for the discussion here.) We subtracted the transform of each projection for the $N = 64$ case from the corresponding transform for the $N = 1024$ case. The latter was assumed to be the true transform, because the projections are oversampled (at least in comparison with

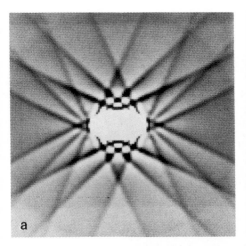
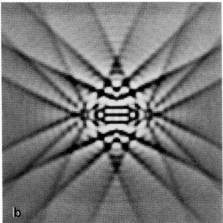

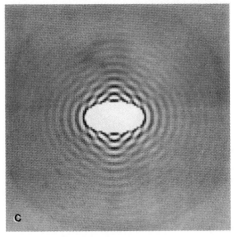

Fig. 24.   (a) Reconstruction of an ellipse with 512 projections and 64 rays in each projection.   (b) Reconstruction from only the aliased frequencies in each projection.   Note that the streaks exactly match those in (a).   (c) Image obtained by subtracting (b) from (a).   This is the reconstruction that would be obtained provided the data for the $N = 64$ case were truly bandlimited.

the $N = 64$ case). The reconstruction obtained from the difference data is shown in Fig. 24(b). Figure 24(c) is the bandlimited image obtained by subtracting the aliased spectrum image of Fig. 24(b) from the complete image shown in Fig. 24(a). Figure 24(c) is the reconstruction that would be obtained if the projection data for the $N = 64$ case were truly bandlimited (i.e., if it did not suffer from aliasing errors after sampling). The aliased-spectrum reconstruction in Fig. 24(b) and the absence of streaks in Fig. 24(c) prove our point that when the number of projections is large, the streaking artifacts are caused by aliasing errors in the projection data.

The reader may have noticed that the thin streaks caused by an insufficient number of projections (see, for example, the image for $N = 512$ and $K = 64$ in Fig. 21) appear broken. This is caused by two-dimensional aliasing due to the display grid being only 128×128. When, say, $N = 512$, the highest frequency in each projection can be 256 cycles per projection length; whereas the highest frequency that can be displayed on the image grid is 64 cycles per image width (or height). The effect of this two-dimensional aliasing is pronounced in the left three images for the $N = 512$ row and the left two images for the $N = 256$ case in Fig. 21. The artifacts generated by this two-dimensional aliasing are also called Moiré patterns.

The thin streaks that are evident in Fig. 21 for the cases of large N and small K (e.g., when $N = 512$ and $K = 64$) are caused by an insufficient number of projections. It is easily shown that when only a small number of filtered projections of a small object are backprojected, the result is a star-shaped pattern. This is illustrated in Fig. 25 where we show in (a) four projections of a point object, in (b) the filtered projections, and in (c) their backprojections.

The number of projections should be roughly equal to the number of rays in each projection. This can be shown analytically for the case of parallel projections by the use of the following argument: By the Fourier slice theorem, the Fourier transform of each projection is a slice of the two-dimensional Fourier transform of the object. In the frequency domain shown in Fig. 26, each radial line, such as $A_1 A_2$, is generated by one projection. If there are M_{proj} projections uniformly distributed over 180°, the angular interval δ between successive radial lines is given by

$$\delta = \pi/M_{\text{proj}}. \tag{90}$$

If τ is the sampling interval used for each projection, the highest spatial frequency W measured for each projection will be

$$W = 1/2\tau.$$

This is the radius of the disc shown in Fig. 26. The distance between the consecutive sampling points on the periphery of this disc is equal to $\overline{A_2 B_2}$ and is given by

$$\overline{A_2 B_2} = W\delta = (1/2\tau)(\pi/M_{\text{proj}}). \tag{92}$$

If there are N_{ray} sampling points in each projection, the total number of independent frequency domain sampling points on a line such as $A_1 A_2$ will also be the same. Therefore, the distance ϵ between any two consecutive sampling points on each radial line in Fig. 26 will be

$$\epsilon = 2W/N_{\text{ray}} = 1/\tau N_{\text{ray}}. \tag{93}$$

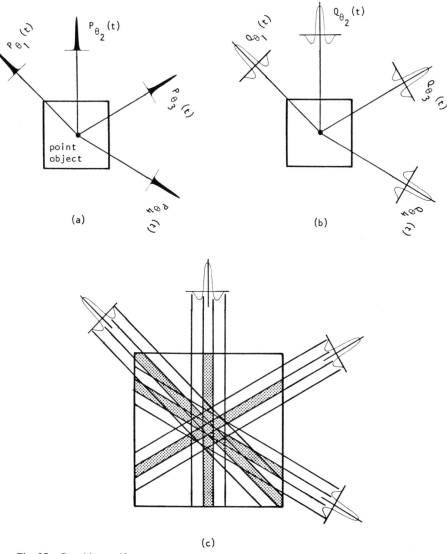

Fig. 25. Streaking artifacts caused by the number of projections being too small. (a) Four projections of a point-like object. (b) The corresponding filtered projections. (c) Backprojections for each of the filtered projections. Sum of these backprojections will form a star shaped pattern.

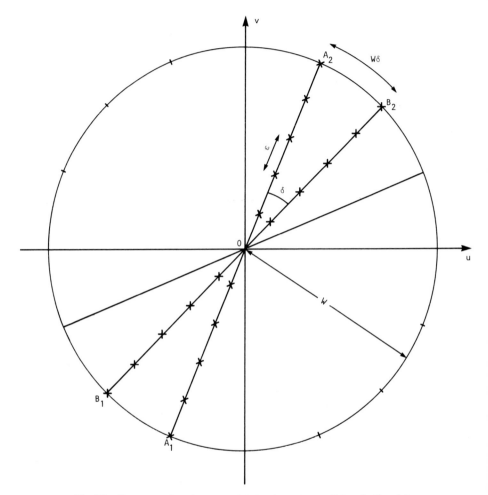

Fig. 26. Frequency domain parameters pertinent to parallel projection data.

Because in the frequency domain the worst-case azimuthal resolution should be approximately the same as the radial resolution, we must have

$$(1/2\tau)(\pi/M_{\text{ray}}) \simeq 1/\tau N_{\text{ray}} , \tag{94}$$

which is obtained by equating Eqs. (92) and (93). Equation (94) reduces to

$$M_{\text{proj}}/N_{\text{ray}} \simeq \pi/2. \tag{95}$$

From the computer simulation and analytical results presented in this section, one can conclude that for a well-balanced $N \times N$ reconstructed image, the number of rays in each projection should be roughly N, the total number of projections should be roughly N, and the total number of projections should also be roughly N.

VI. Bibliographical Notes

The first mathematical solution to the problem of reconstructing a function from its projections was given by Radon [59] in 1917. More recently some of the first investigators to examine this problem theoretically and experimentally (and often independently) include, in roughly chronological order: Bracewell [7], Oldendorf [56], Cormack [16,17], Kuhl and Edwards [48], De-Rosier and Klug [21], Tretiak *et al.* [76], Rowley [63], Berry and Gibbs [6], Ramachandran and Lakshminarayanan [60], Bender *et al.* [5], and Bates and Peters [4]. A detailed survey of the work done in computed tomographic imaging before 1979 appears in [44].

The idea of filtered-backprotection was first advanced by Bracewell and Riddle [8] and later, independently, by Ramachandran and Lakshminarayanan [60]. The superiority of the filtered-backprojection algorithms over the algebraic techniques was first demonstrated by Shepp and Logan [67]. Its development for the fan-beam data was first made by Lakshminarayanan [50] for the equispaced collinear detectors case and was later extended by Herman and Naparstek [37] to the case of equiangular rays. The fan-beam algorithm derivation presented here was first developed by Scudder [66]. Many authors [3,47,49,51,73] have proposed variations on the filter functions discussed in this chapter. The reader is referred particularly to [47,51] for ways to speed up the filtering of the projection data by using binary approximations and/or inserting zeroes in the unit sample response of the filter function.

Images may also be reconstructed by a *direct* two-dimensional Fourier transformation by a method derived first by Bracewell [7] for radioastronomy and later independently by DeRosier and Klug [21] in electron microscopy and Rowley [63] in optical holography. Several workers who applied this method to radiography include Tretiak *et al.* [76], Bates and Peters [4], and Mersereau and Oppenheim [53]. To utilize two-dimensional FFT algorithms for image formation, the direct Fourier approach calls for frequency domain interpolation from a polar grid to a rectangular grid. For some methods to minimize the resulting interpolation error, the reader is referred to [68,69]. Wernecke and D'Addario [78] have proposed a maximum entropy approach to direct Fourier inversion. Their procedure is especially useful if for some reason the projection data are insufficient. The reader is also referred to [64,71] for reconstructions from incomplete and limited projections.

Images may also be reconstructed from fan-beam data by first sorting them into parallel projection data. Fast algorithms for ray sorting of fan-beam data have been developed by Wang [77], Dreike and Boyd [27], Peters and Lewitt [58], and Dines and Kak [24]. The reader is referred to [54] for a filtered-backprojection algorithm for reconstructing from data generated by using very narrow-angle fan-beams that rotate *and* traverse *continuously* around the object.

Aliasing artifacts in image reconstruction have been studied by Brooks *et al.* [11] and Crawford and Kak [18]. In connection with noise in reconstructed images, Shepp and Logan [67] first showed that if one uses filtered-backprojection algorithms, the variance of the noise in a reconstructed image is directly proportional to the area under the square of the filter function. This derivation was based on the assumption that the variance of the measurement noise is same for all the rays in the projection data, a condition that is usually not satisfied; a more general expression (without this assumption) for the noise variance was derived by Kak [43], who has also introduced the concept of *the relative-uncertainty image*. The reader is also referred to [10,13,75] for noise properties of x-ray tomograms, the noise being caused by the quantum nature of x-rays as discussed in [74].

The algebraic approaches to image reconstruction have been studied in great detail by Gordon *et al.* [29–31], Herman *et al.* [36,38], and Budinger and Gullberg [12]. To improve to some extent the convergence properties of the algebraic techniques, a simultaneous iterative reconstruction technique was proposed by Gilbert [28].

The reconstruction implementations that we have presented in this chapter were based on the assumption that uniform sampling is used for projection data and that each projection is complete in the sense that it spans the entire cross section. (Of course, in a sense, an incomplete projection could be considered to be a case of nonuniform sampling also.) The reader is referred to [39,40] for algorithms for nonuniformly sampled projections data and to [9,52,57] for reconstructions from incomplete projections.

The reader is referred to [43] for a survey of medical-imaging applications of reconstructing images from projections. For applications in radio astronomy, in which the aim is to reconstruct the brightness distribution of a celestial source of radio waves from its strip-integral measurements taken with special antenna beams, the reader is referred to [7,8]. For electron microscopy applications, in which one attempts to reconstruct the molecular structure of complex biomolecules from transmissions micrograms, the reader should look up [20,32]. The applications of this technique in optical interferometry, in which the aim is to determine the refractive-index field of an optically transparent medium, are discussed in [6,63,70]. On how the ultrasonic transmission data may be reduced so as to be usable for tomographic reconstructions, the reader is referred to [25,45].

Space limitations have prevented us from discussing algorithms for doing full three-dimensional reconstructions [14,15]. Neither have we presented the circular harmonic transform method of image reconstruction proposed by Hansen [34,35].

Finally, the reader is referred to [44] for some recent advances in computed medical tomography, an area concerned with, and mostly based on, the concepts discussed in this chapter.

Acknowledgments

The author would like to express appreciation to his former graduate students Carl Crawford, Malcolm Slaney, and Anders Andersen for many stimulating discussions on image reconstruction from projections and for providing the computer simulation results shown in this chapter.

References

1. A. H. Andersen and A. C. Kak, Digital ray tracing in two-dimensional refractive fields, *J. Acoust. Soc. Am.* **72**, 1982, 1593–1606.
2. A. H. Andersen and A. C. Kak, Simultaneous algebraic reconstruction technique: A new implementation of the ART algorithm, *Ultrasonic Imaging,* **6**, 1984, 81–94.
3. N. Baba and K. Murata, Filtering for image reconstruction from projections, *J. Opt. Soc. Am.,* **67**, 1977, 662–668.
4. R. H. T. Bates and T. M. Peters, Towards improvements in tomography, *New Zealand J. Sci.* **14**, 1971, 883–896.
5. R. Bender, S. H. Bellman, and R. Gordon, ART and the ribosome: A preliminary report on the three dimensional structure of individual ribosomes determined by an algebraic reconstruction technique, *J. Theor. Biol.* **29**, 1970, 483–487.
6. M. V. Berry and D. F. Gibbs, The interpretation of optical projections, *Proc. R. Soc. London,* **A314**, 1970, 143–152.
7. R. N. Bracewell, Strip integration in radio astronomy, *Aust. J. Phys.* **9**, 1956, 198–217.
8. R. N. Bracewell and A. C. Riddle, Inversion of fan-beam scans in radio astronomy, *Astrophys. J.* **150**, 1967, 427–434.
9. R. N. Bracewell and S. J. Wernecke, Image reconstruction over a finite field of view, *J. Opt. Soc. Am.* **65**, 1975, 1342–1346.
10. R. A. Brooks and G. Dichiro, Statistical limitations in x-ray reconstruction tomography, *Med. Phys.* **3**, 1976, 237–240.
11. R. A. Brooks, G. H. Weiss, and A. J. Talbert, A new approach to interpolation in computed tomography, *J. Comput. Assist. Tomog.* **2**, 1978, 577–585.
12. T. F. Budinger and G. T. Gullberg, Three-dimensional reconstruction in nuclear medicine emission imaging, *IEEE Trans. Nucl. Sci.* **NS-21**, 1974, 2–21.
13. D. A. Chesler, S. J. Riederer, and N. J. Pele, Noise due to photon counting statistics in computer x-ray tomography, *J. Comput. Assist. Tomog.* **1**, 1977, 64–74.
14. M. Y. Chiu, H. H. Barrett, and R. G. Simpson, Three dimensional reconstruction from planar projections, *JOSA* **70**, 1980, 755–762.

15. M. Y. Chiu, H. H. Barrett R. G. Simpson, C. Chou, J. W. Arendt, and G. R. Gindi, Three dimensional radiographic imaging with a restricted view angle, *JOSA* **69**, 1979, 1323–1330.
16. A. M. Cormack, Representation of a function by its line integrals with some radiological applications, *J. Appl. Phys.* **34**, 1963, 2722–2727.
17. A. M. Cormack, Representation of a function by its line integrals with some radiological applications, II, *J. Appl. Phys.* **35**, 1964, 2908–2913.
18. C. R. Crawford and A. C. Kak, Aliasing artifacts in computerized tomography, *Appl. Opt.* **18**, 1979, 3704–3711.
19. C. R. Crawford and A. C. Kak, Multipath artifact corrections in ultrasonic transmission tomography, *Ultrason. Imag.* **4**, 1982, 234–266.
20. R. A. Crowther, D. J. DeRosier, and A. Klug, The reconstruction of a three-dimensional structure from projections and its applications to electron microscopy, *Proc. R. Soc. London* **A317**, 1970, 319–340.
21. D. J. DeRosier and A. Klug, Reconstruction of three dimensional structures from electron micrographs, *Nature* **217**, 1968, 130–134.
22. A. J. Devaney, A filtered backpropagation algorithm for diffraction tomography, *Ultrason. Imag.* **4**, 1982, 336–350.
23. A. J. Devaney, "A computer simulation study of diffraction tomography," *IEEE Trans. Biomed. Engrg.*, **BME-30**, 1983, 377–386.
24. K. A. Dines and A. C. Kak, "Measurement and reconstruction of ultrasonic parameters for diagnostic imaging," Rep. TR-EE 77-4, School of Elec. Eng., Purdue Univ., West Lafayette, IN, 1976.
25. K. A. Dines and A. C. Kak, Ultrasonic attenuation tomography of soft biological tissues, *Ultrason. Imag.* **1**, 1979, 16–33.
26. K. A. Dines and R. J. Lytle, Computerized geophysical tomography, *Proc. IEEE* **67**, 1979, 1065–1073.
27. P. Dreike and D. P. Boyd, Convolution reconstruction of fan-beam reconstructions, *Comput. Graph. Image Proc.* **5**, 1977, 459–469.
28. P. Gilbert, Iterative methods for the reconstruction of three dimensional objects from their projections, *J. Theor. Biol.* **36**, 1972, 105–117.
29. R. Gordon, A tutorial on ART (Algebraic Reconstruction Techniques), *IEEE Trans. Nucl. Sci.* **NS-21**, 1974, 78–93.
30. R. Gordon, R. Bender, and G. T. Herman, Algebraic reconstruction techniques (ART) for three dimensional electron microscopy and X-ray photography, *J. Theor. Biol.* **29**, 1971, 470–481.
31. R. Gordon and G. T. Herman, Three-dimensional reconstruction from projections: A review of algorithms, *International Review Cytology,* **38**, 1971, 111–151.
32. R. Gordon and G. T. Herman, Reconstruction of pictures from their projections, *Commun. Assoc. Comput. Mach.* **14**, 1971, 759–768.
33. R. W. Hamming, "*Digital Filters.*" Prentice-Hall, New York, 1977.
34. E. W. Hansen, Theory of circular image reconstruction, *JOSA* **71**, 1981, 304–308.
35. E. W. Hansen, Circular harmonic image reconstruction: Experiments, *Appl. Opt.* **20**, 1981, 2266–2274.
36. G. T. Herman, A. Lent, and S. Rowland, ART: Mathematics and applications: A report on the mathematical foundations and on applicability to real data of the algebraic reconstruction techniques, *J. Theor. Biol.* **43**, 1973, 1–32.
37. G. T. Herman and A. Naparstek, Fast image reconstruction based on a Radon inversion formula appropriate for rapidly collected data, *SIAM J. Appl. Math.* **33**, 1977, 511–533.
38. G. T. Herman and S. Rowland, Resolution in ART: An experimental investigation of the resolving power of an algebraic picture reconstruction, *J. Theor. Biol.* **33**, 1971, 213–223.

39. B. K. P. Horn, Density reconstructions using arbitrary ray sampling schemes, *Proc. IEEE* **66**, 1968, 551–562.
40. B. K. P. Horn, Fan-beam reconstruction methods, *Proc. IEEE* **67**, 1979, 1616–1623.
41. C. V. Jakowatz, Jr. and A. C. Kak, "Computerized tomography using x-rays and ultrasound," Report TR-EE 76-26. School of Elect. Engrg., Purdue Univ., 1976.
42. S. Kaczmarz, Angenaherte Auflosung von Systemen linearer Gleichungen, *Bull. Acad. Polon. Sci. Lett.* **A**, 1937, 355–357.
43. A. C. Kak, Computerized tomography with x-ray, emission and ultrasound sources, *Proc. IEEE* **67**, 1979, 1245–1272.
44. A. C. Kak (Guest Ed.), Computerized medical imaging, Special Issue of *IEEE Trans. Biomed. Eng.* Feb. 1981.
45. A. C. Kak and K. A. Dines, Signal processing of broadband pulse ultrasound: Measurement of attenuation of soft biological tissues, *IEEE Trans. Biomed. Eng.* **BME-25**, 1978, 321–344.
46. P. N. Keating, More accurate interpolation using discrete Fourier transforms, *IEEE Trans. Acoust. Speech Signal Proc.* **ASSP-26**, 1978, 368–369.
47. S. K. Kenue and J. F. Greenleaf, Efficient convolution kernels for computerized tomography, *Ultrason. Imag.* **1**, 1979, 232–244.
48. D. E. Kuhl and R. Q. Edwards, Image separation radio-isotope scanning, *Radiology* **80**, 1963, 653–661.
49. Y. S. Kwoh, I. S. Reed, and T. K. Truong, A generalized |w|-filter for 3-D reconstruction, *IEEE Trans. Nucl. Sci.* **NS-24**, 1977, 1990–1998.
50. A. V. Lakshminarayanan, "Reconstruction from divergent ray data," Technical Report 92. Dept. of Computer Science, State Univ. of New York at Buffalo, 1975.
51. R. M. Lewitt, Ultra-fast convolution approximation for computerized tomography, *IEEE Trans. Nucl. Sci.* **NS-26**, 1979, 2678–2681.
52. R. M. Lewitt and R. H. T. Bates, Image reconstruction from projections, *Optik,* **50**, Part I: 19–33; Part II: 85–109; Part III: 189–204; Part IV: 269–278, 1978.
53. R. M. Mersereau and A. V. Oppenheim, Digital reconstruction of multidimensional signals from their projections, *Proc. IEEE* **62**, 1974, 1319–1338.
54. D. Nahamoo, C. R. Crawford, and A. C. Kak, Design constraints and reconstruction algorithms for transverse-continuous-rotate CT scanners, *IEEE Trans. Biomed. Eng.* **BME-28**, 1981, 79–97.
55. D. Nahamoo and A. C. Kak, "Ultrasonic diffraction imaging, Technical Report, TR-EE-82-20. Purdue Univ., School of Elect. Engrg., 1982.
56. W. H. Oldendorf, Isolated flying spot detection of radiodensity discontinuities displaying the internal structural pattern of a complex object, *IRE Trans. Biomed. Eng.* **BME-8**, 1961, 68–72.
57. B. E. Oppenheim, Reconstruction tomography from incomplete projections, in *"Reconstruction Tomography in Diagnostic Radiology and Nuclear Medicine,"* (M. M. Ter Pogossian *et al.,* eds.) University Park Press, Baltimore, 1975.
58. T. M. Peters and R. M. Lewitt, Computed tomography with fan-beam geometry, *J. Comput. Assist. Tomog.* **1**, 1977, 429–436.
59. J. Radon, Uber die Bestimmung von Funktionen durch ihre Intergralwerte langs gewisser Mannigfaltigkeiten [On the determination of functions from their integrals along certain manifolds], *Ber. Saechsische Akademie Wissenschaften,* **29**, 1917, 262–279.
60. G. N. Ramachandran and A. V. Lakshminarayanan, Three dimensional reconstructions from radiographs and electron micrographs: Application of convolution instead of Fourier transforms, *Proc. Nat. Acad. Sci.* **68**, 1971, 2236–2240.
61. R. S. Ramakrishnan, S. K. Mullick, R. K. S. Rathore, and R. Subramanian, "Orthogonalization, Bernstein polynomials, and image restoration," *Appl. Opt.* **18**, 1979, 464–468.

62. A. Rosenfeld and A. C. Kak, *"Digital Picture Processing,"* 2nd edition, Vol. 1. Academic Press, New York, 1982.

63. P. D. Rowley, Quantitative interpretation of three dimensional weakly refractive phase objects using holographic interferometry, *J. Opt. Soc. Am.* **59**, 1969, 1496–1498.

64. T. Sato, S. J. Norton, M. Linzer, O. Ikeda, and M. Hirama, Tomographic image reconstruction from limited projections using iterative revisions in image and transform spaces, *Appl. Opt.* **20**, 1980, 395–399.

65. R. W. Schafer and L. R. Rabiner, A digital signal processing approach to interpolation, *Proc. IEEE* **61**, 1973, 692–702.

66. H. J. Scudder, Introduction to computer aided tomography, *Proc. IEEE* **66**, 1978, 628–637.

67. L. A. Shepp and B. F. Logan, The Fourier reconstruction of a head section, *IEEE Trans. Nucl. Sci.* **NS-21**, 1974, 21–43.

68. H. Stark and I. Paul, An investigation of computerized tomography by direct Fourier inversion and optimum interpolation, *IEEE Trans. Biomed. Eng.* **BME-28**, 1981, 496–505.

69. H. Stark, J. W. Woods, I. Paul, and R. Hingorani, Direct Fourier reconstruction, *IEEE Trans. ASSP* 1981, **ASSP-29.**

70. D. W. Sweeney and C. M. Vest, Reconstruction of three-dimensional refractive index fields from multi-directional interferometric data, *Appl. Opt.* **12**, 1973, 1649–1664.

71. K. C. Tam and V. Perez-Mendez, Tomographical imaging with limited angle input, *JOSA* **71**, 1981, 582–592.

72. K. Tanabe, Projection method for solving a singular system, *Numer. Math.* **17**, 1971, 203–214.

73. E. Tanaka and T. A. Iinuma, Correction functions for optimizing the reconstructed image in transverse section scan, *Phy. Med. Biol.* **20**, 1975, 789–798.

74. M. TerPogossian, *"The Physical Aspects of Diagnostic Radiology."* Harper and Row, New York, 1967.

75. O. J. Tretiak, Noise limitations in x-ray computed tomography, *J. Comput. Assist. Tomog.* **2**, 1978, 477–480.

76. O. Tretiak, M. Eden, and M. Simon, Internal structures for three dimensional images, *Proc. 8th Int. Conf. Med. Biol. Eng.* Chicago, IL, 1969.

77. L. Wang, Cross-section reconstruction with a fan-beam scanning geometry, *IEEE Trans. Comput.* **C-26**, 1977, 264–268.

78. S. J. Wernecke and L. R. D'Addario, Maximum entropy image reconstruction, *IEEE Trans. Comput.* **C-26**, 1977, 351–364.

5 Image Data Compression†

Anil K. Jain,
Paul M. Farrelle,‡ and
V. Ralph Algazi

Signal and Image Processing Laboratory
Department of Electrical and Computer Engineering
University of California
Davis, California

† Research supported in part by the Army Research Office, Durham, North Carolina, under grant DAAG29-78-0206 and in part by the Office of Naval Research through a Special Research Opportunity (SRO) project under contract N00014-81-K-0191.

‡ Also Executive Engineer, British Telecom Research Labs, Martlesham Heath, Ipswich, England

DIGITAL IMAGE PROCESSING TECHNIQUES

I. Introduction

Interest in digital image processing has been spurred by technical and commercial activity in image transmission, which goes back to the birth of television in the 1930s. Images contain huge amounts of information and require very large bandwidths. This translates into costly communication channels for transmission and extensive hardware requirements for storage and manipulation of images, whether analog or digital. Television standards for spatial and temporal resolution specify the number of scan lines, the bandwidth, and the number of frames per second. These and other standards for transmission and reconstruction of color images are based to a large extent on acceptable image quality for human observers. The goal of data compression of images is to reduce transmission and storage costs while preserving good image quality. To achieve compression it is necessary to consider representations beyond the simple analog-to-digital conversion of the image into a two-dimensional array of picture elements (pixels). Predictive techniques and orthogonal transformations take advantage of the high correlation between neighboring pixels. For moving images, data are also highly correlated from frame to frame, and several techniques can be used that avoid full-resolution transmission of highly redundant successive images. Because the roots of data compression lie in image broadcast applications, the simplicity and speed of data compression algorithms have, in the past, strongly biased research and development in the field. However, recent evolution of other applications, advances in low-cost and high-speed microelectronic circuits, and, in particular, the surge in the use of computer image storage and manipulation have led to a broader range of techniques and to interest in fundamental principles. Data compression of graphics documents has been motivated by the high cost and low speed of facsimile transmission over telephone lines. Here again, analog techniques are being supplanted by digital equipment.

In this chapter we review and discuss the image coding methods applicable to black and white, still and moving images and to the encoding of graphics. Each section concludes with a summary that outlines the advantages and drawbacks of the methods presented. The emphasis is on algorithms, and several omissions should be noted. Although human vision and image quality have major roles in image coding, they are explicitly discussed only very briefly, because reduction of the known properties of human vision to engineering design practice is still in a rudimentary stage and because a presentation of visual perception would take us too far afield [54]. Note, however, that image quality standards are implicit in the choice of original images to which processed encoded images are compared. Another omission is the coding of color images. The techniques are quite similar to those used for black and white images, except that a color image is usually decomposed into a black and white image (the luminance component) and two color-component images (the chrominance components, which can be represented with greatly reduced spatial resolution)[34].

II. Spatial Domain Methods

In this section we consider digital coding techniques that operate on the data in the spatial domain rather than the transform, or generalized frequency, domain. We further restrict this discussion to within-frame, or intraframe, schemes and defer between-frame, or interframe, schemes until Section V. Methods of pulse code modulation, dither, half toning, statistical coding, delta modulation, and differential pulse code modulation are described.

A. Pulse Code Modulation (PCM)

In a pulse code modulation scheme, the incoming continuous video signal is sampled, usually at the Nyquist rate, and then quantized. Hence, it is nothing more than a digital representation of the original analog signal.

Although the eye's sensitivity to quantizing distortion decreases with increasing luminance level [18], the high gamma of the display tube complements this logarithmic response of the eye and greatly reduces the effect. Consequently, in analog-to-digital conversion of TV images, a uniform quantizer is normally used.

The quantizer usually has N levels, where N is a power of 2 ($N = 2^B$, say), and each sample is represented by a fixed-length binary code word having B bits. The number of bits needed to code a pixel using PCM depends on the type of image. Commonly, 8 bits are sufficient for monochrome broadcast or video-conferencing images, whereas medical images may require 10 to 12 bits to obtain sufficient amplitude resolution.

i. Dither Techniques

For typical TV images, as the number of bits is reduced below five or six, a contouring effect becomes visible in the low detail or flat regions. In a method called pseudorandom noise quantization, these contours are broken up by adding a small amount of broadband, pseudorandom, uniformly distributed noise, called *dither,* to the luminance samples prior to quantization and subtracting it at the receiver. The dither causes the pixels to oscillate about the original quantization levels, which removes the highly visible contours. The proper amount of added dither will reduce contouring while maintaining resolution; usually it will affect the least significant bit of the quantizer. With dither techniques, images subjectively equivalent to 6- or 7-bit PCM can be represented by 3 or 4 bits/pixel [60].

ii. Halftoning

Dither techniques are also used to generate halftone images from gray-level images [52]. Halftone images are binary images that give a gray-scale rendition and are widely used in the printing industry. To each pixel of the original image, often over-sampled, a dither sample is added, and the resulting image is quantized by a one-bit quantizer to produce an array of black and white dots. Compared with a one-bit quantized image, the one-bit halftone has a better visual appearance. The gray-level rendition in halftones is due to local spatial averaging performed by the eye.

iii. Statistical Coding (Entropy Coding)

For signals having nonuniform probability density functions, or histograms, it is possible to use statistical coding, that is a variable-length code such as the Huffman code, which assigns shorter code words to those levels that are more likely to occur. The long-term histogram for TV images is approximately uniform, although the short-term statistics are highly nonstationary. Consequently, statistical coding is not used in PCM coding schemes for images.

B. Predictive Coding

In PCM, successive inputs to the quantizer are treated independently, so there is no exploitation of the significant redundancy present in images.

However, once the signal has been digitized, it is relatively straightforward to exploit this redundancy in the spatial domain by use of predictive coding.

The philosophy underlying predictive coding is to remove the mutual redundancy between successive samples and to quantize only the new information, or innovations. For any pixel, the innovation is the difference between the value of that pixel and its optimum predicted value, which is computed from previously coded pixels. For images whose intensity distribution is non-Gaussian, the optimum mean-square (m.s.) predictor is non-linear, although linear predictors are often used because their design and implementation are easier. Consider a one-dimensional, pth-order, linear predictive coder whose input is a random sequence $u(n)$, and suppose that information in the samples up to $n = k - 1$ has been transmitted somehow (see Fig. 1). Let $u'(n)$ denote the transmitted value of $u(n)$. When $u(k)$ is to be transmitted, a quantity $\bar{u}'(k)$, an estimate of $u(k)$, is predicted from the previously transmitted samples, where

$$\bar{u}'(k) = \sum_{j=1}^{p} a(j)u'(k-j).\tag{1}$$

The design problem is considerably simplified if we assume that the quantization error is sufficiently small so that we have an idealized predictor

$$\bar{u}(k) \triangleq \sum_{j=1}^{p} a(j)u(k-j),\tag{2}$$

from which we determine the prediction weights $a(j)$ to minimize the m.s. prediction error $E\{[u(k) - \bar{u}(k)]^2\}$ by solving the following set of equations

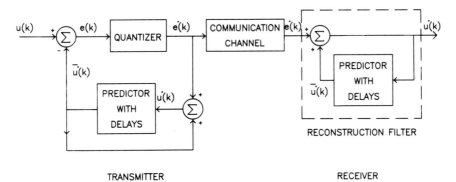

Fig. 1. DPCM. Note that in the absence of transmission errors, the error in reproducing $u(k)$ is merely that introduced when quantizing the prediction error $e(k)$.

(assuming that $u(k)$ is a zero mean stationary process):

$$\rho(l) = \sum_{j=1}^{p} a(j)\rho(l-j), \qquad 1 \le l \le p, \tag{3}$$

where $\rho(l) = E[u(k)u(k-l)]/\sigma_u^2$ is the correlation between $u(k)$ and $u(k-l)$. The m.s. prediction error is then

$$\sigma_e^2 = \sigma_u^2 \left[1 - \sum_{j=1}^{p} a(j)\rho(j) \right]. \tag{4}$$

This analysis is easily extended to higher dimension predictors. For a two-dimensional predictor the design equations become

$$\bar{u}(i,j) = \sum_{(m,n)\in S}\sum a(m,n)u(i-m,j-n), \tag{5}$$

$$\rho(k,l) = \sum_{(m,n)\in S}\sum a(m,n)\rho(k-m,l-n), \qquad k,l \in S \tag{6}$$

$$\sigma_e^2 = \sigma_u^2 \left[1 - \sum_{(m,n)\in S}\sum a(m,n)\rho(m,n) \right], \tag{7}$$

where S is a subset of the previously transmitted pixels, and $\rho(k,l) = E[u(i,j)u(i-k,j-l)]/\sigma_u^2$ is the correlation between $u(i,j)$ and $u(i-k,j-l)$.
As an example, consider a random field whose correlation is

$$\rho(k,l) = a_1^{|k|}a_2^{|l|}. \tag{8}$$

The quantities a_1, a_2 are the one-step correlations of the random field along the i and j axes, respectively. The optimum m.s. predictor is

$$\bar{u}(i,j) = a_1 u(i-1,j) + a_2 u(i,j-1) - a_1 a_2 u(i-1,j-1), \tag{9}$$

and the m.s. prediction error is

$$\sigma_e^2 = \sigma_u^2(1-a_1^2)(1-a_2^2). \tag{10}$$

Once the predictor coefficients are known, the differential pulse code modulation (DPCM) equations for images represented by Eq. (8) are

$$\text{predictor} \qquad \bar{u}'(i,j) = a_1 u'(i-1,j) + a_2 u'(i,j-1)$$
$$- a_1 a_2 u'(i-1,j-1),$$

$$\text{quantizer input} \qquad e(i,j) = u(i,j) - \bar{u}'(i,j),$$

$$\text{reconstructor} \qquad u'(i,j) = \bar{u}'(i,j) + e'(i,j). \tag{11}$$

It should be noted that the validity of the two assumptions, image stationarity and low quantization error (i.e., low compression), will depend on the type of image to be coded and the compression ratio sought. In any case, the above design technique is useful for obtaining initial estimates for the predictor weights $a(m,n)$ which can then be iteratively optimized based on subjective image quality.

Historically, this predictive quantization technique is classified according to the number of quantization levels used. Delta modulation (DM), the simplest implementation, uses only two quantization levels; and differential pulse code modulation (DPCM), the more general case, uses more than two levels.

i. Delta Modulation (DM)

In DM (see Fig. 2) [56], the predictor is a one-step delay function, and a one-bit quantizer is used to achieve a one-bit representation of the signal. An important aspect of this scheme is that it does not require sampling of the input signal. The predictor simply performs integration of the quantizer output signal, which is a sequence of binary pulses. The reconstruction filter of the delta modulator, which is a simple integrator, is unstable. In the presence of transmission errors, the receiver output can accumulate large levels of error. The prediction filter can be stabilized by attenuating the predicted value by a constant $\gamma < 1$ (called leak). Figure 3 shows typical input–output signals of a delta modulator. The primary limitations of

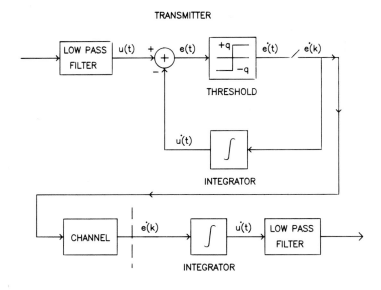

Fig. 2. Delta modulation (DM).

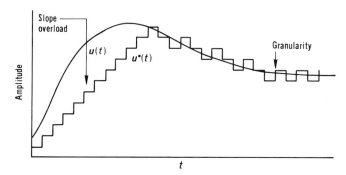

Fig. 3.   Input – output signals for DM.

delta modulation are slope overload, granularity noise, and instability to transmission errors.   Slope overload occurs when there is a large jump or discontinuity in the signal, to which the quantizer can respond only in several delta steps.   Granularity noise is the steplike nature of the output, which cannot be constant, when the input signal is almost constant.   Both of these errors can be compensated to a certain extent by low-pass filtering the input and output signals.   Slope overload can also be reduced by increasing the sampling rate, but that will reduce the achievable compression.   An alternative for reducing granularity while retaining simplicity is to go to a tri-state delta modulation scheme, in which the quantizer has three levels. Most of the pixels are found to be in the "level" or "0" state; thus, run-length coding and statistical coding are appropriate and yield rates around 1 bit/ pixel [47].

Other variations of delta modulation, such as adaptive DM with variable step size, can further improve its performance [33].   However, the increase in complexity has to be measured against the more general case of DPCM coding and other techniques for data compression.   In fact, a large number of the ills of DM can be cured by DPCM, thereby making it often more attractive than DM for data compression.

ii. Differential Pulse Code Modulation (DPCM)

The overall system concept for DPCM is shown in Fig. 1.   The principal components of a DPCM system are its predictor and quantizer.   An important aspect of the scheme is that prediction is based on the output rather than the input samples from the past.   This results in the predictor being in the feedback loop around the quantizer, so that the quantizer noise at a given step is fed back to the quantizer input at the next step.   This has a stabilizing effect that prevents accumulation of errors and dc drift in the reconstructed signal.

For digital data it is possible to design a distortionless or reversible predictive coder for which the predictor is not inside the feedback loop. This may be termed open-loop DPCM. Such an encoder is shown in Fig. 4. A pixel is predicted from the past samples, but the predicted value is rounded to an integer. Then the prediction error also takes integer values and can be coded by a Huffman variable-length, or entropy, coder (see discussion in Section VI). At the receiver, the Huffman code is decoded, and the integer prediction error is added to the integer prediction (formed just as in the coder) to reproduce the original image exactly.

The quantizer should be designed to limit the three types of degradation associated with DPCM. These are granularity, slope overload, and edge busyness. If the inner levels are too coarse, then granularity is visible in flat regions of the image. Slope overload is caused by high-contrast edges when there is a prediction error that exceeds the range, or extreme levels, of the quantizer and manifests itself as a blurred edge. Edge busyness is caused by less-sharp edges when the prediction error is within the range of the quantizer but when the reproduced image is at different quantization levels on adjacent scan lines.

Quantizers can be designed on a statistical basis, or they can take advantage of psycho–visual phenomena. The optimum quantizer for a given number of levels is the Lloyd–Max quantizer, which minimizes the m.s.

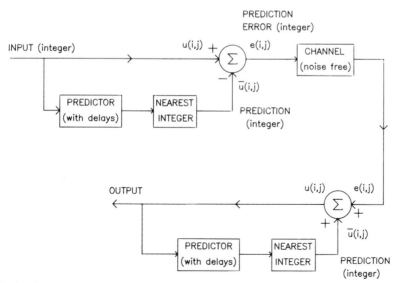

Fig. 4. Open-loop DPCM. Note that in the absence of transmission errors, there is no error in reproducing $u(i,j)$.

quantization error [37]. For a one-dimensional first-order predictor for which the prediction error sequence is modeled as a Laplacian density, the Max quantizer is too *companded,* that is, the inner quantization steps are too small whereas the outer levels are too coarse, resulting in edge busyness. For this reason there have been a number of attempts, and some success, to incorporate psycho–visual phenomena and minimize a subjectively weighted m.s. error [11,44]. To date, there is no recognized best way to incorporate such criteria or even to decide which criterion should be used; thus, at present, such quantizers tend to be optimized iteratively after subjective analysis of the resulting picture quality.

Although the optimum m.s. predictor previously discussed is nonlinear in general, an optimum linear m.s. predictor, which can be one- or two-dimensional, is used in practice. The simplest predictor is the first-order one-dimensional m.s. predictor that weights the previous element on the same line by ρ, the correlation between $u(k)$ and $u(k-1)$. Although this technique usually performs well, sharp vertical or diagonal edges are blurred and exhibit edge busyness. Another shortcoming of this simple predictor is its low tolerance for errors. A transmission error manifests itself as a horizontal streak.

The improvement in m.s. error performance to be gained by two-dimensional processing is dependent on the image statistics and may be fairly small unless the lines are highly correlated. For two-dimensional predictors and DPCM, see [3,15]. The subjective quality of an image and its error tolerance can be improved by use of two-dimensional predictors [12].

The improved subjective quality is most apparent at the edges, which are obvious nonstationarities present in the image. This is to be expected, because in the design of the one-dimensional linear m.s. error predictor it is assumed that the image is stationary. In particular, the one-dimensional predictor is unable to anticipate sharp vertical or diagonal edges; consequently, many "reasonable" two-dimensional predictors will perform better at such discontinuities. However, only certain two-dimensional predictors will have an increased error tolerance.

Assuming a 2:1 interlace system, with adjacent lines on alternate fields, access to the vertical adjacent pixel requires a field of storage. With only one line of storage (512 bytes for 4 MHz TV), the previous line in the same field can be used. Such intrafield predictors using pixels A, B, C, and D (see Fig. 5) have proven useful, and their properties are summarized in Table I. In summary, if the additional hardware complexity is not considered excessive, then the coder should be designed to include a flexible two-dimensional predictor based on pixels A, B, C, and D, which can be subjectively optimized under real operating conditions.

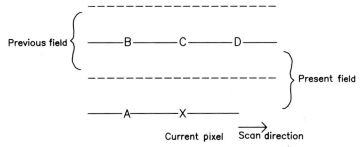

Fig. 5. Pixels used in the two-dimensional prediction of the current pixel X.

iii. Adaptive Techniques

Due to the variation, or nonstationarity, of the image statistics, both within and between frames, it is impossible to design a single predictor or quantizer that is always optimum. Adaptive techniques use a range of quantizing characteristics and/or predictors from which a current optimum is selected according to a decision rule based on local image properties. Any adaptive scheme should limit the amount of extra information to be transmitted, because the ultimate aim is data compression. This is achieved by using only previously coded pixels to determine the mode of operation of the adaptive coder. In the absence of transmission errors, the receiver can keep track of decisions made at the transmitter using the received pixels, which are identical to those used at the transmitter. However, in the presence of transmission errors, a poorly designed adaptation rule could prove catastrophic if the receiver were to lose track of the transmitter operations.

Adaptive predictors provide a much improved subjective image quality, especially at the edges, for reasons already discussed when two-dimensional and one-dimensional predictors [66,67] were compared. A popular technique is to use a number of predictors, each of which performs well if the image is highly correlated in a certain direction. The direction of maximum correlation is computed from previously coded pixels, and the corresponding predictor is chosen.

Adaptive quantization schemes segment the image into different regions according to spatial detail, or activity, and different quantizer characteristics are used for each activity class. The flat regions are quantized more finely than edges or detailed areas. This scheme takes advantage of the fact that noise visibility decreases with increased activity. It has been found that the advantages of adaptation increase as the number of activity classes is increased up to four but that there is little improvement thereafter [45].

TABLE I

Summary of Predictive Coding

Design parameter	Comments
Predictor[a]	
Optimum linear mean square	Prediction coefficients readily determined from correlation properties of image ensemble neglecting quantization errors. Practical predictors use 3 or 4 pixels to form prediction. Error tolerance depends on weights used and may be quite low.
Previous element (differential) γA	Sharp vertical or diagonal edges are blurred and exhibit edge busyness. Low error tolerance; transmission error manifests itself as a horizontal streak.
Averaged prediction (a) $\gamma(A + D)/2$ (b) $\gamma(A + C)/2$ (c) $\gamma[A + (C + D)/2]/2$	Significant improvement over previous element prediction for vertical and most sloping edges. Horizontal and gradually rising edges blurred. (Static blur much less annoying than edge busyness.) The two predictors using pixel D perform equally well and better than $(A + C)/2$ because of better performance on gradual rising edges. Even serious errors result in an error pattern that is hardly visible compared with horizontal streaking (see Fig. 8)
Planar prediction (a) $\gamma[A + (C - B)]$ (b) $\gamma[A + (D - B)/2]$	Better performance than previous element prediction but worse than averaged prediction. Edge busyness on falling edges for predictor (b) and on both rising and falling edges for predictor (a). Transmission errors lead to horizontal and vertical streaks. This leads to an increased visibility of errors when compared with averaged prediction (see Fig. 8).
Leak γ	$0 < \gamma < 1$. As the leak is increased, transmission errors become less visible, but granularity and contouring become more visible.
Quantizer	
Optimum mean square (Lloyd–Max)	Recommended when the compression ratio is not too high ($\leq 3:1$) and a fixed-length code is used. Prediction error can be modeled by Laplacian or Gaussian densities. This quantizer can also be implemented as a compander if a table look-up is not intended.

TABLE I *(Continued)*

Design parameter	Comments
Based on visual criteria	Difficult to design. Possible alternative is to perturb the levels of the Max quantizer to obtain an increased subjective quality.
Uniform	Useful in high compression schemes ($> 3:1$) in which a variable-length code is used, because the quantizer output is entropy coded. Requires a buffer to smooth the variable data rate before transmission over a constant rate channel.

a See Fig. 4 for notation.

iv. Performance

There are two approximate results that relate predictive coder performance to PCM based on the prediction variance reduction ratio $\eta = \sigma_u^2/\sigma_e^2$, where σ_u^2 is the variance of the original signal, and σ_e^2 is the variance of the prediction error signal.

For small but equal distortion in PCM and DPCM, the bit rate difference is approximately

$$\text{rate reduction} \qquad n_{\text{PCM}} - n_{\text{DPCM}} = \tfrac{1}{2} \log_2 \eta.$$

If the PCM quantizer is changed so that the bit rate is the same in PCM and DPCM, then the gain in signal-to-noise ratio of DPCM over PCM is given by

$$\text{SNR improvement} \qquad \text{SNR}' - (\text{SNR}')_{\text{PCM}} = 10 \log_{10} \eta, \qquad (12)$$

where $\text{SNR}' \triangleq 10 \log_{10} \sigma_u^2/e_{ms}^2$ and e_{ms}^2 is the m.s. distortion introduced by the coding scheme. Hence, the data compression ability depends on the variance reduction by prediction, that is, the ability to predict a new pixel, and, therefore, on the mutual dependence of pixels within the original image.

Figure 6 shows the performance characteristics of various predictive coders and PCM. The two-dimensional Gauss–Markov field (curve A) has a correlation given by Eq. (8), $a_1 = a_2 = 0.95$, and the first-order Gauss–Markov field (curve C) is a one-dimensional equivalent with $\rho_k = \rho^k$, $\rho = 0.95$.

Because the entropy of the coder output is usually less than the number of quantizing bits, a variable-length Huffman code could be employed to reduce the average bit rate. However, this would increase the complexity of the encoding–decoding algorithm and would require extra buffer storage at

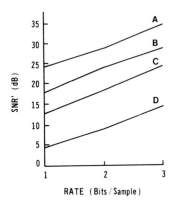

Fig. 6. Performance comparison of PCM and DPCM using a 3-point predictor, $\bar{u}(i,j) = a_1 u(i-1,j) + a_2 u(i,j-1) - a_3 u(i-1,j-1)$. Curve A: two-dimensional, Gauss–Markov field, upper bound, $a_1 = a_2 = 0.95$, $a_3 = a_1 a_2$. Curve B: actual two-dimensional DPCM with 3-point prediction and measured a_1, a_2, a_3. Curve C: line-by-line DPCM, first-order Gauss–Markov, $\rho = 0.95$. Curve D: PCM (Gaussian random variable).

the transmitter and receiver to maintain a constant bit rate over the communication channel. Experimental results on images have shown that for a Laplacian density based 3-bit Max quantizer, the entropy of the quantized output is roughly 2.3 bits. Thus, two-dimensional DPCM techniques achieve a compression for typical 8 bit/pixel raw image data of about 3 for a fixed-length code and roughly 3.5 with a more complex variable-length code. If the quantizer is designed under visual criteria, a further saving of about 1 bit/pixel could be made, which would achieve a compression of 4 or 5 to 1 for acceptable levels of visual distortion [3,44].

Figure 7 shows an encoded image and error images at 2 bits/pixel (a compression of 4:1) using different predictors in the absence of transmission errors. It is interesting to note how the different predictors perform at different edge orientations. These properties are summarized in Table I. Figure 8 shows the effect of a high transmission error rate (10^{-3}) with the same three predictors. Due to ease of hardware implementation, DPCM techniques hold promise for real-time compression of medical images, such as fluoroscopic and radiographic images, that occur in differential radiology. Here, high-resolution images (up to 1024×1024 pixels/frame) arrive at 30 frames/second and have to be stored in real time on a large disc. Data compression offers not only a reduction of memory requirement but also a reduction of speed required to transfer compressed data onto disc. In differential radiology, usually pairs of images from a sequence of motion images are differenced, enhanced, and displayed. Figure 9 shows typical original and compressed images and the enhanced difference image features after DPCM coding with 2:1 compression. Recent studies have indicated that 4:1 compression is achievable by DPCM while preserving the clinically useful information in the enhanced differential images.

Overall, DPCM is a simple and easy to implement online technique that

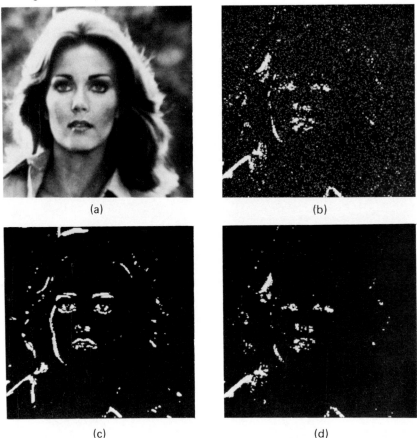

Fig. 7. Two bits/pixel DPCM coding. (a) Reconstructed image using optimum linear predictor. (b) Error image, optimum linear predictor, MSE = 0.5%. (c) Error image, two-point predictor, $\gamma(A + D)/2$, MSE = 1.9%. (d) Error image, three-point predictor, $\gamma(A + C - B)$, MSE = 0.7%

yields favorable compression results. The major drawbacks are its sensitivity to variations in image statistics and its high sensitivity to transmission errors. Table I summarizes the design considerations for predictive coders.

C. Other Spatial Domain Methods

i. Constant Area Quantization (CAQ)

CAQ [7,48] is a one-dimensional spatial differencing technique that performs well down to approximately 1 bit/pixel, for RPV applications [9]. It

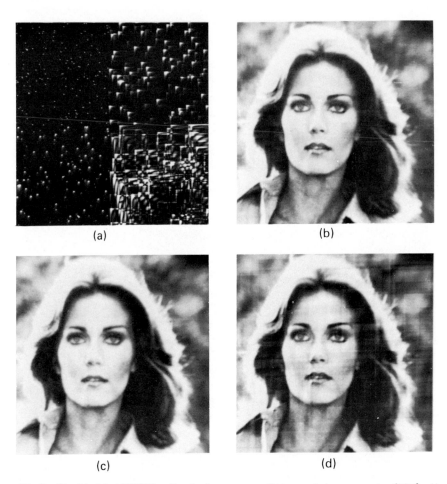

Fig. 8. Two bits/pixel DPCM coding in the presence of a transmission error rate of 10^{-3}. (a) Propagation of transmission errors for different predictors. Clockwise from top left, error location: optimum, three-point, and two-point predictors. (b) Optimum linear predictor. (c) Two-point predictor $\gamma(A + D)/2$. (d) Three-point predictor $\gamma(A + C - B)$.

exploits the psycho–visual property by which the eye sees more detail in high-contrast regions than in low-contrast ones. It is similar to a tri-state DM technique except that it uses a variable step size, which is determined by the separation of nonzero levels or pulses. The product of the step size and pulse separation equals the "constant area"; hence, the step size can be calculated at the receiver without any overhead. Data compression is thus achieved by transmitting lower resolution, fewer pulses, in the low contrast areas.

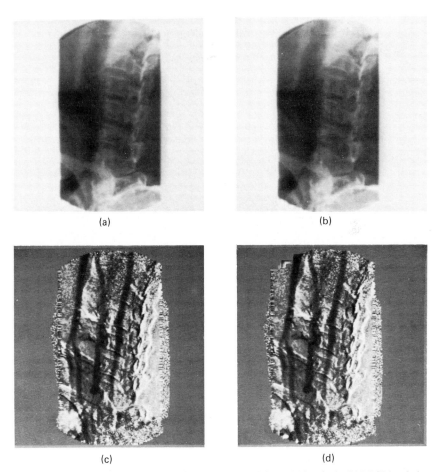

(a)   (b)

(c)   (d)

Fig. 9. DPCM coding of digital radiology images. (a) Original, 10 bits/pixel. (b) DPCM coded, 5 bits/pixel. (c) Uncompressed enhanced difference between post contrast injected and mask images. (d) Enhanced difference after 2:1 compression.

ii. Tree Encoding

Tree encoding [40,41] is a delayed encoding technique that effectively causes the quantizing level representing a given pixel to be based on future pixels. Recent work has shown that two-dimensional tree encoding offers potential advantages over two-dimensional DPCM by reducing the distortion introduced by compression and also by increasing error tolerance. In conventional DPCM, each pixel is quantized to minimize the distortion in that pixel, which can result in unnecessarily large distortion for some future pixel. In delayed encoding, a more global measure of distortion is used, and each pixel is quantized to minimize the distortion in a group of pixels.

III. Transform Coding

An alternative to predictive coding is *transform coding,* which is a special case of *block quantization,* in which a vector or a block of N samples is quantized. An optimum block quantizer could be defined as one that minimizes the average distortion of the quantized elements of the vector for a given number of quantization levels. In predictive coding, the successive inputs to the quantizer are decorrelated by a nonlinear recursive filter. In transform coding, all the samples are first decorreleted jointly and then quantized. The optimal decorrelation filter that minimizes the m.s. distortion in the reconstructed signal turns out to be a linear noncausal filter (as opposed to the causal–predictive filter) known as the Karhunen–Loeve (KL) transform. In practice, the KL transform is substituted by a suboptimal but fast unitary transform.

A. Theory

Suppose a random vector $\mathbf{u}$ with zero mean and covariance $\mathbf{R}$, see Fig. 10(a), is linearly transformed by an $N \times N$ matrix $\mathbf{A}$ to produce a (complex) vector $\mathbf{v}$ such that its components $\{v(k)\}$ are mutually uncorrelated. Each component $v(k)$ is quantized independently. The output vector $\mathbf{v}$ is linearly transformed by a matrix $\mathbf{B}$ to yield a vector $\mathbf{u}$. It is desired to find the optimal decorrelating matrix $\mathbf{A}$, the reconstruction matrix $\mathbf{B}$, and the optimum quantizers such that the overall average mean square distortion,

$$D = \frac{1}{N} E\left\{ \sum_{k=1}^{N} [u(k) - u(k)]^2 \right\}, \tag{13}$$

is minimized. The solution of this problem requires that $\mathbf{B} = \mathbf{A}^{-1}$, that $\mathbf{A}$ be the KL transform of $\mathbf{u}$ (i.e., rows of $\mathbf{A}$ are the orthonormal eigenvectors of $\mathbf{R}$), and that the quantizers be optimum mean square (i.e., Lloyd–Max quantizers). Since the rows of A are orthonormal, it follows that

$$A A^{*\mathrm{T}} = I$$

or

$$A^{-1} = A^{*\mathrm{T}},$$

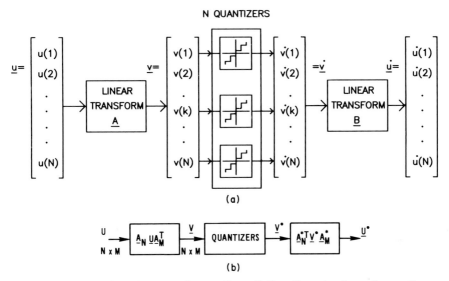

Fig. 10. (a) One-dimensional transform coding. (b) Two-dimensional transform coding.

that is, A is unitary. There are other unitary transforms that also have the energy packing tendency. Hence, the dynamic ranges or, equivalently, the variances of transform coefficients are unequal. For a given quantizer, the compression performance of a transform is completely determined by the distribution of the variances of the transform coefficients. The KL transform distributes them in the most efficient way. It has the property that, for any $n \leq N$, it packs the maximum average energy into some n samples of v. Although the KL transform is optimal, it is often difficult to compute and has no fast algorithm associated with it. For stationary random sequences there are many fast unitary transforms that approach the energy packing efficiency of the KL transform. Examples are the Cosine, Fourier, and Sine transforms. These transforms have been shown to be members of a larger family of sinusoidal transforms [26], all of which have a performance equivalent to that of the KL transform as the size N of the data vector goes to infinity. For first-order stationary Markov processes, the Cosine transform matrix defined in [1] as

$$C_{i,j} = \begin{cases} 1/\sqrt{N}, & i = 1, \quad 1 \leq j \leq N \\ \sqrt{2/N} \cos[(2j-1)(i-1)\pi/2N] & 2 \leq i \leq N, \quad 1 \leq j \leq N \end{cases} \quad (14)$$

has been shown to perform very much like the KL transform when the correlation parameter ρ lies in the interval $(0.5, 1)$, even when N is small [27]. The transformation $\mathbf{y} = \mathbf{C}\mathbf{x}$ can be computed by means of the fast Fourier transform (FFT) in $O(N \log N)$ operations. These properties have made the Cosine transform a useful substitute for the KL transform in image processing, because many images can be modeled by low-order Markov processes with high interpixel correlation.

Nonsinusoidal unitary transforms such as the Hadamard, Haar (which are square-wave transforms), and the Slant transform are also used, because they are computationally faster than the FFT-based fast sinusoidal transforms. Therefore, in transform coding practice, the KL transform is substituted by a suitable fast unitary transform. In two dimensions, Fig. 10(b), one considers the class of transformations of an $N \times M$ image **U** given by

$$V = A_N U A_M^T, \tag{15}$$

where A_N is an $N \times N$ fast unitary transform. This is called a separable transform, because it can be realized by a column-wise transformation $(A_N U)$ followed by a row-wise transformation. Figure 11 shows some of the common transforms of an image.

To make transform coding practical, the given image is divided into small rectangular blocks (typically 16×16), and each block is coded independently. For example, if a 256×256 image is coded in 16×16 blocks rather than a single 256×256 block, then the storage requirement is reduced by a factor of 256, and the speed is improved by a factor of two for a transform that requires $O(N \log_2 N)$ operations for an $N \times 1$ vector.

B. Bit Allocation

If $\sigma_{k,l}^2 \triangleq E[|v(k,l)|^2]$, then for any unitary transform of an $N \times M$ image the mean square distortion becomes

$$D \triangleq \frac{1}{NM} \sum_{k=1}^{M} \sum_{l=1}^{N} E\{[u(k,l) - u'(k,l)]^2\} = \frac{1}{NM} \sum_k \sum_l \sigma_{k,l}^2 f(n_{k,l}), \tag{16}$$

where $f(x)$ denotes the m.s. distortion that would be achieved by an x-bit quantizer if its input were a unity variance random variable, and $n_{k,l}$ is the number of bits allocated to $v(k,l)$, the (k,l)th transform coefficient. If the average desired bit rate per sample is p, then

$$\frac{1}{MN} \sum_k \sum_l n_{k,l} = p. \tag{17}$$

For a fixed value of p, the allocation $n_{k,l} \geq 0$ has to be found so that the distortion D is minimum. For example, if each quantizer is the Shannon optimal block encoder (also called Shannon quantizer), then $f(x) = 2^{-2x}$, and the bit allocation is given by

$$n_{k,l} = n_{k,l}(\theta) = \max[0, \tfrac{1}{2} \log_2(\sigma_{k,l}^2/\theta)], \tag{18}$$

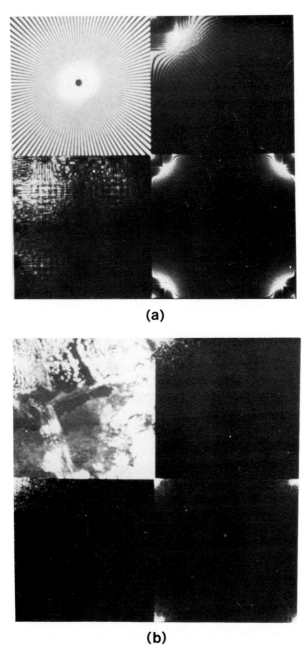

(a)

(b)

Fig. 11. Some common transforms of an image. In both cases, clockwise from top left: original, Cosine, Hadamard, and Fourier transforms.

where θ is determined such that Eq. (17) is satisfied. The minimized distortion at the rate p is given by

$$D_{\min} = \frac{1}{MN} \sum_k \sum_l \min(\theta, \sigma_{k,l}^2). \qquad (19)$$

An algorithm for obtaining the optimal bit allocations and the corresponding rate p versus distortion characteristics for arbitrary zero memory quantizer is given in [4].

The variances $\sigma_{k,l}^2$ can be calculated either directly from an ensemble of transformed blocks of test images or from the knowledge of the power spectral density, or equivalently, from the autocorrelation function of the image. The separable covariance function of Eq. (8) is often used to estimate the transform domain variances. Another function which generally provides a better fit of covariances is given as

$$r(k,l) = \sigma^2 \exp[-(\alpha_1 k^2 + \alpha_2 l^2)^{1/2}]. \qquad (20)$$

Figure 12 shows the bit allocation for Cosine transform coding of a 16×16 block of an image to achieve an average rate of 1 bit/pixel when the image covariance function is modeled by this function. Figure 13 shows the distortion versus rate characteristics for different unitary transforms. Note that the Cosine transform is almost indistinguishable from the KL transform. Figure 14 shows Cosine transform coded images (and the error images) at average rates of 0.5 bit/pixel and 1 bit/pixel, respectively.

j

7	6	5	4	3	3	2	2	2	1	1	1	1	1	0	0
6	5	4	4	3	3	2	2	1	1	1	1	1	0	0	0
5	4	4	3	3	2	2	2	1	1	1	1	1	0	0	0
4	4	3	3	3	2	2	2	1	1	1	1	1	0	0	0
3	3	3	3	2	2	2	1	1	1	1	1	0	0	0	0
3	3	2	2	2	2	2	1	1	1	1	1	0	0	0	0
2	2	2	2	2	2	1	1	1	1	1	0	0	0	0	0
2	2	2	2	1	1	1	1	1	1	1	0	0	0	0	0
2	1	1	1	1	1	1	1	1	1	0	0	0	0	0	0
1	1	1	1	1	1	1	1	1	0	0	0	0	0	0	0
1	1	1	1	1	1	1	1	0	0	0	0	0	0	0	0
1	1	1	1	1	1	0	0	0	0	0	0	0	0	0	0
1	1	1	1	0	0	0	0	0	0	0	0	0	0	0	0
1	0	0	0	0	0	0	0	0	0	0	0	0	0	0	0
0	0	0	0	0	0	0	0	0	0	0	0	0	0	0	0
0	0	0	0	0	0	0	0	0	0	0	0	0	0	0	0

i labels the rows.

Fig. 12. Typical bit allocation for 16×16 block cosine transform coding at approximately 1 bit/pixel.

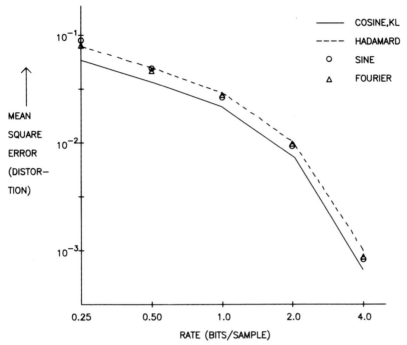

Fig. 13.   Distortion versus rate characteristics for different transforms.

C. Zonal versus Threshold Coding

Examination of the bit allocation pattern of Fig. 12 reveals that a substantial number of transformed samples are allocated zero bits (except at high average rates). Thus, only a small *zone* of the transformed image is transmitted. Denoting by $\mathcal{J}_t$ the address set of transmitted samples,

$$\mathcal{J}_t = \{(k,l);\ n_{k,l} \geqq 1\}, \tag{21}$$

and letting N_t be the number of transmitted samples, we can define a *zonal mask*,

$$\mathcal{H}(k,l) \;=\; \begin{cases} 1, & (k,l) \in \mathcal{J}_t \\ 0, & \text{otherwise} \end{cases} \tag{22}$$

which takes values of unity in the zone of largest N_t variances of the transformed samples. Thus, in transform coding one applies a zonal mask to the transformed image and quantizes only the nonzero elements for transmission or storage. This method is also called *zonal coding*.

(a)

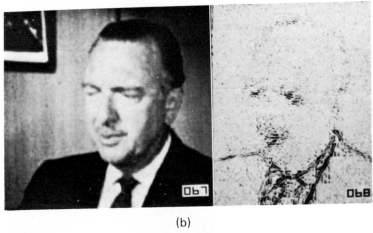

(b)

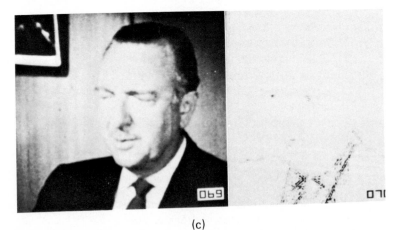

(c)

Fig. 14. Cosine transform coded images and error images. (a) Original, 8 bits/pixel. (b) Intraframe 16 × 16 block coded, 0.5 bit/pixel, SNR = 34.4 dB. (c) Intraframe 16 × 16 block coded, 1.0 bit/pixel, SNR = 40.3 dB.

If instead of transmitting or storing the N_t elements of maximum variance, one considers N_t samples of maximum amplitude (for the given image) in the transform domain, we get what is called *threshold coding*. The address set of transmitted samples is now

$$\mathcal{I}'_t = \{(k,l); |v(k,l)| > \eta\}, \tag{23}$$

where η is a suitably chosen threshold that controls the achievable average bit rate. For a given class of images, because the variances of the transform samples are fixed, the zonal mask remains unchanged from one image to the next (or one block to the next) for a fixed bit rate. However, the *threshold mask* $\mathcal{M}_\eta$ defined as

$$\mathcal{M}_\eta(k,l) \;= \begin{cases} 1, & (k,l) \in \mathcal{I}'_t \\ 0, & \text{otherwise} \end{cases} \tag{24}$$

could change from block to block, because the set $\mathcal{I}'_t$ of largest amplitude samples need not be the same for different blocks. The samples retained after thresholding are typically quantized by a constant word-length quantizer followed by a variable word-length entropy coder.

Although for the same number of transmitted samples (or the number of quantizing bits) a threshold mask would give a better choice of transmission samples (i.e., lower distortion), it can also result in an increased bit rate, because the addresses of the transmitted samples (i.e., the boundary of the threshold mask) would have to be coded for every image block. Typically, a sample line-by-line run-length coding scheme is implemented to code the transition boundaries in the threshold mask. Usually this results in a somewhat more complex scheme than zonal transform coding. However, threshold coding has merits, because it is adaptive in nature and is particularly useful for achieving high compression ratios when image contents change considerably (from block to block) so that a fixed zonal mask is inefficient.

D. Other Transform Coding Methods

i. Recursive Block Coding

In the conventional block-by-block transform coding, each block is processed independently. Therefore, the block size has to be large enough (typically from 16×16 to 64×64) so that interblock redundancy is minimal, although, a large block size will increase the complexity of the hardware. Moreover, at low bit rates (typically less than 1 bit/pixel), the block boundaries start becoming visibly objectionable.

In recursive block coding, redundancy between successive blocks is removed to achieve additional compression while maintaining small block size (8×8 or less). The redundancy between successive blocks can be modeled explicitly using the block boundaries by means of a theory of noncausal representations [24,27]. For example, a block of a stationary Markov sequence $u(n)$, $n = 1,...,N$ has an orthogonal decomposition [24],

$$u(n) = u_o(n) + u_b(n), \qquad n = 1,...,N, \tag{25}$$

where $u_o(n)$ and $u_b(n)$ are orthogonal random sequences such that the KL transform of $u_o(n)$ is the (fast) Sine transform,

$$\Psi_{i,j} = \sqrt{2/(N+1)} \, \sin[ij\pi/(N+1)], \qquad 1 \leq i,j \leq N, \tag{26}$$

and $u_b(n)$ depends only on the block boundary variables $u(0)$ and $u(N+1)$. A recursive block transform coder based on this model is shown in Fig. 15. For each block the boundary variable $u(N+1)$ is first coded. The coded value $u'(N+1)$ and the initial value $u'(0)$ (which is the boundary value of the previous block) are used to generate $u'_b(n)$, an estimate of $u_b(n)$. The differ-

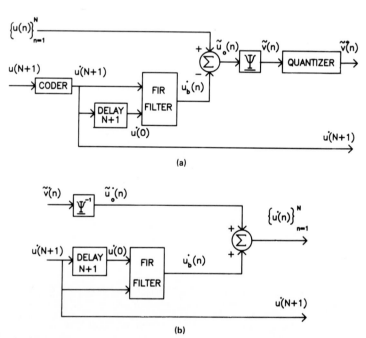

Fig. 15. Recursive block coding of a Markov sequence. (a) Recursive block transform coder (Ψ is the sine transform). (b) Decoder.

ence $\tilde{u}_o(n) \triangleq u(n) - u_b^+(n)$ is transform coded using the Sine transform.† Figure 16(a) compares the rate versus distortion characteristics of small recursive block coders (block size = 4,8) with conventional KLT coding using a
block size of 16.   The recursive block coder with a block size of 8 is more
efficient than the KLT at less than 2% distortion.   (Typically, the distortion
level in transform coding is chosen to be 1% or less.)   In addition to reduced
transform size requirement, such encoders also tend to minimize the block
boundary effects.   Figure 16(b) compares one-dimensional recursive block
coding with cosine transform coding on a 512 × 512 image, and Fig. 16(c)
compares the results for two-dimensional coding [14,28].

ii. Two-Source Coding

Here an image is considered as a composition of smooth and flat areas
that can be represented by low spatial frequencies and edges represented by

† This technique has also been called fast KL transform coding. In recursive block coding,
since the size of the transform can often be quite small, a fast transform is not necessary.

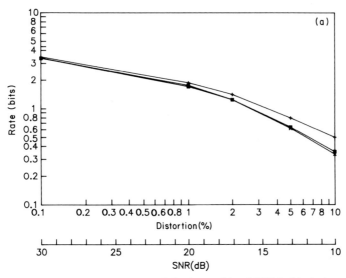

Fig. 16.   (a) Distortion versus rate characteristics for traditional KLT (×,block size = 16) and
recursive block coding RBC(+,blocksize = 4; □,blocksize = 8) of a 1st-order Markov process,
$\rho = 0.95$.   (b) One-dimensional results, block size = 8, entropy = 0.9 bits/pixel. Note that error
images are magnified and enhanced to show coding distortion more clearly.   (c) Two-dimensional results, block size = 8 × 8, entropy = 0.24 bits/pixel. Note that the error images are
magnified and enhanced to show coding distortion more clearly.

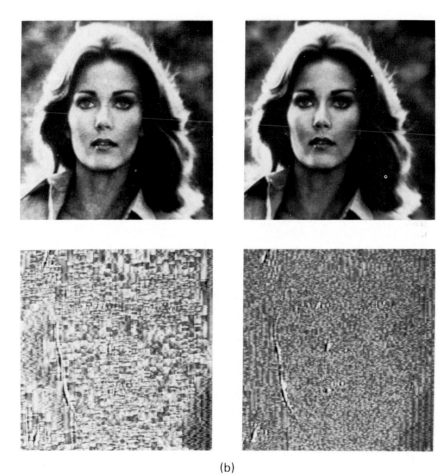

(b)

Fig. 16. (*Continued*)

high spatial frequencies. For encoding, the given image is filtered to separate its low- and high-frequency components. The low-frequency component is transform coded. The addresses of the edges obtained from the
high-frequency component are coded separately. At the receiver, the low-
frequency component is recovered by a transform decoder to which synthe-
sized edges called *synthetic highs* are added at the locations of the edges. A
synthetic high is a signal that is proportional to the second derivative (or
difference) of a step function. In another two-source coding scheme [65] the
high-frequency component is taken to be the sharp changes in the local
mean of the image. The low-frequency component is the difference be-

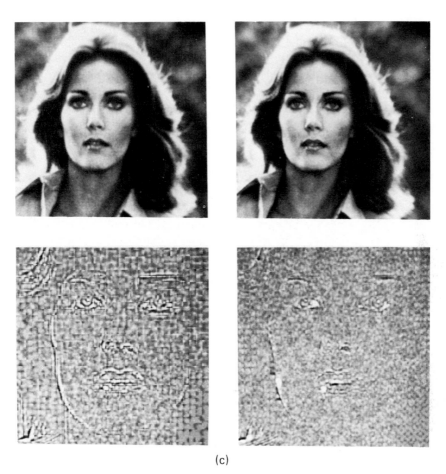

(c)

Fig. 16. (*Continued*)

tween the original image and the sharp changes.   Another method [62] is based on the use of certain stochastic image models to achieve the two-source decomposition.

iii. Transform Coding under Visual Criteria

It is well known that the ordinary m.s. criterion is of limited use for the visual evaluation of images. A weighted m.s. criterion has been found to be useful [36] for transform image coding. Here image contrast, which can be modeled by a cube-root transformation, is considered instead of image intensity. Conceptually, the contrast field is Fourier transformed and multiplied by a frequency weighting function, and the resulting samples are quantized using the usual m.s. criterion. Inverse weighting followed by inverse Fourier transformation gives the reconstructed contrast field.

iv. Adaptive Transform Coding

Performance of transform coding schemes can be improved substantially by adapting them to changes in image statistics. There are essentially three types of adaptation that could be made,

1. Adaptation of transform
2. Adaptation of bit allocation
3. Adaptation of quantizer levels

Theoretically, a change in statistics of the data could require all of these adaptations. Adaptation of the transform basis vectors is the most difficult and expensive, because ideally one should find a new set of KL basis vectors for any change in the statistical parameters. A method of this type was considered by Tasto and Wintz [59]. From a practical standpoint, if one knows the range and type of statistical variations, one can choose a single transform that would be least sensitive to such changes.

A more practical and perhaps more effective method is to adapt the bit assignment to changes in statistics. For example, one could classify an image block into one of several categories and allocate a larger number of bits to blocks that have larger activity and fewer bits to those having lower activity. This results in a variable average rate from block to block but gives a better utilization of the total number of bits over the entire ensemble of image blocks.

Another adaptive scheme is to allocate bits to image blocks so that each block has the same distortion [6]. This results in a uniform degradation of the image and appears less objectionable to the eye. In adaptive quantization schemes, the bit allocation is kept the same but the quantizer levels are adjusted according to changes in the variances of the transform domain samples.

E. Summary of Transform Coding

In summary, transform coding achieves high compression compared with predictive methods. Generally, any distortion due to quantization and transmission errors is distributed, during the inverse transformation, over the entire image. Visually, this is less objectionable than the distortion in predictive coding, which is distributed locally at the source. Predictive coding, compared with transform coding, is quite sensitive to changes in the

TABLE II

Summary of Transform Coding

Design parameter	Comments
Transform	
KL	Statistically optimum but slow. Useful for analysis and performance evaluation.
Cosine	Most efficient fast transform for compression of common images. Hardware transformer available. Difficulties in going to real-time rates.
Sine transform (fast KL)	Useful for recursive block coding. Added complexity of boundary coder minimizes the block effect present in other methods. Uses smaller block size for the same compression as KL.
Hadamard	Easy to implement. No multiplications required. Distortions are more objectionable than sinusoidal transforms.
Slant	Faster and easier to implement than FFT, Sine, Cosine, etc. Performs better than Hadamard.
Haar	Very fast. Gives poor compression but has very good high spatial frequency response.
Slant–Haar	Gives a good compromise between hardware complexity and subjective appeal.
Quantizer	
(a) Optimum mean square (Lloyd–Max)	Recommended for zonal coding if a fixed binary representation of each transform coefficient is used. Transform coefficient probabilities are modeled by Laplacian or Gaussian densities. The dc coefficient is positive and is generally allowed its full dynamic range. This quantizer can also be implemented as a compander if a table lookup is not intended.
(b) Uniform	Useful in threshold coding, because the quantizer output has to be entropy coded.
Zonal coding	Useful for designing fixed-rate coder. Particularly suitable for rates around 1 bit/pixel and in image storage applications, because fixed-length data files offer greater ease and flexibility in database management.
Threshold coding	Most practical adaptive coding technique. Leads to a variable rate coder. Very good performance, but the coder complexity is quite high compared with zonal coding.

statistics of the data, and only some of the adaptive predictive coding algorithms achieve the efficiency of (nonadaptive) transform coding methods. From an implementation point of view, predictive coding has generally a much lower level of complexity, in terms of both memory requirements and number of operations to be performed, but some fast transforms may perform as well with the same number of operations per pixel [4]. However, with the rapidly decreasing cost of digital hardware and computer memory, the hardware complexity of transform coders may not remain a disadvantage for very long. Table II summarizes the design considerations for transform coders.

IV. Hybrid Coding and Vector DPCM

A. Hybrid Coding

Hybrid coding refers to techniques that combine transform and predictive coding techniques [16,30]. Typically, the image is unitarily transformed in one of its dimensions to decorrelate the samples in that direction. Each transform coefficient sequence is then coded independently by a one-dimensional predictive technique such as DPCM in the other direction (see Fig. 17). This technique combines the advantages of hardware simplicity of DPCM and the robust performance of transform coding. The hardware complexity of this method is that of a one-dimensional transform coder and at most N DPCM channels, where N is the size of the transform basis vectors. The number of DPCM channels is significantly less than N, because many elements of the transformed vector are allocated zero bits and, therefore, are not transmitted at all.

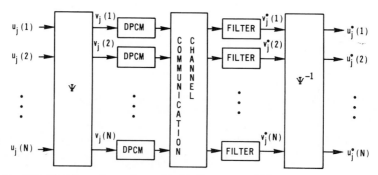

Fig. 17. Hybrid coding of two-dimensional images, where Ψ is a one-dimensional unitary transform.

Consider an $N \times M$ image and let $\mathbf{u}_j$ denote its jth column,

$$\mathbf{u}_j = [u(1,j), u(2,j), \dots, u(N,j)]^T.$$ (27)

A unitary transformation,

$$\mathbf{v}_j = \mathbf{\Psi} \mathbf{u}_j, \qquad 1 \leq j \leq M,$$ (28)

is performed on each vector $\mathbf{u}_j$ such that the elements of $\mathbf{v}_j$ are mutually uncorrelated. Further, for each i, the sequence $\{v_j(i)\}$, is modeled by a suitable autoregressive process such as the first-order Markov model,

$$v_j(i) = \rho_i v_{j-1}(i) + e_j(i), \qquad 1 \leq i \leq N.$$ (29)

The DPCM equations for the ith channel now follow from Section II.B as

$$\begin{aligned}
\text{predictor} && \bar{v}_j(i) &= \rho_i v_{j-1}(i), \\
\text{quantizer input} && \tilde{e}_j(i) &= v_j(i) - \bar{v}_j(i), \\
\text{reconstruction filter} && v_j(i) &= \bar{v}_j(i) + \tilde{e}_j(i).
\end{aligned}$$ (30)

The encoding scheme requires, first, taking the transform $\mathbf{\Psi}$ of each column vector $\mathbf{u}_j$. This is followed by the DPCM channels for predictive coding of successively transformed vectors. The receiver simply reconstructs the transformed vectors according to Eq. (30) and performs the inverse transformation $\mathbf{\Psi}^{-1}$. Because the transform coefficients $v_j(i)$ have unequal variances, the bit rate for each DPCM channel is different. The bit allocation among the various DPCM channels depends on the distribution of prediction error variances $\sigma_e^2(i)$ and can be formulated by a method similar to transform coding [30]. In a typical hybrid coding scheme with $N = 16$ (block size), the bit allocations are (3,3,3,2,2,1,1,1,0,0,0,0,0,0,0,0) for an average rate of 1 bit/pixel. Figure 18 shows a 256×256 image hybrid coded in blocks of 16×256 using the cosine transform.

B. Adaptive Hybrid Coding

The coding scheme of the previous section could be adapted to an image whose spatial statistics vary slowly, by updating the parameters of the model. It is important to consider adaptive schemes that offer reasonable tradeoffs between performance and complexity of the coder.

For a fixed predictor in the feedback loop of a DPCM channel, the variance of the prediction error will fluctuate with changes in the image statistics. A simple method is to update the variance of the prediction error at each step j and use it to adjust the spacing of the quantizer levels in each DPCM channel.

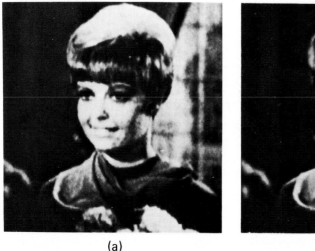

(a)   (b)

Fig. 18. Hybrid encoded images at 1 bit/pixel. (a) Nonadaptive, SNR = 31 dB. (b) Adaptive classification, SNR = 33 dB.

Another adaptation is based on a classification method in which each image column is classified as belonging to one of K predetermined classes that are fixed according to the activity in that image column, as measured by its variance. Quantization bits are allocated according to their dynamic activity. The classification information is communicated by sending an extra $\log_2 K$ bits per column. Figure 18 shows results of adaptive hybrid coding and compares them with the nonadaptive algorithm. Experimentally, it is found that the compression increases by a factor of two for the adaptive techniques [30].

C. Conclusions

In practice, hybrid coders combine the advantages of the relatively simple hardware of DPCM coders and the high performance of transform coders, particularly at moderate bit rates (e.g., 1 bit/pixel for two-dimensional images). In general, hybrid coding performance lies between transform coding and DPCM. It is easily adaptable to coding and filtering of noisy images and to changes in data statistics. It is less sensitive to transmission errors than DPCM but is not as robust as transform coding. Hybrid coders have been implemented for real-time data compression of images acquired by remotely piloted vehicles [38].

V. Interframe Coding

Interframe coding techniques exploit the redundancy that is present between successive frames of a moving sequence of images. This redundancy, or similarity, occurs in regions where there is relatively little motion between frames. The sources of motion are object motion and the associated uncovering of background areas and camera motion, panning, and zooming. Much of the research on interframe image coding has been done recently and is based on spatial–temporal domain techniques. We shall describe four main coding schemes of this type: frame-repeating, resolution exchange, conditional replenishment, and motion compensation techniques. Three-dimensional transform coding and interframe hybrid coding are also discussed.

A. Spatial–Temporal Domain Techniques

i. Frame-Repeating

In broadcast TV systems the picture is displayed at a rate of at least 50 frames/sec to prevent flickering of the display. However, a rate of 30 frames/sec is adequate, and often more than adequate, for displaying motion. This can be likened to the high spatial resolution specification, which is required only in regions containing great detail.

The most widely used technique for obtaining a transmission rate below the flicker frequency is line interlace, whereby each frame is divided into two interleaved fields (odd and even), each containing half the number of lines present in a frame. During a $\frac{1}{60}$-sec interval only one field is transmitted, so that each pixel is displayed once every $\frac{1}{30}$ sec. This is half the flicker rate, but the eye hardly perceives this 30-Hz flicker, because it temporally smooths the displayed signal. Attempts to reduce the data rate by frame-repeating the interlaced pictures do not produce good quality moving images.

An alternative approach is selective replenishment, which requires a frame memory at the receiver only. The picture information is transmitted at a reduced rate according to a fixed, predetermined updating algorithm. At the receiver, any nonupdated data are replaced by corresponding data from previous frames. This technique is simple and produces good quality pictures during periods of little or no movement. However, for more rapid motion, the sampling pattern becomes visible at the edges of the moving regions [42].

ii. Resolution Exchange

It is well known that scenes containing little or no movement can be transmitted and reproduced with good quality using frame-repeating techniques. Furthermore, changing areas of an image can be represented with reduced amplitude and spatial resolution when compared with stationary areas [49]. This resolution exchange phenomenon can be used in straightforward coding schemes to produce good quality images at a constant data rate of 2 to 2.5 bits/pixel. One such scheme segments the image into stationary and moving areas by thresholding the value of the frame-difference sequence. In stationary areas, frame differences are transmitted for every other pixel, and the remaining pixels are repeated from the previous frame. In moving areas, 2:1 horizontal subsampling is used with intervening elements restored by interpolating along the lines. With 5 bits/pixel frame-differential coding, a channel rate of 2.5 bits/pixel is required. The main distortion caused by this system occurs at sharp edges moving with moderate speed [38].

iii. Conditional Replenishment

In this scheme, the image is segmented into stationary and moving areas, and data are transmitted only within the moving areas. This is a more complex technique that requires a frame memory at both the transmitter and the receiver and a buffer at the transmitter to smooth the data, which are generated at a variable rate according to the amount of motion present in the scene. However, it offers much greater compression than selective, or unconditional, replenishment techniques for a comparable picture quality.

A typical interframe predictive coding algorithm requires the following sequence of steps [10,17,35].

1. Segmentation of moving and stationary areas
2. Prediction of a pixel in a given area from pixels in the previous frame(s) and from pixels in the given frame
3. Spatial and temporal resolution exchange used in bit allocation
4. Temporal filtering to reduce jerkiness of motion in the reconstructed image
5. Buffer control

The basis of such a coder is shown in Fig. 19. The movement detector locates significant interframe differences by applying a threshold, which is typically set at 1.5% of the maximum amplitude. The significant pixels tend to occur in clusters along a scan line. Isolated points or very small clusters are ignored to provide an efficient address coding scheme. In this scheme the beginning address of a cluster is transmitted, followed by the quantized

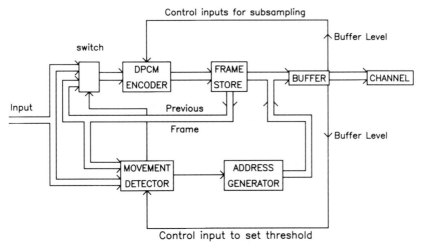

Fig. 19. Conditional replenishment.

amplitudes of the significant interframe differences, and then a cluster terminator code. At the receiver, the data replenish the moving area pixels identified by the addressing information.

The channel capacity needed by such a coder depends on the amount of movement and the buffer size. The larger the buffer the closer the capacity will be to the average data rate. The averge data rate is less than 1 bit/pixel for a typical visual telephone conversation. However, peak data rates are an order of magnitude higher and last for up to a few seconds, that is, more than a hundred frames. It is impractical to try to use a buffer to smooth out these large peaks, because it would be too large and would introduce intolerable delay for face-to-face communication. (It has been found that a delay exceeding $\frac{1}{3}$ sec is distracting.) The practical alternative is progressively to reduce the spatial and amplitude resolution within the moving areas as the activity in the scene increases.

To predict the value of a pixel in a moving area, the methods discussed in Section II are applicable. A useful approach is to send line-to-line differences of frame-to-frame differences. This predictor does not use pixels from the same line or previous field. Hence, horizontal subsampling or field-repeating do not degrade the performance of the predictor. Adaptive prediction is also possible and is called motion compensation. This technique is discussed in the next subsection.

The buffer fullness has been found to be a useful measure of activity, and Fig. 20 shows a typical buffer control scheme with different control levels [10,17,35].

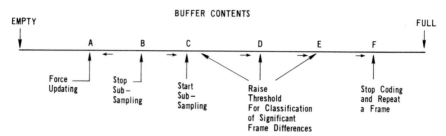

Fig. 20.   Buffer control for conditional replenishment interframe coding.

In the beginning, the first three lines of the first frame are force updated, that is, transmitted at 8 bits/pixel.   In subsequent frames, the next group of three lines is force updated except when the coder is in a buffer overflow state.   At this rate, a complete frame is refreshed every 85th frame (for a 256-line image frame) or approximately every 3 sec for 30 frames/sec. Hence, errors at the receiver are periodically cleared.   If the contents of the buffer fall below state A, force update is continued.   Beyond the state C, the frame differences in a cluster are subsampled by transmitting every other frame difference.   At the receiver, the missing samples are linearly interpolated.   This continues until the buffer contents fall below B.   Beyond C, D, and E, the threshold is increased to lower the number of significant changes.   Beyond the point F, coding is stopped for one frame period, and subsampling is continued for the next frame period.

Because the buffer control plays a leading role, the preceding method is sensitive to available buffer capacity.   Often the buffer length is taken to be the average number of bits per frame.

Simulation studies [25,29] have shown that a 1 bit/pixel rate could be achieved conveniently with an average SNR of about 33 dB.   The buffer overflow occurs about 7% of the time.   Lowering the rate to $\frac{1}{2}$ bit/pixel degrades the performance substantially to 59% buffer overflow rate and 28 dB SNR.   Figures 21(a) and 21(b) show encoded images and the error images for a typical frame at rates of $\frac{1}{2}$ bit/pixel and 1 bit/pixel.   A high rate of buffer overflow results in jerky movement at the receiver's display, as evidenced by Fig. 21(a).   Recent work has shown that sophisticated prefiltering of the source video prior to encoding can decrease the noticeable visual degradations due to mode changes (e.g., subsampling) in a conditional replenishment coder.

iv. Motion Compensation

A recent improvement over conditional replenishment is motion compensation, which adaptively varies the predictor, usually according to an

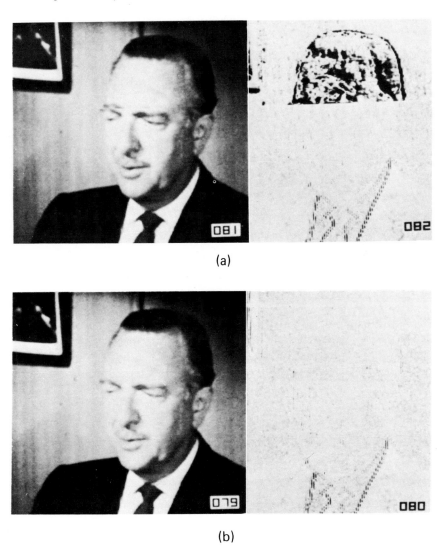

(a)

(b)

Fig. 21. Conditional replenishment interframe encoded images (left) and error images (right). (a) 0.5 bit/pixel, SNR = 27.9 dB. (b) 1.0 bit/pixel, SNR = 33 dB.

estimate of the translational motion of objects [25,29,46]. If the estimate is accurate, the prediction is improved by use of appropriately spatially dis-placed pixels from the previous frame. For video-conference images quan-tized to 8 bits/pixel, motion compensation might save 20–70% of the total bit rate required by a conditional replenishment coder. The saving clearly depends on the amount of motion and the accuracy of the motion estimation algorithm. Future coders will undoubtedly incorporate some form of motion compensation to capitalize on this further compression.

B. Three-Dimensional Transform Coding

In many applications (e.g., multispectral imaging, interframe video imaging, biomedical cineangiography, and computed tomography), one has to work with data of three or more dimensions. Transform coding schemes are possible for compression of such data. The basic ideas of Section III are extended in this development.

A three-dimensional (separable) transform of an $L \times M \times N$ sequence $u(i,j,k)$ defined as

$$v(l,m,n) = \sum_{i=1}^{L} \sum_{j=1}^{M} \sum_{k=1}^{N} u(i,j,k) a_L(l,i) a_M(m,j) a_N(n,k), \qquad (31)$$

where $1 \leq l \leq L$, $1 \leq m \leq M$, $1 \leq n \leq N$, and $\{a_L(i,j)\}$ are the elements of an $L \times L$ unitary matrix $\mathbf{A}_L$. In higher dimensions, one simply takes the $\mathbf{A}$ transform with respect to each index. For an arbitrary $\mathbf{A}$, the number of operations is $LMN(L + M + N)$. If $\mathbf{A}$ is a fast transform, such as the DFT, Sine, and Cosine, then the operation count is generally $LMN(\log_2 LMN)$. The storage requirement for the data is LMN. To reduce the online storage and computation requirements, one often partitions the available data into smaller blocks (e.g., $16 \times 16 \times 16$), and each block is processed independently. The coding algorithm after transformation is the same as before except that one is working with triple-indexed variables.

C. Interframe Hybrid Coding

Hybrid coding is particularly useful in interframe image data compression of motion images. A two-dimensional $M \times N$ block of the kth frame, denoted $\mathbf{U}_k$, is first transformed to give $\mathbf{V}_k$. For each (m,n) the sequence $\{v_k(m,n), \quad k = 1,2,...\}$ is considered a one-dimensional random process and is coded independently by a suitably designed DPCM method. The receiver performs the two-dimensional inverse transform of the sequence $\{v_k(m,n)\}$. Because motion is characterized by deterministic variations along the temporal axis, various motion compensation schemes can be used in the DPCM coder [57]. Adaptive hybrid coding with motion compensation based on trajectory estimation and frame skipping can yield high compression ratios (64:1 for 8-bit images) [25,29]. Such adaptations are not feasible in 3-dimensional transform coding. Thus, with motion compensation the adaptive hybrid coding method performs better than either adaptive predictive coding or adaptive three-dimensional transform coding [25,29].

VI. Coding of Graphics

A. Overview

Among the types of image data most amenable to compression, one has to include graphics and two-tone images such as sketches, diagrams, letters, maps, or newsprint. The need for digital storage and transmission of such images is growing rapidly, and for some applications, commercial products that incorporate compression techniques are already common. Telecopy or facsimile transmission of documents over telephone and data lines is widespread and will become pervasive with the increased use of digital and computer technology in society.

Examples of two-tone documents are shown in Fig. 22. These documents are among those chosen by the CCITT† as test documents for development of data compression techniques in telecopy. These documents have been scanned at a resolution of 200 pixels/in. horizontally and 100 lines/in. vertically; thus, an 8-$\frac{1}{2}$ × 11 in. document contains approximately 2 million bits. Reducing this huge number of bits without loss of quality is worthwhile for either storage or transmission applications. On the documents, one can readily observe properties of the data that are exploited in data compression. The data occur in large contiguous white blocks that can be represented more compactly than by simply listing all the pixels within the image. Some compression algorithms have been devised on the basis of heuristic models; others make use of formal probabilistic models and measurements.

i. Characterization of Graphics Data

Two-tone representations of documents require high sampling densities. For business letters, densities of 100 points per inch (ppi) result in legible documents but with apparent staircase effects at the boundaries between black and white. The CCITT standard of 200 ppi and 100 or 200 lines per inch (lpi) results in acceptable quality in the reproduction of a variety of documents. For newspaper pages that contain text material and halftone images, sampling densities as high as 1000 ppi have been used to avoid Moiré patterns in the halftone images. For business documents, the use of

---

† CCITT: Comité Consultatif International de Téléphonie et Télégraphie

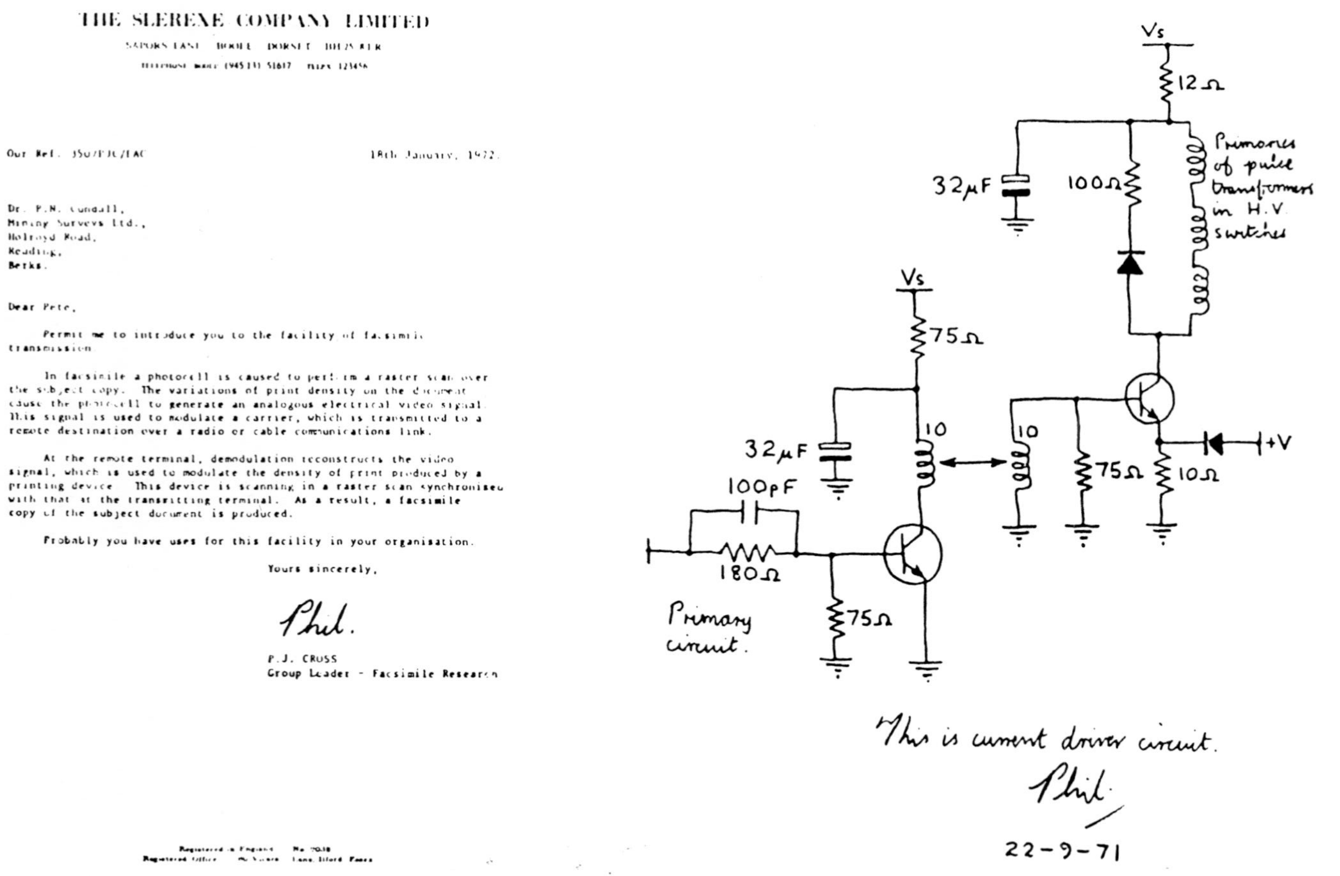

THE SLERENE COMPANY LIMITED

SAPORS LANE · BOOLE · DORSET · BH25 8ER

TELEPHONE BOOLE (945 13) 51617 · TELEX 123456

Our Ref. 350/PJC/EAC 18th January, 1972.

Dr. P.N. Cundall,
Mining Surveys Ltd.,
Holroyd Road,
Reading,
Berks.

Dear Pete,

Permit me to introduce you to the facility of facsimile transmission.

In facsimile a photocell is caused to perform a raster scan over the subject copy. The variations of print density on the document cause the photocell to generate an analogous electrical video signal. This signal is used to modulate a carrier, which is transmitted to a remote destination over a radio or cable communications link.

At the remote terminal, demodulation reconstructs the video signal, which is used to modulate the density of print produced by a printing device. This device is scanning in a raster scan synchronised with that at the transmitting terminal. As a result, a facsimile copy of the subject document is produced.

Probably you have uses for this facility in your organisation.

Yours sincerely,

Phil.

P.J. CROSS
Group Leader - Facsimile Research

Registered in England: No. 2038
Registered Office: 60 Vicars Lane, Ilford. Essex.

(a) (b)

Fig. 22. Examples of graphics documents. (a) Document 1. (b) Document 2.

gray-scale information allows some reduction in sampling density with an acceptable quality of reproduction. We shall assume that only two levels are used and that the sampling density is 200 ppi unless noted otherwise. Methods and algorithms discussed are thus relevant to the data compression of drawings and text but not of halftone images.

ii. Heuristic Models

Two properties of the data are exploited for data compression: the amount of black in an image is generally a small fraction of the total image; and this black information appears in regular two-dimensional patterns such as characters, lines, and symbols. Three types of coding schemes are suggested by these properties:

1. White areas are skipped, and only the portions of images with substantial fraction of black are transmitted in detail.

2. The boundaries of the two-dimensional black objects over a white background are transmitted or stored.

3. Specific patterns such as symbols or characters are identified, and the resulting identification is transmitted or stored.

By contrast to the coding of gray-scale images, the coding of graphics is generally required to be reversible; and thus, no image degradation in the encoding process is allowed. Another substantial difference is in the use of variable rate coding schemes. Because the time used in the transmission or decoding of a document does not have to be the same for all documents, the statistical properties of the data can be used to increase data compression. Further, because no degradation of data quality is allowed, a fixed-rate constraint would limit algorithms and performance quite drastically. Graphics coding algorithms and techniques have to be considered in terms of tradeoff between simplicity, which in general implies lower equipment costs or higher speed, and performance as measured by the number of bits per pixel needed to represent a document. These considerations lead naturally to a separate discussion of one-dimensional techniques, which only use information on a single scan line, and two-dimensional techniques, which exploit image structure and correlations from line to line.

B. One-Dimensional Codes

On a single scan line, a two-tone image consists of sequences of white pixels (0) that alternate with sequences of black pixels (1). A sequence of pixels of the same brightness is called a *run;* thus, the scan line consists of alternate black and white runs. As discussed earlier, there will generally be a preponderance of long white runs and much shorter black runs.

Data	0000	0000	1100	0001	0100	0000
Block number	1	2	3	4	5	6
WBS codewords	0	0	11100	10001	10100	0
Run lengths	8W, 2B, 5W, 1B, 1W, 1B, 6W					

Fig. 23. One-dimensional data.

i. White Block Skipping

White block skipping (WBS) is a very simple coding scheme [21]. A block of N white pixels is encoded as 0. If not all pixels within the block are white, then $N + 1$ bits, consisting of a 1 followed by the binary pattern for the black and white pixels is used to encode the block. An example is shown in Fig. 23 for $N = 4$. Thus, for each block of N pixels we use a 1-bit codeword if the block is all white and an $(N + 1)$-bit codeword if the block is not all white. If the probability of white blocks of N pixels is p_N, then the average number of bits per pixel for the one-dimensional WBS code is

$$b_{\text{WBS},N} = \frac{1}{N}[p_N + (1 - p_N)(N + 1)] = \left(1 - p_N + \frac{1}{N}\right) \quad \text{bits/pixel.} \quad (32)$$

Although the performance of the WBS code depends on the block size, it has been determined [21] that the choice of N is not critical and that $N = 10$ is suitable for a wide range of documents. The number of bits per pixel ranges from 0.11 to 0.32 for a reference set of CCITT documents which are predominantly white and are 0.11 for Document 1 and 0.15 for Document 2 shown in Fig. 22.

ii. Theoretical Basis for One-Dimensional Codes

The heuristic WBS code takes advantage of the likely occurrence of long white runs in the data but does not explicitly consider the transitions from black to white or white to black (i.e., the change of color) as the information to be transmitted. Instead, run-length codes directly address the representation and coding of change of color information. We show, in Fig. 23, the white and black run lengths. Consider now the run lengths as the output of an information source. The source symbols are no longer binary, but there are substantially fewer run lengths than pixels in the original data. Assume that run length l_k with probability p_k is encoded with r_k bits. The average number of bits needed to encode the run lengths is, therefore,

$$b = \sum_k r_k p_k. \quad (33)$$

There are a large number of methods for choosing the binary code for the run lengths. A simple one is to use a binary representation of the run length.   For the example of Fig. 23, a 3-bit code word is sufficient; thus, 21 bits are needed to represent the run lengths for the original sequence of 24 pixels.

A question of theoretical and practical interest is the least number of bits needed to represent the output of a discrete information source.   A fundamental result due to Shannon [55] establishes the entropy H of the source as the lower limit to the average number of bits required per source symbol,

$$H = -\sum_k p_k \log_2 p_k. \tag{34}$$

As noted earlier, black runs are shorter than white runs.   If we evaluate separately the probabilities of white run length and black run length, we can obtain a better bound as

$$H = \tfrac{1}{2}(H_B + H_W), \tag{35}$$

where H_B and H_W are entropies of the black or white run lengths evaluated by Eq. (34).   If the average run lengths are R_B and R_W, the lower bound on the number of bits per pixel is given by

$$b_r = (H_B + H_W)/(R_B + R_W). \tag{36}$$

The lower bounds evaluated for the two documents of Fig. 22 are 0.05 for Document 1 and 0.045 for Document 2, respectively.   As compared with the WBS code, these bounds indicate that further compression by a factor of two may be possible.

The coding method due to Huffman [22] assigns codewords on the basis of the symbol probability p_k and is an optimum technique.   The code is thus adapted to the run-length probabilities and to the specific document. Huffman codes systematize the heuristic rules of assigning short codewords to the most frequent symbols and result in a table that lists codewords for all possible run lengths.   For graphics coders, run lengths can be as large as 1700; thus, large memories and look-up tables are needed to implement Huffman codes.   Several other codes have been devised that require smaller memory.   Some of these codes are algorithmic in that a calculation can be performed at the encoder to construct each codeword.   Others, such as Huffman codes, are nonalgorithmic and require a precalculated codebook stored in memory and a table look-up encoding procedure.

iii. Some Efficient One-Dimensional Codes

The B_1 *code* [39] is an algorithmic code that is quite efficient for negative power distributions, in which the probabilities of the run length decrease rapidly with the length.   The code is constructed as follows: Bit 0 is assigned to white and bit 1 to black.   Color bits alternate with the bits used in the

Run length	Binary representation	Codeword white run	Codeword black run
1	0	00	10
2	1	01	11
3	00	0000	1010
4	01	0001	1011
5	10	0100	1110
6	11	0101	1111
7	000	000000	101010

Fig. 24. B_1 codes.

representation of the run length.   The run length is represented in binary form by the shortest possible binary word.   An example is shown in Fig. 24. Codewords always have even length, and the first bit of each of the 2-bit sequences always indicates the color of the run.   Encoding and decoding are thus very simple.

The *truncated Huffman code* assigns Huffman codewords for white run lengths up to L_W and black runs up to L_B with different tables for white and black run length.   All longer run lengths, which have low probabilities, are assigned a single binary prefix followed by a fixed-length binary representation of the specific run length.

In a specific scheme [43] proposed to the CCITT, $L_W = 47$, $L_B = 15$, and specific codewords for black and white Huffman codes based on average statistical properties of documents have been determined.   Thus, the coding method is simple to use, because it requires two small, known, look-up table codes and two fixed-length algorithmic codes.

In the *modified Huffman code*, Huffman codewords are used for run lengths up to 63.   For run lengths greater than 63, the run length is characterized by two numbers less than 63, the quotient and the remainder of the division of the run length by 64.   The quotient is encoded by a *make up* codeword and the remainder by a *terminating* codeword.   Again, specific codeword tables have been proposed that are optimized for a typical run-length distribution.

Other one-dimensional codes have been reported [53] that are comparable in performance and complexity to the algorithms we have just described.

C. Two-Dimensional Codes

Two-dimensional codes exploit the two-dimensional structure of graphics data to provide additional data compression.   White block skipping (WBS), discussed earlier, can be extended to two dimensions by representing a two-dimensional $M \times N$ white block of pixels by a single codeword, 0. If the $M \times N$ block contains some black element, then a 1 prefix followed by the MN color bits of the block are used [13]. The choice of block size will affect performance; and for a given document, it is worthwhile to have the option of using several possible block sizes. Several adaptive WBS codes have been proposed and are discussed in a review article by Huang [13,20].

Several two-dimensional codes consider only two adjacent scan lines. Because of the two-dimensional structure of the image, run lengths will be highly correlated from line to line.

i. Adjacent Line Coders

The *prediction differential quantization (PDQ)* [19] coding scheme encodes changes in corresponding color runs on successive lines. This can be done by encoding changes in color transition locations, changes in run lengths, or changes in transition location for one color and run length for the other. Special provisions must be made for the cases of a color run ending or starting.

Relative address coding (RAC) [64] is one of the most efficient coding schemes that uses only two adjacent lines. We refer to Fig. 25 for one explanation of the algorithm. In RAC the changes of color in the current line are referred either to the last change of color on the same line or to the nearest change of the same color on the previous line. Assume that the change Q is to be encoded. The two candidate reference changes are P on the current line and Q' on the previous line. Distance PQ is 5 and distance QQ' is -2; therefore, the RAC distance of the change of color at Q is -2. Note that the RAC distance is negative, because Q on the current line is to the left of the previous line reference. The RAC distances of the color changes on the current line are shown in Fig. 25. Note that the RAC distances will generally be small. The encoding of the RAC distances is done by a code similar to the B_1 code discussed earlier, except for the choice of the reference line and for very short distances $(0, +1, -1)$, which are given special code-words based on their probability of occurrence, as for a Huffman code. The RAC technique requires 0.06 bit/pixel for document 1 and 0.041 bit/pixel for

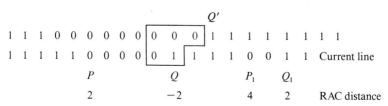

Fig. 25. RAC technique.

document 2 [43]. This is substantially better than any of the one-dimensional codes and is also lower than the entropy of the one-dimensional run lengths.

Another two-dimensional encoding scheme, which slightly outperforms the RAC scheme, uses probabilistic models more explicitly [31,50]. This Technical University of Hanover (TUH) method considers the four closest neighbors of the current pixel as characterizing the state of the data source at the time the current pixel occurs. See, for instance, the four neighbors of Q that are outlined in Fig. 25. Based on the state of the data source, a prediction of the current pixel is made. Prediction errors are indicated by a 1 and correct predictions by a 0. Thus, the original data are used to build a two-dimensional array of prediction errors. This array has sparse errors and can be encoded by run-length encoding techniques. Alternatively, the runs between prediction errors are encoded separately for each state of the source.

ii. Pattern Recognition Coding

The basic idea of using pattern recognition coding for graphics is simple and appealing especially for the case of printed characters. If a character is identified on the basis of a complete two-dimensional pattern of pixels, then only the character and not the pattern needs to be stored or transmitted. Even if one requires additional information for the position or the type of character, the potential data compression is very high. For the specialized case of characters, an efficient encoding technique based on pattern recognition has been proposed [8]. For documents containing other graphics information as well as characters, an interesting method based on pattern recognition has been described [51]. The reference or template patterns that serve in the pattern recognition phase are obtained from the received data. As a new pattern is received, it is compared to all the stored patterns; and if no identification is possible, the pattern is stored and used as part of the reference set. There are a number of distinct cases to consider for documents that consist of graphs, diagrams, and so on, as well as characters. Thus, pattern recognition schemes quickly become complex and memory intensive.

D. Susceptibility to Errors

As in gray-scale image coding, an important consideration in the choice of encoding technique for graphics is the susceptibility of the code to transmission errors. Note that if the error rate is fixed, fewer transmission errors will occur during the transmission of a document when data compression is used. When errors do occur, their effect on the quality of the reproduced document will differ greatly for different coding schemes. Steps can be taken to limit the effect of transmission errors to as small an area of the image as possible.

i. One-Dimensional Coding

An important advantage of all one-dimensional codes is the ease in the introduction of end-of-line (EOL) signals. By providing for EOL signals, one can limit the disturbance of the image due to a single error to a single line. Whenever an error in a line is detected, a simple method that reduces the degradation of the image is to substitute the previous line for the line in error. This method is useful for scanning at 200 lpi where the data are highly redundant from line to line. It is also feasible to determine the position of the error on the line and, therefore, to do only a partial substitution from other lines by correlating decoded data from the current line to the previous one and keeping the portions of the line with a good correlation as essentially correct [53].

ii. Two-Dimensional Coding

Two-dimensional codes are inherently more susceptible to errors than one-dimensional codes. In the case of the RAC or TUH codes, errors can destroy several scan lines and produce intolerable degradation of quality. One method that provides some protection from errors is to change from a two-dimensional to a one-dimensional code every several lines. This prevents vertical propagation of errors, but at the expense of a loss of compression. Experimental results [43] on burst errors occasionally encountered on telephone lines indicate that two-dimensional codes would probably need to be supplemented by error correction to operate reliably in such an environment.

E. Summary of Graphics Coding

The evaluation of graphics coders is based on their performance for different types of documents at high and low sampling densities, on the complexity of implementation, and on the degradation caused by transmission errors. See Table III.

At high resolution, 200 ppi and 200 lpi, and for the most efficient codes, the two-dimensional RAC or TUH codes, the bits per pixel range from a low of 0.035 for Document 2 of Fig. 22, a sparse circuit diagram, to 0.185 for a page of solid text in English or in Japanese [43]. Thus, the performance may

TABLE III

Summary of Graphics Coders

Design parameter	Comments
One-dimensional codes	Simple implementation, good performance, principally for graphs. Limited degradation due to transmission errors.
(a) B_1 code[a]	Simple algorithmic code.
(b) Modified Huffman and truncated Huffman codes	Table look-up and algorithmic. Up to 20% better performance than B_1. Can be adapted to type of document.
Two-dimensional codes: RAC and TUH	Two scan line processing, fairly complex. Very high performance, principally for graphs. Substantial, 2 to 6 lines degradation due to transmission errors. Table look-up and algorithmic codes.
Pattern recognition codes	Excellent performance for limited set of patterns, such as for text. Complex implementation.

[a] See Section VI.B.3.

vary in the ratio 5:1 depending on the document encoded. The effect of the encoding scheme is not as large. The simple one-dimensional B_1 code (see Section VI.B.3) has twice the rate of the best two-dimensional scheme for Document 2 and only 1.5 times the lowest rate for Japanese text. As the vertical sampling rate decreases from 200 to 100 lpi, the performance of one-dimensional codes does not change, but the performance of two-dimensional codes will decrease by 20% or more. Thus, performance and implementation considerations indicate that two-dimensional codes should only be considered at high sampling densities. With respect to the effect of transmission errors in the absence of error protection, one-dimensional codes are preferable.

VII. Applications

In this section we shall discuss a number of application areas in which data compression is or could be widely used to enable efficient transmission and/or storage of images. Rather than discussing particular systems that have already been implemented, we shall adopt a more general approach and discuss the merits of various coding schemes in the following areas: archival storage, medical image data compression, video conferencing, remotely piloted vehicles, and broadcast television [2].

A. Archival Storage

Currently, one of the most important applications is storage, because of the rapid development of very large digital databases. The types of database are diverse and include documents, speech files, personnel files, maps, fingerprints, parts manuals, medical images, and satellite images. A distinction should be drawn between computer storage and bulk storage of archival data. In the former case, speed of recall is important and, hence, fixed-rate coding techniques should be used to allow a fixed storage format. Furthermore, intraframe coding schemes should be employed for optimal recall speed rather than interframe schemes, which would necessitate reconstruction of the whole sequence up to the point of interest. However, the speed of recall is not as important in archival storage applications; hence, more-efficient variable-rate techniques that are fairly close to information preserving should be used.

B. Medical Image Data Compression

Despite the high cost of computer tomography systems, they are being used in many hospitals throughout the world. Physicians would like to store the x-ray images and maintain an image database for each patient. Currently, there is no compression of these images; they are stored directly on computer tape. Physicians are concerned that data compression might introduce distortions that would remove important information or introduce spurious information, either of which could lead to incorrect diagnosis. Consequently, error-free (or extremely low-error) techniques should be considered. One possibility is open-loop DPCM which is an error-free encoding technique [5].

C. Video-Conferencing

Unlike broadcast television, there is rarely a large amount of movement in the pictures used for video-conferencing. This allows conditional replenishment coding to be used effectively, which has been demonstrated by hardware realization of complete systems [23,58,61].

D. Remotely Piloted Vehicles

Remotely piloted vehicles (RPV) are small aircraft, not much larger than model planes, that are used by the military in reconnaissance operations. In a military environment, transmission must be protected from jamming by the enemy. This is achieved by use of spread spectrum techniques in which the information bandwidth is spread and transmitted over a wideband signalling structure, and the information is retrieved at the receiver by means of correlation techniques. Consequently, the original information bandwidth must be kept as small as possible to allow a number of RPVs to operate simultaneously using the available bandwidth [9].

Any compression system used in this application must be compact, light, and require very little power; also, because an RPV is to some extent expendable, it must be a low-cost system. Possibly, the most suitable coding technique is a hybrid scheme that does not require a frame memory in the RPV and achieves high compression without the full complexity of a two-dimensional transform coder. Furthermore, hybrid or transform schemes are more flexible than DPCM, since they can operate at switchable data rates by selecting different quantizers. This may be useful, because different stages of a mission require differing spatial and temporal resolutions.

E. Broadcast Television

In broadcast television [32,63] there is large movement between frames because of scene changes and camera motion, zooming, and panning. Under such conditions, which are not infrequent, conditional replenishment is rather inefficient. A promising alternative is motion compensation, an adaptive interframe technique that is currently receiving a great deal of attention.

Bibliography

Data compression has been a topic of immense interest in digital image processing. Several special issues and review papers have been devoted to this field. For details, see

Davisson, L. D., and Gray, R. M. (Eds.) *Data Compression,* Benchmark Papers in Electrical Engineering and Computer Science. Dowden, Hutchinson & Ross, Inc., Stroudsberg, Pennsylvania 1976.
Huang, T. S. "PCM Picture Transmission," *IEEE Spectrum,* **12,** No. 12, pp. 57–60, December 1965.
Huang, T. S., and Tretiak, O. J. (Eds.) *Picture Bandwidth Compression.* Gordon and Breach, New York, 1972.
Jain, A. K. "Some New Techniques in Image Processing," *In Image Science Mathematics,* C. O. Wilde, E. Barrett (Eds.) pp. 201–233, West Periodical Co., North Hollywood, California 1977.
Jain, A. K. "Image Data Compression: A Review," *Proc. IEEE,* **69,** No. 3, pp. 349–389, March 1981.
Jain, A. K. "Fundamentals of Digital Image Processing," book (to appear).

Netravali, A. N., and Limb, J. O. "Picture Coding: A Review," *Proc. IEEE,* **68,** No. 3, pp. 366–406, March 1980.

Pratt, W. K. Digital Image Processing. John Wiley, New York, pp. 591–710, 1978.

Pratt, W. K. (Ed.) *Image Transmission Techniques,* Academic Press, NY. 1979.

Special Issue on Digital Communications, *IEEE Commun. Tech.,* **COM-19,** 6, Part I, December 1971.

Special Issue on Digital Encoding of Graphics, *Proc. IEEE,* **68,** No. 7, July 1980.

Special Issue on Image Bandwidth Compression, *IEEE Trans. Commun.,* **COM-25,** 11, November 1977.

Special Issue on Image Processing, *Proc. IEEE,* **69,** No. 5, May 1981.

Special Issue on Picture Communication Systems, *IEEE Trans. Commun.,* **COM-29,** December 1981.

Special Issue on Redundancy Reduction, *Proc. IEEE,* **55,** No. 3, March 1967.

References

1. N. Ahmed, T. Natarajan, and K. R. Rao, Discrete cosine transform, *IEEE Trans. Computers* (corresp.), **C-23,** January 1974, 90–93.

2. V. R. Algazi (ed.), "Proceedings of NSF Workshop on Image Coding," SIPL Report 79–5, University of California, Davis, Dec. 1977.

3. V. R. Algazi and J. T. DeWitte, Theoretical performance of entropy-encoded DPCM, *IEEE Trans. Commun.* **COM-30**(5), May 1982, 1088–1095.

4. V. R. Algazi and B. J. Fino, Performance and computation ranking of fast unitary transforms in applications, *Proc. ICASSP* 1982, 32–35.

5. V. R. Algazi and A. K. Jain, "Noise Analysis and Measurement and Data Compression Study for a Digital X-Ray System." Sign. & Image Proc. Lab., Dept. of Elect. & Comput. Engr., Univ. of Calif., Davis, CA, Nov. 1981.

6. V. R. Algazi and D. J.-Sakrison, Encoding of a Counting Rate Source with Orthogonal Functions, in "Computer Processing in Communications," pp. 85–100. Polytechnic Institute of Brooklyn, New York, 1969.

7. J. F. Arnold and M. C. Cavenor, Improvements to the CAQ bandwidth compression scheme, *IEEE Trans. Commun.* **COM-29**(12), December 1981, 1818–1822.

8. R. N. Ascher and G. Nagy, A means for achieving a high degree of compaction on scan-digitized printed text, *IEEE Trans. Comput.* **C-23**(11), Nov. 1974, 1174–1179.

9. P. Camana, Video bandwidth compression: A study in tradeoffs, *IEEE Spectrum* **16**(6) June 1979, 24–29.

10. J. C. Candy *et al.,* Transmitting television as clusters of frame-to-frame differences, *Bell Syst. Tech. J.* **50,** August 1971, 1889–1917.

11. J. C. Candy and R. H. Bosworth, Methods for designing differential quantizers based on subjective evaluations of edge busyness, *Bell Syst. Tech. J.* **51**(7), September 1972, 1495–1516.

12. D. J. Connor, R. F. W. Pease, and W. G. Scholes, Television coding using two-dimensional spatial prediction, *Bell Syst. Tech. J.* **50,** March 1971, 1049–1061.

13. F. De Coulon and O. Johnsen, Adaptive block scheme for source coding of black-and-white facsimile, *Elec. Lett.* **12,** February 1976, 61–62.

14. P. M. Farrelle, "Recursive block coding techniques for data compression," M. S. Thesis, U. C. Davis, 1982.

15. A. Habibi, Comparison of the nth-order DPCM encoder with linear transformations and

block quantization techniques, *IEEE Trans. Commun. Tech.* **COM-19,** December 1971, 948–956.

16. A. Habibi, Hybrid coding of pictorial data, *IEEE Trans. Commun.* **COM-22,** May 1974, 614–626.

17. B. G. Haskell *et al.,* Interframe coding of video-telephone pictures, *Proc. IEEE* **60,** July 1972, 792–800.

18. S. Hecht, The visual discrimination of intensity and the Weber-Fechner law, *J. Gen. Physiol.* **7,** 1924, 241.

19. T. S. Huang, Runlength coding and its extensions, in "Picture Bandwidth Compression," (T. S. Huang and O. J. Tretiak, eds.). Gordon and Breach, New York, 1972.

20. T. S. Huang, Coding of two-tone images, *IEEE Trans. Commun.* **COM-25,** November 1977, 1406–1424.

21. T. S. Huang and A. B. S. Hussian, Facsimile coding by skipping white, *IEEE Trans. Commun.* **COM-23**(12), Dec. 1975, 1452–1466.

22. D. A. Huffman, A method for the construction of minimum-redundancy codes, *Proc. IRE* **40,** 1952, 1098.

23. K. Iinuma *et al.,* "NETEC-6: Interframe Encoder for Color Television Signals," No. 44, pp. 92–96. NEC Res. Devel., Jan. 1977.

24. A. K. Jain, A fast Karhunen-Loéve transform for a class of stochastic processes, *IEEE Trans. Commun.* **COM-24,** September 1976, 1023–1029.

25. J. R. Jain, Ph.D. dissertation, Dept. of Electr. Engr. SUNY Buffalo, September 1979."Interframe adaptive data compression techniques for images."

26. A. K. Jain, A sinusoidal family of unitary transforms, *IEEE Trans. Pattern Anal. Mach. Intelligence* **PAMI-1**(4), October 1979, 356–365.

27. A. K. Jain, Advances in mathematical models for image processing, *Proc. IEEE* **69,** March 1981, 502–528.

28. A. K. Jain and P. M. Farrelle, Recursive block coding, *Sixteenth Annual Asilomar Conf. on Circuits, Systems and Computers,* Nov. 1982.

29. J. R. Jain and A. K. Jain, Displacement measurement and its application in interframe image coding, *IEEE Trans. Comm.* **COM-29,** Dec 1981, 1799–1808.

30. A. K. Jain and S. H. Wang, "Stochastic Image Models and Hybrid Coding," Final Report, NOSC contract N00953-77-C-003MJE. Dept. of Electrical Engineering, SUNY Buffalo, New York, October 1977.

31. N. S. Jayant (ed.), "Waveform Quantization and Coding," pp. 368–372. IEEE Press, New York, 1976.

32. T. Koga *et al.,* Statistical performance analysis of an interframe encoder for broadcast television signals, *IEEE Trans. Commun.* **COM-29**(12), Dec. 1981, 1868–1876.

33. T. R. Lei, N. Scheinberg, and D. L. Schilling, Adaptive delta modulation systems for video encoding, *IEEE Trans. Commun.* **COM-25**(11), November 1977, 1302–1314.

34. J. O. Limb, C. B. Rubinstein, and J. E. Thompson, Digital coding of color video signals — A review, *IEEE Trans. Commun.* **COM-25**(11), November 1977, 1349–1385.

35. J. O. Limb *et al.,* Combining intraframe and frame-to-frame coding for television, *Bell Syst. Tech. J.* **53,** August 1974, 1137–1173.

36. J. L. Mannos and D. J. Sakrison, The effects of a visual fidelity criterion on the encoding of images, *IEEE Trans. Inf. Theory* **IT-20,** July 1974, 525–536.

37. J. Max, Quantizing for minimum distortion, *IRE Trans. Inf. Theory* **IT-6,** March 1960, 7–12.

38. R. W. Means, E. H. Wrench, and H. J. Whitehouse, "Image Transmission via Spread Spectrum Techniques," ARPA Quarterly Technical Report, ARPA-QR6. Naval Ocean Systems Center, San Diego, CA, Jan.-Dec. 1975; also see ARPA-QR8, Annual Report, Jan.-Dec. 1975.

39. H. Meyr, H. G. Rosdolsky, and T. S. Huang, Optimum run-length codes, *IEEE Trans. Commun.* **COM-22,** June 1974, 825–835.
40. J. W. Modestino and V. Bhaskaran, Robust two-dimensional tree encoding of images, *IEEE Trans. Commun.* **COM-29**(12), December 1981, 1786–1798.
41. J. W. Modestino *et al.,* Tree encoding of images in the presence of channel errors, *IEEE Trans. Inf. Theory* **IT-27**(6), November 1981, 677–697.
42. F. W. Mounts, Low-resolution TV: An experimental digital system for evaluating bandwidth-reduction techniques, *Bell Syst. Tech. J.* **46,** January 1967, 167–198.
43. H. G. Musmann and D. Preuss, Comparison of redundancy reducing codes for facsimile transmission of documents, *IEEE Trans. Commun. Technol.* **COM-25,** Nov. 1977, 1425–1432.
44. A. N. Netravali, On quantizers for DPCM coding of picture signals, *IEEE Trans. Inf. Theory* **IT-23,** May 1977, 360–370.
45. A. Netravali and B. Prasada, Adaptive quantization of picture signals using spatial masking, *Proc. IEEE* **65,** April 1977, 536–548.
46. A. N. Netravali and J. D. Robbins, Motion compensated television coding, Part I, *Bell Syst. Tech. J.* **58**(3) March 1979, 631–670.
47. I. M. Paz, G. C. Collins, and B. H. Batson, A tri-state delta modulator for run-length encoding of video, *Proc. National Telecomm. Conf.,* Dallas, Texas, Vol. I, pp. 6.3-1–6.3-6, November 1976.
48. J. J. Pearson and R. M. Simonds, Adaptive, Hybrid, and Multi-Threshold CAQ Algorithms, *Proc. SPIE Conf. Adv. Image Transmis. Tech.* **87,** August 1976, 19–23.
49. R. F. W. Pease and J. O. Limb, Exchange of spatial and temporal resolution in television coding, *Bell Syst. Tech. J.* **50,** January 1971, 191–200.
50. D. Preuss, Comparison of two-dimensional facsimile coding schemes, *Int. Conf. on Communications,* San Francisco, USA, 1975, pp. 7/12-7/16.
51. W. K. Pratt *et al.,* Combined symbol matching facsimile data compression system, *Proc. of the IEEE* **68**(7), July 1980, 786–796.
52. P. G. Roetling, Halftone method with edge enhancement and Moiré suppression, *J. Opt. Soc. Am.* **66,** 1976, 985–989.
53. V. Rothgordt, G. Aaron, and G. Renelt, One dimensional coding of black and white facsimile pictures, *Acta Elec.* **21**(1), 1978, 21–38.
54. D. J. Sakrison, On the role of the observer and a distortion measure in image transmission, *IEEE Trans. Commun.* **COM-25**(11), November 1977, 1251–1266.
55. C. E. Shannon, The mathematical theory of communication, Parts I and II, *Bell Syst. Tech. J.* **27,** 1948, 379, 623.
56. R. Steele, "Delta Modulation Systems. Wiley, New York, 1975.
57. J. A. Stuller and A. N. Netravali, Transform domain motion estimation, *Bell Syst. Tech. J.,* **58**(7), September 1979, 1673–1702, also see pp. 1703–1718 for application to coding.
58. K. Takikawa, "Simplified 6.3 Mbit/s codec for video conferencing," *IEEE Trans. Commun.* **COM-29**(12), Dec. 1981, 1877–1882.
59. M. Tasto and P. A. Wintz, Image coding by adaptive block quantization, *IEEE Trans. Commun. Technol.* **COM-19**(6), December 1971, 956–972.
60. J. E. Thompson, A 36-Mbit/s television coder employing pseudorandom quantization, *IEEE Trans. Commun. Technol.* **COM-19**(6), December 1971, 872–879.
61. J. E. Thompson, European collaboration on picture coding research for 2Mbit/s transmission, *IEEE Trans. Commun.* **COM-24**(12), Dec. 1981, 2003–2004.
62. S. H. Wang, "Applications of stochastic models for Image Data Compression", Ph.D. dissertation, Dept. of Elec. Engr., State Univ. of New York, Buffalo, September 1979.
63. H. Yamamoto *et al.,* 30 Mbit/s codec for NTSC color TV signal using an interfield-intrafield adaptive prediction, *IEEE Trans. Commun.* **COM-29**(12), Dec. 1981, 1859–1867.

64. Y. Yamazaki, Y. Wakahara, and H. Teramura, "Digital Facsimile Equipment 'Quick-FAX' Using a New Redundancy Reduction Technique," NTC '76, pp. 6.2-1/6.2-5, 1976.
65. J. K. Yan and D. J. Sakrison, "Encoding of images based on a two component source model", *IEEE Trans. Commun.* **COM-25,** November 1977, 1315–1322.
66. H. Yasuda and H. Kawanishi, Predictor adaptive DPCM, *Proc. SPIE Conf. Applications of Digital Image Processing,* **149,** August 1978, 189–195.
67. W. Zschunke, DPCM picture coding with adaptive prediction, *IEEE Trans. Commun.* **COM-25**(11), November 1977, 1295–1302.

6 Image Spectral Estimation

S. W. Lang
T. L. Marzetta

Schlumberger–Doll Research
Ridgefield, Connecticut

I. Introduction

Many scientific theories and system design formalisms involve the modeling of signals as random processes. These signals are often functions of space, time, or both. Homogeneous random processes possess statistical descriptions that are independent of an absolute origin in space and time. Random processes of this important class consist of superpositions of complex exponentials, or plane waves, of various temporal frequencies and spatial wavevectors, with random amplitudes and with a deterministic distribution of power as a function of frequency and wavevector. This power distribution is described by the *power spectrum,* a nonnegative function of frequency and wavevector. The power spectrum is an important partial description of a homogeneous random process.

A common problem is that of *estimating* the power spectrum from samples of the random process. Power spectral estimation is important in many of the sciences, as a means of suggesting and testing theories that involve

227

DIGITAL IMAGE PROCESSING TECHNIQUES

random process models for physical phenomena. For example, power spectral estimation has been used to test theories that predict particular models for ocean wave spectra, internal gravity wave spectra, and cloud brightness distribution spectra [2,49,54].

Power spectral estimation is important in the design and implementation of many signal processing systems. The incoherent imaging problems and the bearing estimation problems encountered in radio-astronomy, radar, and sonar applications are often equivalent to problems of power spectral estimation [Fig. 1(a)] [17,26]. Spectral estimation is used to construct codes for efficient image transmission systems, to classify textures in pattern recognition systems, and to design image restoration systems [Fig. 1(b)] [24,47,50].

This chapter contains descriptions of various methods of power spectral estimation that are applicable to the multidimensional problems encountered in image and array processing applications. The assumptions that each method makes about the random process are examined, as well as the type of data required by each method and the description of the power spectrum that each method produces. The relationships and differences between the one-dimensional theory of time series spectral estimation and multidimensional spectral estimation theory are noted.

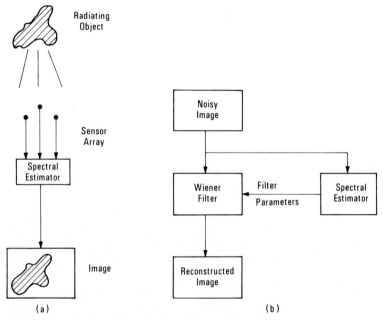

Fig. 1.   Two applications of power spectral estimation.

Section II contains a brief presentation of the random process theory that is necessary to understand and design spectral estimation methods. A homogeneous random process, its correlation function, and its power spectrum are defined. Spectral supports and random process sampling are discussed, as well as extendibility of correlation function samples.

Section III contains a classification of multidimensional spectral estimation methods in accordance with the assumptions that each method makes about the random process, the type of data required by each method, and the type of estimate that each method produces. Each resulting class of methods is discussed, and some particular methods are examined in detail.

Section IV contains a summary of the chapter. The area of multidimensional spectral estimation is of current research interest and, as a result, is undergoing rapid evolution. The direction of current research in this area and important unsolved problems will be mentioned.

II. Background

This section is a brief introduction to random process theory and to the problem of spectral estimation in multiple dimensions. The theory of multidimensional random processes is similar in many respects to that of ordinary temporal random processes, but there are several important differences, which we are careful to emphasize. A homogeneous multidimensional random process is defined, and the correlation function is defined in terms of the expected value of products of random process samples. The power spectrum is implicitly defined by expressing the correlation function as the inverse Fourier transform, over the spectral support, of the power spectrum. Ergodicity and its implications for the spectral estimation problem are briefly discussed. The problem of multidimensional spectral estimation is introduced, and the issues of the extendibility of correlation function samples and the ill-posedness of the spectral estimation problem are discussed.

A. Homogeneous Random Processes

A multidimensional *random process,* sometimes known as a random field, is a random function of a vector-valued argument. It is a generalization of the familiar temporal random process, which is a random function of a single variable, time. As in the case of temporal random processes, multidimensional random processes can be real or complex, scalar- or vector-valued functions. In this chapter, attention is restricted to the class of complex scalar-valued multidimensional random processes, denoted $a(\mathbf{x})$, where $\mathbf{x}$ is a vector-valued argument. What constitutes the elements of the argument $\mathbf{x}$ depends on the problem at hand. In image processing, a picture might be modeled as a multidimensional random process with $\mathbf{x}$ being a two-dimensional vector of spatial position. A wide variety of geophysics, radar, sonar, and radio-astronomy problems involve fields that are functions of both space and time, in which case the argument $\mathbf{x}$ is of dimension four, three spatial dimensions and one temporal dimension.

A complete statistical characterization of a random process $a(\mathbf{x})$ consists of a specification of the joint probability density for the values of any set of random process samples $\{a(\mathbf{x}_1), \ldots, a(\mathbf{x}_N)\}$ for all finite values of N and for all arguments $\{\mathbf{x}_1, \ldots, \mathbf{x}_N\}$ [25]. However, such a general probabilistic model is seldom used in practice; it is more usual to work with special cases of this model. One important special case is the class of homogeneous random processes, for which the joint probability density for the values of any set of random process samples depends only on the relative separations between samples, that is, the statistical description of a homogeneous process is independent of an absolute origin in space and time. Homogeneity corresponds to the temporal random process property of stationarity.

The restriction to homogeneous random processes is too strong for most of the results to be presented. A generalization of the class of homogeneous random processes is the class of *wide-sense* homogeneous random processes, characterized by a mean that is a constant with respect to the space–time argument and by an autocorrelation function that depends only on the vector difference between the two arguments. In this chapter we assume that the mean is equal to zero. The autocorrelation function, or more simply, the correlation function, is a function of a single variable and is denoted $r(\delta)$,

$$r(\delta) = E[a^*(\mathbf{x})a(\mathbf{x} + \delta)]. \tag{1}$$

The concept of wide-sense homogeneity corresponds to the property of wide-sense stationarity for the one-dimensional case [63]. In terms of the first and second moments of the random process, it is immaterial whether the process is homogeneous or merely wide-sense homogeneous, and the terms can be used interchangeably. An important property of a wide-sense homogeneous random process is that it can be represented as a linear superposition of complex exponentials,

$$\exp(j\mathbf{k} \cdot \mathbf{x}), \tag{2}$$

each of a particular wavevector $\mathbf{k}$, with uncorrelated complex random amplitudes. Just as the vector $\mathbf{x}$ may contain both temporal and spatial compo-

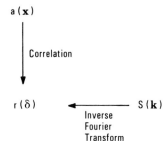

Fig. 2. The correlation function r is the expected value of samples of the random process a as well as the inverse Fourier transform of the power spectrum S.

nents, the wavevector **k** may contain both temporal frequency and spatial wavenumber. Roughly speaking, the power density spectrum of the random process is the variance of the random amplitude as a function of wavevector; its units are that of power density.[†]

The correlation function of a wide-sense homogeneous random process is conjugate-symmetric, and it is related to the power density spectrum $S(\mathbf{k})$ by a multidimensional inverse Fourier transform,

$$r(\delta) = \int_K \exp(j\mathbf{k}\cdot\delta)S(\mathbf{k})\,dv(\mathbf{k}). \tag{3}$$

The *spectral support* K is the set of wavevectors at which the power spectrum is nonzero (Fig. 2). The power density spectrum is always real-valued and nonnegative.[‡] The integration is done with respect to some fixed measure v which allows Eq. (3) to be interpreted as a surface or volume integral, possibly weighted, over the spectral support in multidimensional wavevector space.

The issue of spectral support is usually not important for one-dimensional random processes, for which it is generally taken to be some finite interval. However, there are some freedoms in the selection of a multidimensional spectral support that do not occur in the one-dimensional case. The spectral support can have a shape associated with a particular sampling lattice [41,45]. The shape of the spectral support in the processing of photographic images may be dictated by considerations of camera resolution or film grain size. In sensor array applications, there are often natural physical constraints on the spectral support, such as dispersion relations that

---

† For a precise treatment of the multidimensional spectral representation the reader is referred to Yaglom [64].

‡ In the signal processing literature, the term *power spectrum* is used in both the random process sense, as we use it in this chapter, and in the deterministic sense, that is, as the magnitude-squared of the Fourier transform of a deterministic function. It is important to distinguish between the two uses.

limit the region of wavevector space in which power may be present. Additional constraints on the spectral support are associated with other known properties of the wavefield sources, of the medium, or of the array sensors. For example, a finite temporal bandwidth may result from known source characteristics, from attenuation in the medium, or from finite sensor temporal bandwidth. A source of known spatial extent, or an array of directive sensors, will further constrain the spectral support. Thus, the selection of an appropriate spectral support is a more complex problem in multidimensions than in one dimension; it is also a mechanism by which prior knowledge can be incorporated into the multidimensional power spectral estimation problem [31].

At this point it is appropriate to comment on the validity of the homogeneous multidimensional random process model. Not all multidimensional signals are well modeled as homogeneous random processes, although there is often a strong temptation to do so on account of the simplicity and elegance of the formulation. One class of problems for which the homogeneous assumption is very good is in optical and radio astronomy, in which there are sources of incoherent radiation far enough away so that the Fraunhoffer approximation is valid [19]. The sources of radiation are frequently thermal and are well modeled as stationary temporal processes. Because the sources are far away, power at different spatial wavevectors comes from sources that are far apart, justifying the assumption of uncorrelated spectral amplitudes required for wide-sense homogeniety.

Seismic waves are sometimes modeled as homogeneous random processes [7], but the independence of spectral components with different wavevectors does not always hold in geophysical problems, particularly in experiments in which a single source excites several components of the field at different wavevectors. In these situations, spectral components are correlated with each other unless there happens to be some mechanism that separately randomizes the excitation of each component [51,58].

Sometimes a homogeneous random process model may be adopted because a particular signal processing scheme is to be used. For example, it may be desired for the sake of computational efficiency to implement an image noise reduction filter by segmenting a large image and applying a single identical filter to each segment. Although there may be no good stochastic model for the original image, the segments might be modeled as independent realizations of a homogeneous random process. The power spectrum of this homogeneous random process, for this one original image, may be estimated from the segments and used to design the noise reduction filter.

An important property of a homogeneous random process property is *ergodicity*. Roughly speaking, a random process is ergodic if, by taking enough samples from a single realization of the process, statistics of the random process can be estimated with arbitrary accuracy. As an example of a nonergodic random process, consider a complex-valued temporal random process consisting of a single complex exponential plus white noise,

$$a(t) = c \exp(j\omega_0 t) + v(t), \tag{4}$$

where ω_0 is a nonrandom constant, and c is a zero-mean complex random variable uncorrelated with the white noise $v(t)$. This random process is homogeneous, but it is not ergodic. A single realization of this random process is equivalent to only a single realization of the random variable c. Although it is possible to estimate the frequency ω_0 or the complex amplitude c of a single realization, the variance of c and, hence, the power spectrum cannot be reliably estimated from a single realization.† Estimating the power spectrum of a nonergodic random process requires multiple independent realizations of the process.

It is impossible to determine whether a random process is ergodic from a finite number of samples of a single realization: it is always possible to fit a finite number of sinusoids to the samples, which can be interpreted as a realization of a nonergodic random process with a spectrum consisting of a finite number of impulses. Consequently, in attempting to estimate a power spectrum from a single realization, one must make the a priori assumption that the process is ergodic.

B. The Spectral Estimation Problem

The multidimensional spectral estimation problem is one of estimating the power density spectrum of the process from a finite number of samples of one or more realizations of the random process. Let X be a set of sample points in D-dimensional space, an *array*

$$X = \{\mathbf{x}_1, \ldots, \mathbf{x}_N\}. \tag{5}$$

Here X is the set of points at which random process samples are available for one or more realizations. Let the *coarray* Δ be the difference set of X,

$$\Delta = \{\mathbf{x}_i - \mathbf{x}_j\}. \tag{6}$$

Thus, the coarray Δ is a finite subset of R^D of the form

$$\Delta = \{\mathbf{0}, \pm\boldsymbol{\delta}_1, \ldots, \pm\boldsymbol{\delta}_M\}. \tag{7}$$

---

† Estimating the complex amplitudes and frequencies of sinusoidal components of a nonergodic random process from a single realization is an important problem in its own right [55], but it will not be treated in this chapter.

This coarray is the set of vector separations for which correlation estimates can be made by averaging products of random process samples over repeated separations or over independent realizations.

As pointed out previously, a valid multidimensional power density spectrum is real-valued and nonnegative over its support. Because many estimation methods start with estimated samples of the correlation function, there is an existence question of some importance. Given a set of estimated samples of the correlation function on Δ, is there any nonnegative power density spectrum such that Eq. (3) is satisfied? If so, the set of correlation samples is said to be *extendible.*†

Let R be the $N \times N$ correlation matrix for samples of a homogeneous random process on X, that is, $R_{ij} = r(\mathbf{x}_i - \mathbf{x}_j)$. R has some special structure. In common with all correlation matrices, R must be Hermitian and nonnegative definite. Furthermore, elements of R corresponding to equal vector separations must be equal, because R describes a homogeneous random process. For example, R must be Toeplitz in the *time series case.*‡ These conditions are clearly necessary for a correlation matrix to be extendible. However, they are almost never sufficient except in the time series case [15]. Extendibility is, as a result, a more interesting property in multidimensional problems than it is in the one-dimensional time series case. Futhermore, it is important in the understanding and implementation of many spectral estimation methods [13,31,39].

The spectral estimation methods to be discussed treat each dimension of a multidimensional problem in an equivalent manner. It may so happen, however, that one dimension in a multidimensional problem is special. A good example of this is an array processing problem in which there are few sensors but in which many time samples are taken from each sensor. If the data samples are visualized as points in a multidimensional space, the distribution of points is much longer along the time axis than along any of the spatial axes. In a problem with a special dimension, it is often convenient to treat that dimension in a special manner.§ Rather than trying to form one global estimate of spectral density as a function of temporal frequency and spatial wavevector, one might find it more convenient to form separate estimates of spectral density as a function of spatial wavevector alone, for various temporal frequencies. This can be done, for example, by passing the output of each sensor through an identical quadrature demodulator and low-pass filter for each temporal frequency to be examined.

In some applications, particularly those in which samples of one realiza-

---

† The correlation function associated with such a spectrum *extends* the correlation samples to a function of all separations.

‡ $D = 1$, $K = [-\pi,\pi]$, $X = \{0, 1, \ldots, M\}$, and $\Delta = \{0, \pm 1, \ldots, \pm M\}$.

§ This is discussed under the topic of separable spectral estimation by McClellan [35].

tion are available on a uniform sampling lattice, it may be convenient to divide the realization into many, possibly overlapping, segments. If the segments are large and the overlap is small compared with the correlation length of the process, then it is often possible to treat each segment as an independent, albeit smaller, realization of the random process [57]. The assumption of a short correlation length is equivalent to an assumption that the random process is ergodic, as discussed in Section II.A.

The spectral estimation problem is inherently *ill-posed:* it is impossible to estimate an unknown function reliably for each value of its argument from a finite number of random process samples. Likewise, even if a finite set of actual correlation samples were available, the problem of determining the spectrum would still be ill-posed, because an infinite number of spectra are generally consistent with any finite set of correlation samples.

III. Techniques

Spectral estimation has been an active area of research for a number of years. As a result there has been a proliferation of new techniques, for the multidimensional case as well as the time-series case. Nevertheless, classical techniques continue to be the most commonly used spectral estimation methods, because of their ease of implementation and interpretation. The well-known limitations of classical methods have stimulated much research on new high-performance techniques. The volume of literature in this area can be intimidating to the beginner. Likewise, it may be difficult to recognize from the literature that even the new techniques have their limitations and that, regardless of the technique used, spectral estimation remains an ill-posed problem in the absence of a priori knowledge about the power spectrum.

This section consists largely of a detailed description of various types of multidimensional spectral estimation methods. We have chosen to emphasize the theoretical aspects of the methods rather than numerical examples. It is our feeling that numerical experiments in which various spectral estimation techniques are tested against one another on the same data seldom allow one to say anything conclusive, simply because the results of such experiments are heavily dependent on the nature of the data and the choice of performance criteria. Instead, our approach is to discuss each method with emphasis on the prior knowledge about the spectrum that each method assumes (either explicitly or implicitly), the type of data required by each method, the description of the power spectrum that each method produces, and on the existing analytical evidence that favors one method over another.

It has not been possible to include every existing method of multidimensional spectral estimation because of space limitations. In particular, we have deliberately excluded any techniques for which there is little or no theoretical motivation.

A. Classifications of Spectral Estimation Methods

Ideally, a spectral estimate should be based directly on a set of random process samples, avoiding any possible loss of information as a result of a preliminary data reduction step. Methods that do this are called *data-direct*. On the other hand, it is simpler to base a spectral estimate on estimated samples of the correlation function; the stochastic side of the problem has been eliminated, and there is a simple linear relationship, the inverse Fourier transform, Eq. (3), between the correlation samples and the power density spectrum. Such methods are called *correlation-based*. See Fig. 3.

The primary method of estimating samples of the multidimensional correlation function is to compute a weighted sum of products of random process samples. One such technique requires that the random process samples occur on a regular multidimensional grid, so that repeated separa-

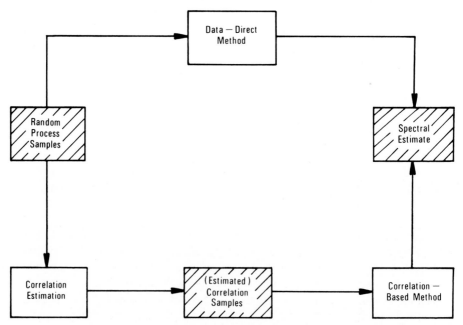

Fig. 3. Data-direct versus correlation-based spectral estimation.

tions are available to be averaged over. The details are discussed later in this section. Another technique depends on there being multiple independent realizations of the random process, so that averaging can be done over the different realizations. For example, suppose that L statistically independent realizations $\{a_1, \ldots, a_L\}$ of an N-dimensional vector consisting of N samples of the random process are available. Several spectral estimation methods, including the classical methods and the maximum likelihood method (MLM), can be obtained from an estimate for the correlation matrix of a of the form

$$\hat{R} = \frac{1}{L} \sum_{l=1}^{L} a_l a_l^\dagger. \tag{8}$$

This estimate is guaranteed to be Hermitian nonnegative-definite. If the true correlation matrix is positive-definite, and if L is greater than or equal to N, then the estimate is positive-definite with probability one [7]. Under the mild condition of finite fourth-order moments, the estimate can be shown to converge in probability to the true correlation matrix as L goes to infinity. Moreover, for a Gaussian random process, the expression in Eq. (8) is a sufficient statistic for the data [20]. However, the above correlation matrix estimate does not have the special structure possessed by the correlation matrix of a homogeneous random process: elements of the correlation matrix estimate corresponding to the same vector separation are generally not equal. Some spectral estimation methods require only that the estimated correlation matrix be Hermitian positive-definite, whereas others require estimates for actual samples of the correlation function, or equivalently they require that the correlation matrix estimate have a particular structure. The estimation of correlation matrices with some particular structure has received attention only recently [6].

In some array processing problems, the time duration of the data may be very long with respect to the correlation time of the space–time random process, and it is useful to reduce the data to an estimate for the spectral density matrix for the spatial data as a function of temporal frequency [7]. This can be done by segmenting the data in time into L segments of equal duration. Each data segment, consisting of an N-dimensional vector corresponding to the data from the N array elements, is Fourier transformed in time to yield an N-dimensional vector as a function of temporal frequency. The estimate for the spectral density matrix is obtained as in Eq. (8) by taking the average of the outer products of the L transform vectors.

Some spectral estimation problems are most naturally formulated in terms of a correlation-based spectral estimate, for instance, when many independent realizations are available, allowing very stable correlation estimates to be made. In such situations, the primary cause of uncertainty in the spectral estimate is not the uncertainty in the correlation estimates but rather the finite sampling of the correlation function. Although the primary cause of uncertainty may be the finite sampling of the correlation function, some correlation-based methods do allow for the incorporation of confidence intervals on the correlation estimates, and variance analyses are available for others [3,8]. These features help take into account the stochastic nature of the correlation estimates.

Though the combination of a correlation estimation procedure with a correlation-based spectral estimation method yields an estimate that is ultimately based on random process samples, a data-direct spectral estimator is usually considered to be a method that does more than this. In particular, a data-direct method should use some more complicated probabilistic model of the random process to produce spectral estimates in a more optimal manner. In practice, the distinction between data-direct and correlation-based methods is not always clear; some spectral estimation methods can be thought of as either data-direct or correlation-based.

Ideally, a spectral estimation method produces an estimate of the power density spectrum at each wavevector. Such an estimate is the most detailed possible estimate, implicitly providing estimates of any function of the power density spectrum. As mentioned in Section II.B, this is an ill-posed problem in the absence of additional prior knowledge about the spectrum, because only a finite number of observed random variables are available from which to estimate an infinite number of unknowns (power density versus wavevector).

Rather than attacking the ill-posed problem of estimating spectral density at single frequencies, one can use the so-called *window methods* to provide estimates of the total power in a particular spectral window. This problem is not so ill-posed. Window methods include the classical methods and more modern methods such as the MLM. Another class of techniques, known as *parametric methods,* assume a priori a structure for the spectrum in terms of a finite and relatively small number of unknown parameters, which can be estimated from the given random process samples. The assumption of such a parametric model generally converts the ill-posed spectral estimation problem into a well-posed parametric estimation problem. Examples of parametric methods are the maximum entropy method (MEM) and Pisarenko's method. Potentially, parametric methods are extremely powerful; but in practice, the usefulness of results obtained from them depends critically on the correctness of the parametric model assumed.

When a parametric model for the spectrum is available in addition to a probabilistic model, then a theoretically motivated data-direct technique for estimating the unknown parameters, such as maximum likelihood estimation [56], is ideal. In practice, such estimation problems almost always require the solution of difficult optimization problems, and usually a simplified problem is tackled, yielding a suboptimal solution. Such is the case for the various one-dimensional methods of autoregressive model fitting [27].

B. Classical Window Methods

The classical methods of multidimensional spectral estimation produce estimates of the power through certain classical spectral windows. Most often, the power estimate is computed and plotted for a range of window positions, holding the window shape fixed. For illustrative purposes we first discuss a two-dimensional example. Assume that a set of complex-valued samples of a two-dimensional Gaussian random process, taken on a regular grid, $\{a(m_1,m_2): 0 \le m_1, m_2 \le M - 1\}$, is available. Further assume that the spectral support is the Nyquist square $[-\pi,\pi]^2$; hence,

$$r(\delta) = \int_{-\pi}^{\pi} \int_{-\pi}^{\pi} \exp(j\mathbf{k} \cdot \delta)S(\mathbf{k}) \frac{dk_1\,dk_2}{(2\pi)^2}. \tag{9}$$

This is a simple generalization of the well-known one-dimensional time-series problem.

The modified periodogram approach to classical spectral estimation begins with the Fourier transform of the windowed random process samples:

$$F(\mathbf{k}) = \sum_{m_1=0}^{M-1} \sum_{m_2=0}^{M-1} \{w_1(m_1,m_2) \exp[-j(m_1k_1 + m_2k_2)]\}a(m_1,m_2). \tag{10}$$

The *modified periodogram* $I(\mathbf{k})$ is the magnitude-squared of $F(\mathbf{k})$. Its mean value at a particular frequency is equal to the spectral power seen through a spectral window $|W_1(\mathbf{k})|^2$,

$$E[I(\mathbf{k})] = \int_{-\pi}^{\pi} \int_{-\pi}^{\pi} |W_1(\mathbf{k} - \mathbf{u})|^2 S(\mathbf{u}) \frac{du_1\,du_2}{(2\pi)^2}, \tag{11a}$$

where

$$W_1(\mathbf{k}) = \sum_{m_1=0}^{M-1} \sum_{m_2=0}^{M-1} w_1(m_1,m_2) \exp[-j(m_1k_1 + m_2k_2)]. \tag{11b}$$

Classical spectral windows are designed to have a *main lobe* region where the window is large and a *side lobe* region where the window is small (Fig. 4). Then the expected power through a spectral window with sufficiently small side lobes can be attributed, to a good approximation, to power in the

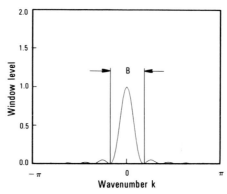

Fig. 4. A classical spectral window with main-lobe region *B*.

main-lobe region of the spectral support. The requirement that the side lobes be small leads to a frequency resolution limit. If the side lobes of a classical window are to remain significantly smaller than the main lobe, the main-lobe region cannot be made too small, hence, the classical or Rayleigh resolution limit. Tabulated designs for one-dimensional window functions exist that allow main-lobe width (and hence, resolution) to be traded off against side-lobe level [22]. Such one-dimensional designs can be used to produce two-dimensional window designs using the same methods, such as transformation [34,40], by which one-dimensional filter designs are used to produce two-dimensional filter designs. Alternatively, two-dimensional Chebyshev filter design techniques [21] can be applied to window design, as has been done for one-dimensional window design [10]. Finally, some tabulated designs for two-dimensional windows do exist [52].

For a complex Gaussian random process, it can be shown that the variance of a periodogram sample is equal to the square of its mean. Hence, the periodogram does not provide, by itself, reliable estimates of the windowed power [44,57]. If multiple independent realizations of the random process are available, the periodograms corresponding to the different realizations can be averaged together to give statistically reliable estimates of the windowed power.

In the absence of multiple independent realizations, the only way to obtain statistically reliable power estimates is to smooth the periodogram. For a Gaussian random process, samples of the periodogram corresponding to spectral windows having disjoint main lobes are approximately independent, provided the spectrum is sufficiently smooth. Under these conditions, reliable estimates of the windowed power are obtained from a single realization of the random process by smoothing the periodogram. The variance of the smoothed periodogram is inversely proportional to the number of independent periodogram samples appearing in the smoothing window. The expectation of the smoothed periodogram is given by

$$E[I_S(\mathbf{k})] = E\left[\int_{-\pi}^{\pi} \int_{-\pi}^{\pi} W_2(\mathbf{k} - \mathbf{u})I(\mathbf{u})\frac{du_1\,du_2}{(2\pi)^2}\right] \tag{12a}$$

$$= \int_{-\pi}^{\pi} \int_{-\pi}^{\pi} \left[\int_{-\pi}^{\pi} \int_{-\pi}^{\pi} W_2(\mathbf{k} - \mathbf{u} - \mathbf{v})|W_1(\mathbf{v})|^2\frac{dv_1\,dv_2}{(2\pi)^2}\right]$$

$$\times S(\mathbf{u})\frac{du_1\,du_2}{(2\pi)^2}. \tag{12b}$$

Samples of the smoothed periodogram provide an estimate of the power through an effective spectral window composed of the convolution of $W_2(\mathbf{k})$ and $|W_1(\mathbf{k})|^2$. The mean and variance of the estimates obtained in this way, for Gaussian processes, are discussed by Nuttall and Carter [42]. Qualitatively, a large smoothing window reduces the variance of the windowed power estimate but at the price of increasing the window main-lobe width. This is the well-known tradeoff between the resolution and statistical stability of the smoothed periodogram.

The periodogram method of classical spectral estimation is equivalent to the so-called Blackman–Tukey method, which involves an intermediate step of correlation estimation. It is straightforward to verify that the smoothed modified periodogram can be written, using the array–coarray notation of Section II, in the form

$$I_S(\mathbf{k}) = \sum_{\delta \in \Delta} w_e(\delta)\exp(-j\mathbf{k}\cdot\delta)\hat{r}(\delta), \tag{13}$$

where $w_e(\delta)$ is an effective window,

$$w_e(\delta) = w_2(\delta)\sum_{\mathbf{x}_j-\mathbf{x}_i=\delta} w_1^*(\mathbf{x}_i)w_1(\mathbf{x}_j) \tag{14}$$

and where $\hat{r}$ is an estimate for the correlation function of the form

$$\hat{r}(\delta) = \left[\sum_{\mathbf{x}_i-\mathbf{x}_j=\delta} w_1^*(\mathbf{x}_i)a^*(\mathbf{x}_i)w_1(\mathbf{x}_j)a(\mathbf{x}_j)\right]\Bigg/\left[\sum_{\mathbf{x}_j-\mathbf{x}_i=\delta} w_1^*(\mathbf{x}_i)w_1(\mathbf{x}_j)\right]. \tag{15}$$

Note that $\hat{r}$ is an unbiased correlation estimate, $E[\hat{r}(\delta)] = r(\delta)$.† Although Eq. (13) implies that there are many different w_1, w_2 window pairs that result in smoothed modified periodograms of the same expected value, these periodograms actually have different variances [42].

---

† Alternatively, the normalization can be removed and $I_s(\mathbf{k})$ expressed in terms of a biased correlation estimate.

Thus, classical methods produce estimates of the power through windows, chosen a priori. If the spectrum is assumed to be smooth over the main-lobe region, the power through the window, properly normalized, provides an estimate of the power spectral density in the main-lobe region. The variance of these power estimates can be reduced by averaging over independent realizations or over wavevector. The estimates are simple quadratic functions of data, based on the periodogram. When the data are produced by sampling on a uniform lattice, the periodogram can be efficiently computed using fast Fourier transform algorithms.

Some complication is introduced if the samples are nonuniformly spaced or if a different spectral support or a weighting function is introduced. The definition of the modified periodogram can be generalized to

$$I(\mathbf{k}) = \left| \sum_{\mathbf{x} \in X} w_{\mathbf{k}}(\mathbf{x}) a(\mathbf{x}) \right|^2, \tag{16}$$

hence,

$$E[I(\mathbf{k})] = \int_K |W_{\mathbf{k}}(\mathbf{u})|^2 S(\mathbf{u}) \, dv(\mathbf{u}), \tag{17}$$

with

$$W_{\mathbf{k}}(\mathbf{u}) = \sum_{\mathbf{x} \in X} w_{\mathbf{k}}(\mathbf{x}) \exp(-j\mathbf{u} \cdot \mathbf{x}) \tag{18}$$

representing a window centered around **k**. The smoothed periodogram can be similarly generalized. However, a serious practical problem is that suitable window designs are not available for arbitrary arrays/spectral-support/weighting combinations. Indeed, these combinations vary from application to application; so tabulated designs, as are available for the time-series case, would not be suitable. What is needed are algorithms for automatic window design. Although some work has been done in this direction [10], the problem deserves further study.

To summarize, classical multidimensional spectral estimation for uniformly sampled data is a straightforward extension of one-dimensional classical spectral estimation. The chief advantages of classical methods are their ease of implementation and their relative ease of interpretation. The chief disadvantage is their limited resolution. The lack of window designs for nonuniformly sampled data is also a serious impediment to their application.

C. Data Adaptive Window Methods

One of the drawbacks of classical spectral estimation is its limited frequency resolution. If the side lobes of a classical window are to remain significantly smaller than the main lobe, the main-lobe region cannot be made too small, hence, the classical or Rayleigh resolution limit. Two data adaptive window methods of spectral estimation, the data adaptive spectral estimator (DASE) of Davis and Regier [14] and the MLM of Capon [7], a special case of DASE, represent attempts to overcome the limited resolution of classical methods.

Both MLM and DASE produce estimates of windowed power in terms of an estimate for a correlation matrix for a particular set of random process samples. The methods require only that the correlation matrix estimate be Hermitian positive-definite; no other structure is assumed. In what follows, the problem of estimating the correlation matrix is ignored, and the true correlation matrix is assumed to be available. A variance analysis of MLM is available for the case when the correlation matrix is estimated as an average of outer products of independent realizations of a Gaussian random process [8]. The $N \times N$ correlation matrix for the vector of random process samples **a** is denoted by R. The correlation matrix is related to the power spectrum by the formula

$$R = \int_K \gamma_k \gamma_k^\dagger S(\mathbf{k}) \, dv, \qquad (19)$$

where the vector γ_k, the steering vector, is

$$\gamma_k^\dagger = [\exp(-j\mathbf{k} \cdot \mathbf{x}_1), \ldots, \exp(-j\mathbf{k} \cdot \mathbf{x}_N)]. \qquad (20)$$

A general class of power spectral estimators are quadratic forms in terms of the correlation matrix R. Alternatively, they can be represented as the integral of the product of the power density spectrum and a nonnegative spectral window:

$$\hat{P}(\mathbf{k}) = \mathbf{h}_k^\dagger R \mathbf{h}_k = \int_K W_k(\lambda) S(\lambda) \, dv(\lambda), \qquad (21)$$

where

$$W_k(\lambda) = |\mathbf{h}_k^\dagger \gamma_\lambda|^2. \qquad (22)$$

If R is computed as an average of outer products of independent random process realizations, Eq. (21) is equivalent to an average of modified periodograms, Eq. (16), with $w_k(\mathbf{x})$ identified with $\mathbf{h}_k$. In the case of MLM and DASE, the weight vector $\mathbf{h}_k$ and, consequently, the spectral window depend on the correlation matrix, whereas for the classical spectral estimator, the weight vector is chosen a priori. In classical spectral estimation, the weight vector is chosen so that the spectral window has a main-lobe peak in the neighborhood of wavenumber **k** and low side lobes elsewhere. If the side lobes are sufficiently low, we are assured that the estimated power corresponds, for the most part, to spectral power within the main lobe of the window.

Instead, consider choosing a region B of wavevector space, which will be analog to the main-lobe region of a classical method, and designing a window so that, if the spectral density is constant over B, and there is no spectral power outside of B, the windowed power will be equal to the total power in B. This consideration leads to a constraint on the window of the form

$$\mathbf{h}_k^\dagger \Gamma \mathbf{h}_k = 1, \tag{23}$$

where

$$\Gamma = \frac{1}{v(B)} \int_B \gamma_k \gamma_k^\dagger \, dv, \tag{24}$$

and $v(B)$ is the multidimensional volume of B. In the DASE method, the weight vector $\mathbf{h}_k$ is chosen subject to this constraint to minimize the power estimate Eq. (21), thereby minimizing the influence of spectral power outside of B. This constrained minimization problem results in a weight vector $\mathbf{h}_k$ that is proportional to the eigenvector of $R^{-1}\Gamma$ having the largest eigenvalue, and in a windowed power equal to the inverse of that eigenvalue. The MLM power estimate can be obtained as a special case of the DASE estimate by shrinking the region B to the single point $\mathbf{k}$, in which case the power estimator has an explicit formula,

$$\hat{P}(\mathbf{k}) = 1/\gamma_k^\dagger R^{-1} \gamma_k. \tag{25}$$

DASE and MLM differ from classical window methods in that the window shape changes with wavevector and depends explicitly on the estimated correlation matrix.

When the size of the region B is made sufficiently small, the DASE estimate is a high-resolution spectral estimate in the sense that it may resolve peaks in the spectrum too close together to be resolved by classical methods. This property has been extensively studied for the special MLM case [7,11]. The high-resolution property of MLM can be understood by considering the extreme case in which the power spectrum consists of $N - 1$ or fewer impulses, where N is the dimension of the correlation matrix. In this case, the correlation matrix is singular, so the MLM power estimate is zero everywhere except at isolated wavevectors. The wavevectors for which the MLM power estimate is nonzero include the wavevectors where there are impulses in the true spectrum and, except in pathological cases, no others. Thus, for this special case, the MLM spectral estimator resolves all of the impulses in the spectrum. If the true spectrum consists of a small but nonzero continuous component in addition to the N impulses, then by a perturbation argument it can be seen that the MLM spectral estimate should contain peaks at the correct wave vectors, provided the continuous spectral component is sufficiently small.

The chief drawback of these high-resolution methods is that, in general, peaks appearing in the spectral estimate may not actually be present in the true spectrum. Such artifacts, for the case of MLM, have been reported by Davis and Regier [14].

Unlike classical windows, the adaptive windows of MLM and DASE need not have definite main-lobe and side-lobe regions; indeed they cannot if, as is usually the case, B is smaller than the classical resolution limit. Therefore, the windowed power cannot be attributed to a well-defined main-lobe region as in classical methods; it is impossible to determine whether the estimated power is due to spectral power in B or to the leakage of spectral power from outside of B (Fig. 5).

The MLM spectral estimate has been used extensively, but the more general DASE estimate has received little attention. Apart from the special MLM case, it may well be that the primary advantage of DASE over the classical techniques is not in resolution but rather in the automatic design of the spectral window that it provides. Although there are no tabulated classical window designs for nonuniformly spaced data, DASE takes care of the window design problem automatically. If the main-lobe region B is chosen large enough, the window level should be small enough outside of B so that the DASE power estimate can be simply interpreted as due to power in the region B alone.

It has been shown [38,39] that the DASE power estimate can be interpreted as an *upper bound* on spectral power in the region B. The DASE procedure implicitly assumes that the correlation matrix is of the form

$$R = p\Gamma + Q, \tag{26}$$

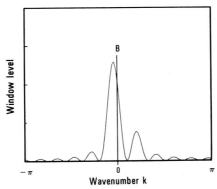

Fig. 5.   A data-adaptive (MLM) spectral window.

where Q is the correlation matrix corresponding to spectral power outside of the region B. The object is to estimate the signal power p, a problem complicated by the fact that Q is unknown. As the problem stands, it is ill-posed, because there is a continuous range of values of p and Q such that Eq. (26) is satisfied. Clearly, a trivial lower bound on p is zero, because power must be nonnegative. A nontrivial upper bound on p is imposed by the requirement that Q be nonnegative-definite; it can be shown that this upper bound is identical to the DASE power estimate. Consequently, MLM and DASE can be interpreted as giving nontrivial upper bounds on spectral power in a particular region. The fact that MLM and DASE give only upper bounds on power explains why the methods can give false peaks. In practice, only an estimate is available for the correlation matrix, so strictly speaking, the power estimates are bounds subject to the assumption that the correlation matrix estimate is correct.

A natural question to ask is whether tighter upper and lower bounds on spectral power exist. If one assumes once again that the true correlation matrix is available, the answer is yes. In particular, for DASE one assumes only that the matrix Q is nonnegative-definite. In fact, because it corresponds to spectral power in a particular region, it must also satisfy the stronger constraint of extendibility. Imposing the extendibility constraint results in a tighter upper bound on power than the DASE upper bound, as well as a nontrivial lower bound. However, these extendibility-derived bounds still make the a priori assumption that the spectrum is constant in the region B.

Cybenko has studied the related problem of bounding windowed spectral power, given the true correlation matrix, for the case of an arbitrary window with no assumptions made about the shape of the spectrum [12]. The tightest possible upper and lower bounds on windowed power can be obtained by solving semi-infinite linear programming problems. The upper and lower bounds can be plotted as a function of the window position, and the convolution of the true spectral density with the window must lie between the bounds. The quality of the bounds is indicated by how close the upper and lower bounds are to each other [39].

The MLM and DASE procedures have been used, for the most part, in frequency-wave vector spectral estimation for sensor arrays. As described earlier in this section, the available space–time data are reduced to an estimate for the spectral density matrix, which is a function of temporal frequency, for the sensor array. The MLM and DASE procedures can then be applied to this matrix to obtain a high-resolution spectral estimate with respect to wavenumber. As noted previously, these procedures assume only that the estimated spectral density matrix is Hermitian positive-definite. In contrast, the tighter bounds based on the extendibility constraint require both that the spectral density matrix have a particular structure and that it be extendible, properties not guaranteed by any existing correlation matrix estimator. A possible way to ameliorate this problem is discussed by Marzetta and Lang [39]. The ultimate practicality of extendibility-based bounds will depend largely on the development of reliable and efficient algorithms for their computation.

To obtain a MLM or a DASE estimate from a single realization of a random process, one is faced with the difficult choice of the size of the correlation matrix to be estimated. Generally, the correlation samples at larger separations cannot be estimated as reliably as those at smaller separations, so there is a tradeoff between resolution and the statistical stability of the spectral estimate. In this respect, DASE and MLM have the same drawback as the classical window methods.

To summarize the advantages of the MLM and DASE spectral estimators: they are straightforward to compute, they can be applied to correlation matrices for nonuniform sampling, and they have higher resolution than classical methods. They can be interpreted either as giving the power through certain data adaptive windows or, more simply, as upper bounds on the power in selected regions of the spectral support. The price that one pays for high resolution is that, in general, it cannot be determined whether peaks in the spectral estimate are also present in the true spectrum or whether they are merely artifacts. Also, the requirement of a positive-definite estimate for the correlation matrix can be difficult to satisfy in practice [51]. If only a single realization of the random process samples is available, then an assumption of spectral smoothness with respect to one or more of the wave-vector components is necessary, so that some sort of averaging can be employed to estimate the correlation matrix.

D. Parametric Methods

Parametric approaches to spectral estimation are based on an a priori model for the spectrum in terms of a finite and relatively small number of unknown parameters. In contrast to window methods, in which the object is to estimate windowed power, in parametric methods the unknown parameters themselves are estimated. The spectral density estimate can then be obtained by evaluating a formula in terms of the estimated parameters, although the parameters themselves are often of primary interest.

When a trustworthy parametric model for the spectrum is available, estimates of the model parameters can be obtained using standard parametric estimation techniques such as maximum likelihood [56]. However, such techniques often result in difficult nonlinear optimization problems; more practical algorithms may result from approximations based on the period-ogram [4,32] or from approximations based on fitting the model to estimated correlation samples.

Parametric methods have great potential when a good parametric model for the spectrum is available. However, aside from these cases, a group of parametric methods have been proposed for general use, regardless of the validity of the particular model. Among these are Burg's maximum en-tropy method (MEM), autoregressive (AR) modeling, and Pisarenko's method.

MEM is an example of a correlation-based parametric method. The method starts with an extendible set of correlation samples, and it produces a spectral density that is consistent with the given correlation samples. In general, there are an infinite number of such spectra, and MEM chooses the spectrum with the maximum entropy†, where the entropy H is defined by

$$H = \int_K \log S(\mathbf{k}) \, dv. \tag{27}$$

It can be shown that this constrained maximization problem generally has a solution in the form of a parametric model for the maximum entropy spec-trum, specifically,

$$S(\mathbf{k}) = 1 \Big/ \sum_{\delta \in \Delta} p(\delta) \exp(-j\mathbf{k} \cdot \delta), \tag{28}$$

where the set Δ contains the separations associated with the given correla-tion samples, and the set of model parameters $\{p(\delta) : \delta \in \Delta\}$ are determined by the correlation-matching constraint.‡ For the time series case, an alge-braic solution for the model parameters in terms of the correlation samples is available. For the general multidimensional case, there is no algebraic solution; however, reliable iterative computational algorithms are available [30,36].

Another correlation-matching spectral estimator is Pisarenko's method, which was originally developed for the time series case [46] and later ex-tended to the multidimensional case [29,31]. Of all the spectra that are consistent with a given set of correlation samples, Pisarenko's method chooses the spectrum with the largest possible component of known spec-tral shape, but with unknown power. It can be shown that this maximiza-tion problem leads to a parametric model for the spectrum consisting of the spectral component of known shape plus a sum of $2M$ impulses, where M is

† Undoubtedly, some of the popular appeal of MEM is due to the word *entropy* with its suggestion of thermodynamics and information theory.

‡ There are problems in some multidimensional cases regarding the existence of the MEM estimate in the form Eq. (28) [30,36].

related to the size of the coarray, as in Eq. (7). The model parameters consist of the power of the spectral component of known shape plus the powers and locations of the impulses. The computation of the model parameters from the correlation samples involves the solution of an eigenvalue problem in the time series case. The multidimensional estimate is computed as the solution to a semi-infinite linear programming problem.

Both MEM and Pisarenko's method can be viewed as *extension* methods, in that they produce spectral estimates that are each, by means of the inverse Fourier transform, one possible extension of the given correlation samples.† Such spectral estimates can be difficult to interpret, because they provide only one of an infinite number of spectra that are consistent with the data; they provide no information about the range of equally consistent spectra. The primary objection to the use of a parametric method, such as MEM or Pisarenko's method, when the model is not known a priori to be correct is that there is no reason to expect one particular extension to be closer to that true spectrum than any other extension. Furthermore, when only a single realization of the random process is available, one is faced with the difficult problem of choosing the size of the correlation matrix to be estimated, just as with MLM or DASE. In contrast, window methods provide measures of a property, windowed power, of the entire range of consistent spectra.

Although the general purpose use of parametric methods such as MEM or Pisarenko's method must be viewed with skepticism, these methods may be valuable in certain applications in which the model is known to be appropriate or in which the estimate exhibits certain desirable behavior. For example, MEM has attracted considerable attention because of its ability to resolve closely spaced sinusoids in background noise [33,37], a feature that is important in problems such as bearing estimation [26].

Autoregressive (AR) modeling is a third parametric estimation method, applicable to random processes sampled on uniform grids. A random process is autoregressive if it satisfies an all-pole, recursive, constant-coefficient difference equation driven by white noise. A considerable number of different methods, mostly of the data-direct type, have been proposed for fitting one-dimensional AR models [27]. Generally, these methods can be shown to yield approximate maximum likelihood estimates of the AR parameters for Gaussian random processes. AR models for two-dimensional random processes, as well as more general interpolative models, have been investigated [16,18,53,59,61]. A good survey of this work is available in the book by Willsky [60].

† Conceptually, at least, both MEM and Pisarenko's method could be converted to data-direct methods by use of statistical estimation techniques to estimate the model parameters. For the Gaussian case, however, the resulting likelihood estimation problems appear to be intractable.

Although AR modeling has been successfully applied to a number of one-dimensional spectral estimation problems, such as speech vocoding [43], it has received little attention in the general multidimensional case. The general-purpose use of AR modeling in either the one-dimensional or the multidimensional case is not recommended. Not only are the resulting estimates difficult to interpret, but also one is faced with a difficult model order determination problem [27]. In the one-dimensional case, AR modeling has been used successfully to estimate the frequencies of sinusoidal components of a single realization of a random process. In this application, the AR modeling algorithms are not used to estimate a power spectrum (because the process is nonergodic), but rather are used to estimate the sinusoidal frequencies. Recently, however, new frequency estimation algorithms have been proposed for this purpose whose performance is superior to the AR methods [55].

To summarize, if a good parametric model is available, then a parametric estimation method has much to recommend it. Even when a parametric model is not valid, parametric modeling may still be practically useful in certain problems, such as the resolution of closely spaced sinusoids. However, the general-purpose use of parametric methods such as MEM or Pisarenko's method should be viewed with skepticism. Although they may provide one particular spectrum that is consistent with the data, they provide no information about the range of equally consistent spectra and so can be difficult to interpret.

IV. Summary

Power spectral estimation is important in the design and implementation of many signal processing systems, as well as in the testing of models for physical processes. An introduction has been given to the theory of multidimensional homogeneous random processes, those random processes that can be partially described by a power spectrum.

Multi-dimensional spectral estimation problems often differ from time series spectral estimation problems in more than just dimensionality. Most one-dimensional problems deal with uniformly sampled random processes, whereas many multidimensional problems deal with nonuniformly sampled processes. Further complications result from the fact that even though an image may be sampled on a regular lattice, the boundary of the sampled region may be irregular. In uniformly sampled one-dimensional problems, the spectral support is almost always the Nyquist interval associated with the uniform sampling interval. In multidimensional problems, particularly sensor array problems, a priori information such as known source characteristics, sensor directionality and temporal bandwidth, and dispersion relations impose useful constraints on the spectral support that are independent of sampling.

Various methods of power spectral estimation have been described that are applicable to the multidimensional problems encountered in image and array processing. The methods have been classified according to whether they work directly from random process samples or start with estimates of correlation function samples. They have been further classified according to the type of estimates they produce. Some provide estimates of windowed power, whereas others provide actual power spectral density estimates by making use of a spectral model with a finite number of parameters.

The simplest and best understood spectral estimation techniques, multi-dimensional as well as one-dimensional, are classical window based methods. These methods are appropriate for data sampled on a regular lattice with regular boundaries and an associated Nyquist spectral support. These methods can deal with estimated correlation samples or with samples of one or more realizations of the random process. They produce estimates of windowed power for classical polynomial type windows and are the basis of approximate maximum likelihood parameter estimation methods for model fitting. When appropriate, these methods provide computationally simple estimates whose reliability may be simply assessed. Their main disadvantage is their limited resolution and the lack of tabulated window designs for nonuniformly spaced data. An important practical contribution might be the development of fast and reliable automatic window design algorithms.

Capon's data adaptive window method, MLM, provides an upper bound on the power at a single wave vector. MLM has the advantage of being computationally simple, although the bounds that it produces are not as tight as possible. A more general data adaptive window method, DASE, allows a main-lobe region to be specified. Its chief advantage over classical methods may not be so much in resolution as in the provision of an automatic window design.

Methods have been proposed that produce tight upper and lower bounds on windowed power for nonclassical windows, based on an extendibility constraint. These techniques are promising partly because the formulation of the bound as the solution to a convex optimization problem allows for the easy incorporation of application specific a priori constraints on the power spectrum. However, the utility of such bounding techniques remains to be shown and will depend in large part on the development of efficient and reliable algorithms for their computation. Further work also needs to be done on the incorporation of the stochastic aspect of the spectral estimation problem into bounding techniques.

If a good parametric model is available, then a parametric estimation method has much to recommend it. Even when a parametric model is not valid, parametric modeling may still be useful in certain problems, such as the resolution of closely spaced sinusoids. However, the general-purpose use of parametric methods such as MEM or Pisarenko's method should be viewed with skepticism.

V. Bibliographical Notes

A classic reference in spectral estimation is Blackman and Tukey [3]. Modern surveys of spectral estimation include Kay and Marple [27] and McClellan [35].

The issues of sampling and spectral support for multidimensional random processes are discussed by Peterson and Middleton [45], Mersereau and Speake [41], and Lang and McClellan [31].

Dickinson [15] first pointed out that for the multidimensional case a set of correlation samples with a positive-definite correlation matrix need not be extendible. The extendibility problem was discussed by Lang and McClellan [31] and by Cybenko [13]. An extendibility test based on whether or not a Pisarenko-type extension can be obtained was proposed by Lang and McClellan [31].

A recent discussion of one-dimensional classical spectral estimation is that of Nuttall and Carter [42]. Classical multidimensional spectral estimation for uniformly sampled data is a straightforward extension of the one-dimensional case. The multidimensional case of nonuniformly sampled data needs further study. An important aspect of this problem is the design of windows for nonuniformly sampled data.

The maximum likelihood of spectral estimation was introduced by Capon [7], and its high-resolution properties were investigated extensively by Cox [11]. The more general DASE method was devised by Davis and Regier [14]; it has received comparatively little attention in the literature. Marzetta [38] demonstrated that the MLM power estimator is an upper bound on the class of nonunique maximum likelihood estimates for an ill-posed estimation problem. A purely deterministic interpretation for the DASE power estimate, as an upper bound, was obtained by Marzetta and Lang [39]. Cybenko considered the problem of bounding windowed spectral power, for an arbitrary window, given a set of correlation samples, with no assumptions made about the power spectrum. The properties of the Cybenko bounds were studied by Marzetta and Lang [39].

General parametric estimation techniques are discussed by Van Trees [56]. Approximate one-dimensional maximum likelihood parametric estimation techniques that are based on the periodogram and are capable of generalization to multiple dimensions are discussed by Levin [32] and Brillinger [4].

Burg's thesis [5] is still a useful reference for the one-dimensional maximum entropy method. A number of theoretical problems associated with the multidimensional MEM problem were discussed by Lang and McClellan [30,36], as well as reliable computational algorithms based on the dual problem and a survey of previously proposed computational algorithms. Some properties of the multidimensional MEM estimate have been studied by Lim and Malik [33,37].

Pisarenko's method [46] was generalized to the multidimensional case by Lang and McClellan [29,31].

The survey paper of Kay and Marple [27] is a useful reference for one-dimensional AR modeling. Useful background material concerning multidimensional AR modeling is contained in Willsky [60] and Ekstrom and Marzetta [16].

References

1. A. B. Baggeroer, Sonar Signal Processing, in *"Applications of Digital Signal Processing"* (A. V. Oppenheim, ed.), pp. 169–237. Prentice-Hall, New Jersey, 1978.
2. T. H. Bell, Jr., J. M. Bergin, J. P. Dugan, Z. C. B. Hamilton, W. D. Morris, B. S. Okawa, and E. E. Rudd, Internal waves: Measurements of the two-dimensional spectrum in vertical-horizontal wave number space, *Science* **189**, 1975, 632–634.
3. R. B. Blackman and J. W. Tukey, *"The Measurement of Power Spectra,"* Dover, New York, 1959.
4. D. R. Brillinger, Fourier analysis of stationary processes, *Proc. IEEE* **62**, 1974, 1628–1643.
5. J. P. Burg, "Maximum Entropy Spectral Analysis," Ph.D. thesis, Stanford University, May 1975.
6. J. P. Burg, D. G. Luenberger, and D. L. Wenger, Estimation of structured covariance matrices, *Proc. IEEE* **70**, 1982, 963–974.
7. J. Capon, High-resolution frequency-wavenumber spectrum analysis, *Proc. IEEE* **57**, 1969, 1408–1418.
8. J. Capon and N. R. Goodman, Probability distributions for estimators of the frequency-wavenumber spectrum, *Proc. IEEE,* **58,** 1970, 1785–1786. (Correction in *Proc. IEEE* **59**, 1971, 112.)
9. D. G. Childers (ed.), *"Modern Spectrum Analysis."* IEEE Press, New York, 1978.
10. J. W. Cooley and S. Winograd, On the use of filter design programs for generating spectral windows, *Proc. ICASSP 81* **1**, 1981, 394–396.
11. H. Cox, Resolving power and sensitivity to mismatch of optimum array processors, *J. Acoust. Soc. Am.* **54**, 1973, 771–785.
12. G. Cybenko, Affine minimax problems and semi-infinite programming, *Math. Programming.,* in press.

13. G. Cybenko, Moment problems and low rank Toeplitz approximations, *Circuits, Systems, Signal Proc.,* **1,** 1982, 345–366.
14. R. E. Davis and L. A. Regier, Methods for estimating directional wave spectra from multi-element arrays, *J. Marine Res.* **35,** 1977. 453–477.
15. B. W. Dickinson, Two-dimensional Markov spectrum estimates need not exist, *IEEE Trans. Inf. Theory* **IT-26,** 1980, 120–121.
16. M. P. Ekstrom and T. L. Marzetta, Fundamentals of multidimensional time-series analysis, in *"Identification of Seismic Sources—Earthquake or Underground Explosion,"* (E. S. Husebye and E. Mykkeltveit, eds.), pp. 615–647. Reidel, New York, 1981.
17. E. B. Fomalont, Fundamentals and deficiencies of aperture synthesis, in *"Image Formation from Coherence Functions in Astronomy"* (C. van Schooneveld ed.), pp. 3–18. Reidel, 1979.
18. D. Goodman and M. Ekstrom, Multidimensional spectral factorizations and unilateral autoregressive models, *IEEE Trans. Autom. Control* **AC-25,** 1980, 258–262.
19. J. W. Goodman, *"Introduction to Fourier Optics."* McGraw-Hill, New York, 1968.
20. N. R. Goodman, Statistical analysis based on a certain multivariate complex Gaussian distribution (an introduction), *Ann. Math. Stat.* **34,** 1963, 152–177.
21. D. B. Harris and R. M. Mersereau, A comparison of algorithms for minimax design of two-dimensional linear phase FIR digital filters, *IEEE Trans. Acoust. Speech, Signal Process* **ASSP-25,** 1977, 492–500.
22. F. J. Harris, On the use of windows for harmonic analysis with the discrete Fourier transform, *Proc. IEEE* **66,** 1978, 51–83.
23. S. Haykin (ed.), *"Nonlinear Methods of Spectral Analysis."* Springer-Verlag, New York, 1979.
24. B. R. Hunt, Digital image processing, in *"Applications of Digital Signal Processing"* (A. V. Oppenheim, ed.), pp. 169–237. Prentice-Hall, New Jersey, 1978.
25. A. H. Jazwinski, *"Stochastic Processes and Filtering Theory."* Academic Press, New York, 1970.
26. D. H. Johnson, The application of spectral estimation methods to bearing estimation problems, *Proc. IEEE* **70,** 1982, 1018–1028.
27. S. M. Kay and S. L. Marple, Jr., Spectrum analysis—A modern perspective, *Proc. IEEE* **69,** 1981, 1380–1419.
28. R. T. Lacoss, Data adaptive spectral analysis methods, *Geophysics* **36,** 1971, 661–675.
29. S. W. Lang and J. H. McClellan, The extension of Pisarenko's method to multiple dimensions, *Proc. ICASSP 82* **1,** 1982, 125–128.
30. S. W. Lang and J. H. McClellan, Multi-dimensional MEM spectral estimation, *IEEE Trans. Acoust. Speech, Signal Process.* **ASSP-30,** 1982, 880–887.
31. S. W. Lang and J. H. McClellan, Spectral estimation for sensor arrays, *IEEE Trans. Acoust. Speech, Signal Process.,* **ASSP-31,** 1983, 349–358.
32. M. J. Levin, Power spectrum parameter estimation, *IEEE Trans. Inf. Theory* **IT-11,** 1965, 100–107.
33. J. S. Lim and N. A. Malik, A new algorithm for two-dimensional maximum entropy power spectrum estimation, *IEEE Trans. Acoust. Speech, Signal Process.,* **ASSP-29,** 1981, 401–413.
34. J.H. McClellan, The design of 2-D digital filters by transformation, *Proc. 7th Annual Princeton Conf. Inf. Sci. Syst.* pp. 247–251, 1973.
35. J. H. McClellan, Multi-dimensional spectral estimation, *Proc. IEEE* **70,** 1982, 1029–1039.
36. J. H. McClellan and S. W. Lang, Duality for multi-dimensional MEM spectral estimation, *IEE Proc.* **130,** Pt.F, 1983, 230–235.

37. N. A. Malik and J. S. Lim, Properties of two-dimensional maximum entropy power spectrum estimates, *IEEE Trans. Acoust. Speech, Signal Process.* **ASSP-30,** 1982, 788–798.

38. T. L. Marzetta, A new interpretation for Capon's maximum likelihood method of frequency-wavenumber spectral estimation, *IEEE Trans. Acoust. Speech, Signal Process..* **ASSP-31,** 1983, 445–449.

39. T. L. Marzetta and S. W. Lang, Power spectral density bounds, *IEEE Trans. Inf. Theory,* **IT-30,** 1984, 117–122.

40. R. M. Mersereau, W. F. G. Mecklenbrauker, and T. F. Quatieri, Jr., McClellan transformations for 2-D digital filtering: I — Design, *IEEE Trans. Circuits Syst.* **CAS-23,** 1976, 405–414.

41. R. M. Mersereau and T. C. Speake, The processing of periodically sampled multi-dimensional signals, *IEEE Trans. Acoust. Speech, Signal Process.* **ASSP-31** 1983, 188–194.

42. A. H. Nuttall and G. C. Carter, Spectral estimation using combined time and lag weighting, *Proc. IEEE* **70,** 1982, 1115–1125.

43. A. V. Oppenheim, Digital processing of speech, in *"Applications of Digital Signal Processing"* (A. V. Oppenheim, ed.), pp. 117–168. Prentice-Hall, New Jersey, 1978.

44. A. V. Oppenheim and R. W. Schafer, *"Digital Signal Processing."* Prentice-Hall, New Jersey, 1975.

45. D. P. Peterson and D. Middleton, Sampling and reconstruction of wave-number-limited functions in N-dimensional euclidean spaces, *Inf. Control* **5,** 1962, 279–323.

46. V. F. Pisarenko, The retrieval of harmonics from a covariance function, *Geophys. J. R. Astr. Soc.* **33,** 1973, 347–366.

47. W. K. Pratt, *"Digital Image Processing."* Wiley, New York, 1978.

48. L. R. Rabiner and B. Gold, *"Theory and Application of Digital Signal Processing,"* Prentice-Hall, New Jersey, 1975.

49. L. A. Regier and R. E. Davis, Observations of the power and directional spectrum of ocean surface waves, *J. Marine Res.* **35**(3), 1977, 433–451.

50. A. Rosenfeld and J. S. Weszka, Picture recognition, in *"Digital Pattern Recognition"* (F. S. Fu, W. D. Keidel, and H. Wolter, eds.), pp. 135–166. Springer, New York, 1976.

51. M. Schoenberg, T. Marzetta, J. Aron, and R. Porter, Space-time dependence of acoustic waves in a borehole, *J. Acoust. Soc. Am.* **70,** 1981, 1496–1507.

52. T. C. Speake and R. M. Mersereau, A note on the use of windows for two-dimensional FIR filter design, *IEEE Trans. Acoust. Speech, Signal Process.* **ASSP-29,** 1981, 125–127.

53. D. Tjostheim, Statistical spatial series modelling, *Adv. Appl. Prob.* **10,** 1978, 130–154.

54. L. D. Travis, "Nature of the atmospheric dynamics on Venus from power spectrum analysis of Mariner 10 images," *J. Atmos. Sci.* **35,** 1978, 1584–1595.

55. D. W. Tufts and R. Kumaresan, Estimation of frequencies of multiple sinusoids: Making linear prediction perform like maximum likelihood, *Proc. IEEE* **70,** 1982, 975–989.

56. H. L. Van Trees, *"Detection, Estimation, and Modulation Theory, Part I."* Wiley, New York, 1968.

57. P. D. Welch, The use of the fast Fourier transform for the estimation of power spectra: A method based on time averaging over short, modified periodograms, *IEEE Trans. Audio Electroacoust.* **AU-15,** 1967, 70–73.

58. W. White, Angular spectra in radar applications, *IEEE Trans. Aerospace Elec. Syst.* **AES-15,** 1979, 895–899.

59. P. Whittle, On stationary processes in the plane, *Biometrika* **41,** 1954, 434–449.

60. A. S. Willsky, *"Digital Signal Processing and Control and Estimation Theory."* MIT Press, Cambridge, Massachusetts, 1979.

61. J. W. Woods, Two-dimensional discrete Markovian fields, *IEEE Trans. Inf. Theory* **IT-18,** 1972, 232–240.
62. J. W. Woods, Two-dimensional Markov spectral estimation, *IEEE Trans. Inf. Theory* **IT-22,** 1976, 552–559.
63. J. M. Wozencraft and I. M. Jacobs, *"Principles of Communication Engineering."* Wiley, New York, 1965.
64. A. Yaglom, *"An Introduction to the Theory of Stationary Random Functions."* Dover, New York, 1962.

7 Image Analysis

Azriel Rosenfeld

Center for Automation Research
University of Maryland
College Park, Maryland

I. Introduction

A. Image Analysis and Its Applications

This chapter deals with *image analysis,* the goal of which is to produce a description of a given image. The description depends on the domain of application. In *character recognition,* the input is an image of printed or written characters, and the desired output is a sequence of digital codes for the characters (e.g., ASCII). The input in cytology might be an image of a blood smear on a microscope slide, and the output might then be the num-

The support of the National Science Foundation under Grant MCS-79-23422 is gratefully acknowledged, as is the help of Janet Salzman in preparing this paper.

DIGITAL IMAGE PROCESSING TECHNIQUES

ber of blood cells of each type in the smear. The input in radiology might be a chest x ray, and the output might be measurements of the shape of the heart, the texture of the lung tissue, or the positions of spots on the lungs that appear to be tumors. The input in remote sensing might be a satellite image of terrain, and the output might be a map showing crop classes, land uses, or terrain types. The output in industrial automation might be the positions of defects in a manufactured article or the positions and identifications of parts on a belt or in a bin. These are just a few of the major fields in which image analysis is extensively used.

The applications of image analysis are numerous and diverse, and the specific steps involved vary widely from one application to another. However, certain basic subtasks, in one form or another, are involved in nearly all these applications. In Section I.B we outline a general paradigm for the image analysis process, and in Section I.C we identify a set of general subtasks that will be treated in the body of the chapter.

B. A Simplified Paradigm

Figure 1 shows a simplified diagram of the steps in a typical image analysis process. The nature of the simplifications will be discussed, but first we describe the steps and illustrate the forms that they might take in various applications.

Image descriptions nearly always refer to parts (regions, features, or objects) of the given image. Thus, the initial step in nearly every image analysis process is *segmentation,* in which the image is divided into parts. With reference to the typical applications listed in Section I.A, segmentation processes are used to distinguish characters, blood cells, or manufactured parts from their background; to detect defects or tumors; to distinguish the heart, lungs, ribs, and so forth; and to subdivide a satellite image into land-use classes.

The initial results of segmentation do not always correspond to the desired parts. Segmentation may yield a large set of parts, for example, characters or blood cells. The parts may touch or overlap, or they may be fragmented; so it may be necessary to split them into pieces or to assemble them into groups. It may also be desirable to simplify them in various ways for purposes of description. Thus, resegmentation of the parts obtained from the initial image segmentation is often necessary. The techniques used for region segmentation are usually very different from those used for the initial image segmentation and, in fact, are usually not applicable to an unsegmented image.

Once the desired parts have been obtained, they can be described in terms of various types of geometric properties (number, size, shape, etc.), as well as

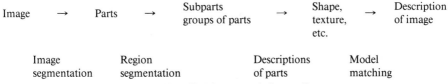

Fig. 1. Simplified image analysis paradigm.

in terms of their colors, textures, and so forth. This level of description is sufficient for many applications; for example, the measured properties (shapes of characters, blood cell nuclei, or industrial parts; texture of lung tissue; etc.) can be used as features to identify or classify the parts using standard statistical pattern classification techniques. In other cases, however, we need to identify arrangements or configurations of the parts. A common method of doing this is to construct a labeled graph or similar data structure, possibly hierarchical, in which the nodes represent the parts, labeled with their properties, and the arcs represent relationships among the parts (adjacency, surroundedness, relative position, etc.). Recognition of the configuration of parts can then be accomplished by comparing this data structure with stored models, that is, with generic data structures representing given types of configurations. Because this level of recognition involves abstract data structures rather than images, it will not be treated in detail in this chapter.

The image analysis paradigm just described is simplified in many respects. It assumes that the image is ready for segmentation, but in practice it may be desirable to enhance or restore the image (e.g., by noise cleaning) before segmentation. More important, our paradigm treats the image on a strictly two-dimensional level, ignoring the fact that it is a projection of a three-dimensional scene. In many applications, including character recognition and analysis of satellite images, two-dimensional treatment is reasonable; but in others, notably industrial parts recognition, it is not, and the analysis process must attempt to relate the image to a three-dimensional representation of the scene, for example, by inferring three-dimensional shape from shading or texture gradients, occlusions from edges, and so forth. Our paradigm also deals only with single images in which the values represent scene brightnesses or colors; it does not consider the possibility of obtaining explicit depth information about the scene (e.g., using stereo, range finding, or tomography), nor does it consider time sequences of images.

A final simplification of our paradigm is that it treats image analysis as a sequential process that involves successive stages of segmentation, description, and so forth. Often, however, feedback between stages is necessary;

for example, if the parts do not satisfy certain criteria, one may need to resegment. In general, models for the class of images being analyzed should be used as a guide in choosing segmentation and property measurement techniques, in adjusting their parameters, and in evaluating their results, in an attempt to obtain an image description consistent with the models.

C. The Basic Tasks

In accordance with our simplified paradigm, we shall deal in the body of this chapter with the following set of basic image analysis subtasks. For each subtask we shall describe appropriate algorithms, discuss their domains of applicability and computational aspects, and give selected examples of their performance.

Image segmentation, the most fundamental subtask, can be performed in many different ways; it will be treated in Section II. One approach is to use pattern recognition techniques to classify the image pixels into populations based on their brightnesses or colors (Section II.A); this is the standard segmentation method used in character recognition, cytology, and remote sensing. Pixel classification on the basis of local properties (computed over a neighborhood of each pixel) can also be used to extract various types of local features (edges, curves, etc.) from an image (Section II.B). It is also used to segment an image into differently textured regions, and statistics of local pixel properties are used to describe textures (Section II.C). Another class of segmentation techniques extracts global shapes or patterns from an image, for example, by template matching (Section II.D). Still another approach to segmentation is based on partitioning the image into a set of homogeneous connected regions (Section II.E).

Region segmentation and the measurement of geometric properties of regions are closely related topics; they are both treated in Section III. Section III.A deals with region connectedness and segmentation into connected parts. Section III.B deals with the extraction of significant parts from regions, such as their borders and skeletons. Section III.C treats size and shape properties of regions. Finally, in Section III.D, we briefly discuss some general aspects of image description in terms of parts.

It should be pointed out that few of the tasks treated in this chapter can be regarded as well-posed problems that have general solutions. There is no canonical segmentation task applicable to all images; the type of segmentation desired depends on the domain of application, and the success of a segmentation technique depends ultimately on the correctness of the descriptions that are obtained from it.

References for individual tasks and techniques will be cited in each section. The most comprehensive treatment of image analysis can be found in [13], from which most of the figures in this chapter are taken. An extensive bibliography on image processing and analysis appears annually in the journal *Computer Vision, Graphics and Image Processing.*

II. Image Segmentation

A. Pixel Classification into Subpopulations

The parts into which it is desired to segment an image can often be distinguished on the basis of gray level or color properties of the individual pixels that compose these parts. For example, in an image of printed or written characters, the pixels belonging to the characters are (usually) darker than those belonging to the background; in an image of a white blood cell, the pixels belonging to the nucleus tend to be darker than those belonging to the cell body, which in turn are darker than those belonging to the background. In a multispectral image of terrain, the pixels representing different crop types or land-use classes often have characteristic spectral signatures (i.e., combinations of brightness values in the different spectral bands). In these and similar situations, it is reasonable to segment the image by assigning its pixels to classes on the basis of their gray levels or colors.

The simplest type of pixel classification into subpopulations is *thresholding,* in which each pixel is classified as light or dark depending on whether or not its gray level exceeds a threshold. In effect, the threshold partitions the gray-level axis into two intervals, and the pixels belonging to each interval constitute a class. For color or multispectral imagery, the analogous process is more complicated; we partition the color space into regions, for example, using a set of (hyper)planes or quadric surfaces, and the pixels belonging to each region constitute a class. The question of how to define the partition (or select the threshold) belongs to the general subject of pattern classification and will not be treated here in detail; we give only a few brief heuristic remarks. In some cases, the classes can be characterized on an absolute basis; for example, if we know the illumination and the distribution of reflectivities of ink and paper, we can define a threshold that will segment ink from paper with minimum error. In the absence of such knowledge, we examine the observed distribution of brightness or color values (the *histogram* of gray levels present in the image or the *scatter plot* of spectral signatures) and look for densely populated subregions (peaks on the histogram or clusters in the scatter plot). It is reasonable to assume that such peaks or clusters correspond to significant subpopulations of pixels; this suggests that it should be possible to segment the image usefully by choosing thresholds that separate the peaks or surfaces that separate the clusters.

The computational cost of thresholding is negligible (it requires only comparison of each pixel's gray level to the threshold), and the cost of color classification using hyperplanes or quadric surfaces is also quite low (it involves verifying a set of linear or quadratic inequalities for the color components of each pixel). These costs tend to be dominated by the time required to read-in the image from auxiliary storage and read-out the results of the classification (a symbol for each pixel representing the class to which it has been assigned). We do not consider here the cost of determining the classification criteria (thresholds or separating surfaces). Creating a histogram or scatter plot also requires reading-in the image; but once a histogram is created, analyzing it is usually not a costly process, because a relatively small amount of data is involved. (Higher-dimensional scatter plots, on the other hand, can present problems as regards their storage and manipulation.)

Figure 2 shows an image of a white blood cell, its histogram, and the results of segmenting it into nucleus, cell body, and background using thresholds that separate the three peaks on the histogram. Figure 3 shows the red, green, and blue components of a color image of a house; the scatter plot of its color values projected on the red–green, green–blue, and blue–red planes; and the results of segmenting it into sky, sunlit brick, shadowed brick, grass, and bushes by partitioning the color space (in fact, just its red–blue projection) so as to separate major clusters in the scatter plot.

Thresholding is the standard segmentation technique in character recognition; for a brief review of thresholding techniques, including the use of thresholds that vary in different parts of an image, see [16]. The method of selecting thresholds that separate histogram peaks was introduced in [12]. Color space partitioning is the standard method of pixel classification in multispectral imagery; for a general treatment of this subject see [15]. An

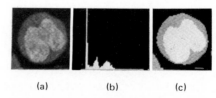

(a)   (b)   (c)

Fig. 2. Thresholding applied to an image of a white blood cell. (a) Image. (b) Histogram. (c) Results of thresholding at 9 and 19; the classes are displayed as white, gray, and black, respectively.

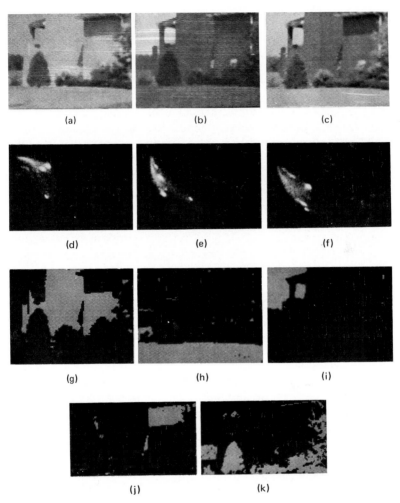

Fig. 3. Pixel classification in color space applied to an image of a house. (a – c) Red, green, and blue components of the image. (d – f) Projections of the color scatter plot on the red – green, green – blue, and blue – red planes. (g – k) Results of partitioning the red – blue color space into the following regions: (1) $15 \leq red \leq 20$, $7 \leq blue \leq 12$ or $21 \leq red \leq 28$, $11 \leq blue \leq 17$; (2) $13 \leq red \leq 19$, $13 \leq blue \leq 23$, or $20 \leq red \leq 22$, $18 \leq blue \leq 23$, or $18 \leq red \leq 26$, $24 \leq blue \leq 33$; (3) $37 \leq red \leq 44$, $22 \leq blue \leq 31$; (4) $27 \leq red \leq 39$, $37 \leq blue \leq 44$; and (5) $37 \leq red \leq 41$, $46 \leq blue \leq 49$. The pixels in each of these classes are displayed as white in parts (g–k).

introductory treatment of statistical pixel classification and a variety of threshold selection techniques can be found in [13, Section 10.1].

B. Local Feature Detection

Often the parts that one wants to extract from an image are local features, such as edges and curves, that are defined by local patterns of gray levels. Examples of applications in which the desired parts are curves include the detection of strokes in handwriting, of roads in remote sensor imagery, or of cracks in materials. Extraction of edges is often used as an aid in segmenting an image into regions in cases in which the pixel subpopulations are too varied to produce well-defined peaks on the image's histogram, but in which there are still significant changes in gray level from one region to the next. Edge information can also be used in threshold selection; for example, one can pick a threshold for which the border(s) of the above-threshold region(s) best coincide with edges [8].

The most common class of edge detection techniques is based on the application of a local difference operator at every pixel; evidently, the output of such an operator will have high magnitude at pixels where the gray levels are changing rapidly, that is, at pixels that lie on edges. (Analogous remarks apply to edge detection based on color differences; we shall consider here only the gray-level case.) In analogy with the definition of the *gradient* in calculus, most such operators are based on taking differences of gray levels in two orthogonal directions and then using the rms (or, for simplicity, the maximum of absolute values) of the results as an estimate of edge steepness and the arc tangent of their quotient as an estimate of edge direction. Note that the differences (taken at every pixel) are convolution operators, but their combination to yield magnitude and direction is a nonlinear process. The operators are usually defined on a 2×2 or 3×3 neighborhood; for example, in the *Roberts* operator the image is convolved with

$$\begin{array}{cc} 0 & 1 \\ -1 & 0 \end{array} \quad \text{and} \quad \begin{array}{cc} 1 & 0 \\ 0 & -1 \end{array},$$

and in the *Sobel* operator it is convolved with

$$\begin{array}{ccc} -1 & 0 & 1 \\ -2 & 0 & 2 \\ -1 & 0 & 1 \end{array} \quad \text{and} \quad \begin{array}{ccc} 1 & 2 & 1 \\ 0 & 0 & 0 \\ -1 & -2 & -1 \end{array}.$$

(This responds to an edge in two positions, but it is less sensitive to the noise, because it combines differencing across the edge with averaging along the edge.) A digital version of the *Laplacian,* in which the image is convolved with, for example,

$$
\begin{matrix}
0 & -1 & 0 \\
-1 & 4 & -1, \\
0 & -1 & 0
\end{matrix}
$$

is also sometimes used as an isotropic difference operator; but it responds more strongly to isolated points than it does to lines or edges (of equal contrast) and so is not a good choice when the image is noisy. (Note that for a ramp edge it has nonzero magnitude at the top and bottom of the ramp but is zero on the ramp itself.)

One way of defining difference operators for edge detection purposes is to fit a polynomial surface to the gray levels in the neighborhood of the given pixel and take the gradient (magnitude and direction) of this polynomial as an estimate of the image gradient. The x and y partial derivatives of the polynomial can be expressed directly in terms of the gray levels in the neighborhood, so that actual surface fitting is not necessary. For example, if we fit a plane by least squares to a 2×2 neighborhood, the magnitude of the gradient of that plane turns out to be the same as the (rms) magnitude of the Roberts operator; and if we fit a quadric surface to a 3×3 neighborhood, the x and y partial derivatives are proportional to the results of convolving the image with

$$
\begin{matrix}
-1 & 0 & 1 \\
-1 & 0 & 1 \\
-1 & 0 & 1
\end{matrix}
\qquad \text{and} \qquad
\begin{matrix}
1 & 1 & 1 \\
0 & 0 & 0, \\
-1 & -1 & -1
\end{matrix}
$$

the *Prewitt* operator. The results of applying several difference operators to the blood cell image are shown in Fig. 4. The final edge detection decision is based on thresholding the edge mangitudes; unfortunately, their histograms are usually unimodal, so that it is difficult to define a "natural" threshold.

Because it involves only a few small-support convolution operations, the computational cost of edge detection is low and is dominated by the cost of image read-in and read-out. An alternative but very similar approach to edge detection is to convolve the image with a set of four or eight patterns (masks) representing step edges in the principal directions; the maximum of the magnitude of the results at a given pixel is then taken as the edge magnitude, and the orientation of the mask giving this maximum defines the edge direction. This is basically a template matching approach (see Section II.D), in which we detect steplike patterns in given orientations in the image by convolving it with second differences of these patterns. A somewhat more costly approach is to determine the best-fitting step function to the gray levels in the neighborhood of each pixel and to use the height and direction of this step as the edge magnitude and direction at that pixel. Generally

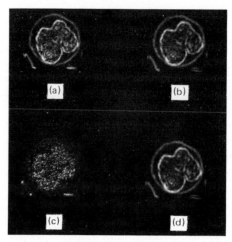

Fig. 4. Edge detection using difference operators, applied to the blood cell image. (a) Roberts operator. (b) Sobel operator. (c) Prewitt operator. (d) Laplacian. In (a–c), the output is the maximum of the absolute values; in (d), it is the absolute value.

speaking, the results obtained using these more elaborate approaches are not significantly (if at all) better than those obtained using simple difference operators. For a survey of edge detection techniques see [4], and for a detailed textbook treatment see [13, Section 10.2].

Template matching is the standard approach to curve detection; one convolves the image with a set of masks representing second differences of linelike patterns in various orientations, for example,

$$
\begin{array}{ccc}
-1 & 2 & -1 \\
-1 & 2 & -1, \\
-1 & 2 & -1
\end{array}
\qquad
\begin{array}{ccc}
-1 & 2 & -1 \\
-1 & 2 & -1, \\
-1 & 2 & -1
\end{array}
\quad \ldots ,
$$

and other rotations, and takes the maximum of the resulting magnitudes at a given pixel as a measure of the *line strength* at that pixel. This process will detect smooth curves (if they are thin), because locally a digital smooth curve resembles a straight line. Note, however, that these operators also respond to non-linelike patterns, such as edges and points. To define operators that are specific to lines or curves, one can use *gated* templates in which the response is set to zero unless certain conditions are satisfied, for example, for the vertical direction, requiring that in the neighborhood

$$
\begin{array}{ccc}
A & B & C \\
D & E & F \\
G & H & I
\end{array}
$$

we have $B > A$, $B > C$, $E > D$, $E > F$, $H > G$, and $H > I$. A comparison of

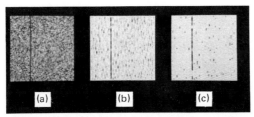

Fig. 5. Vertical line detection using gated and ungated templates. (a) Input image (the line has gray level 50 on a 0 – 63 gray scale; the noise is Gaussian with $\mu = 32$, $\sigma = 9$. (b) Positive values of ungated template output; note streaky responses to noise. (c) Values of gated template output; note gaps at points on the line where the conditions are not all satisfied because of the noise.

the results obtained using gated and ungated templates for vertical line detection is shown in Fig. 5. For a more detailed discussion of curve detection, see [13, Section 10.3]. Analogous templatelike operators can be designed for the detection of other types of local patterns, such as spots and corners. These methods can be extended to larger types of features, for example, thick curves, by scaling up the operators appropriately.

C. Texture Analysis

In edge and curve detection we are classifying pixels (as to whether their neighborhoods have high rates of change of gray level or are steplike or linelike, etc.) by thresholding the values of local properties computed in the neighborhood of each pixel. Pixel classification on the basis of (sets of) local property values can also be used to segment an image into regions that differ in texture. Textural properties of regions, for example, as measured by statistics of local property values, are also important in image description. For example, such properties are used in cytology as an aid in classifying white blood cells (based on the texture of the nucleus), in radiology to detect black lung disease, and in remote sensing to characterize land use classes; they also have applications in materials inspection.

Several classes of properties have been traditionally used to describe and classify textures. These include values of the Fourier power spectrum in various bands, second-order statistics of pairs of gray levels at various separations, and first-order statistics of absolute gray-level differences at various separations. The latter approaches are less costly computationally, because the necessary statistics can be computed in a single scan of the image. More refined approaches have used statistics of gray levels (or local property values) in the vicinities of local features detected in the image or statistics of gray levels and geometric properties of microregions extracted from the image. A general review of texture analysis can be found in [7]; see also [13, Section 12.1.5].

Segmentation of an image into regions based on texture is not a trivial task; reliable measurement of textural properties requires large samples of pixels, and the larger the samples used, the less likely they are to belong to a single region. Nevertheless, good results can often be obtained using local properties computed in small neighborhoods of each pixel (e.g., 3×3 or 5×5), particularly if the values obtained are then smoothed (by local averaging, median filtering, etc.) to decrease their variability. As an example, Fig. 6 shows the results of segmenting a black-and-white version of the house picture using two local properties of each pixel: its gray level and its local busyness (measured by the average of the absolute differences of all pairs of adjacent pixels in the 3×3 neighborhood of the given pixel). One can also detect texture edges (where the texture of the image changes abruptly) by comparing local property statistics in pairs of adjacent neighborhoods. These processes are computationally relatively inexpensive, because they typically involve only a few tens of arithmetic operations per pixel.

D. Global Feature Detection and Template Matching

A more computationally costly class of segmentation tasks is that of detecting specific global shapes or patterns in an image, for example, straight lines, circles, rectangles, or more complex configurations. Such tasks have obvious applications in the analysis of images containing man-made objects (remote sensor imagery of cultural features or recognition of industrial parts). Finding matches to a piece of one image in another image is also used extensively for registration of successive images of a scene or for stereomapping.

The brute force approach to this type of task is *template matching;* classically, this is convolution (or cross-correlation) of a copy of the desired pattern with the image. For large patterns, this requires many operations per pixel (particularly if the orientation and scale of the pattern are not known, so that one must use templates having many orientations and sizes); it is best done in the Fourier domain, by taking the inverse transform of the product of the transforms of the image and pattern, in accordance with the Fourier convolution theorem. Because images are usually quite correlated, a copy of the desired pattern is usually not the ideal matched filter; it is better to cross-correlate the first differences of the pattern and the image with each other, or the second difference of the pattern with the image. (The use of second-difference templates was illustrated in Section II.B.) Intuitively, matching differences yields more sharply localized matches and is also less sensitive to gray-scale differences between the image and pattern [1,2].

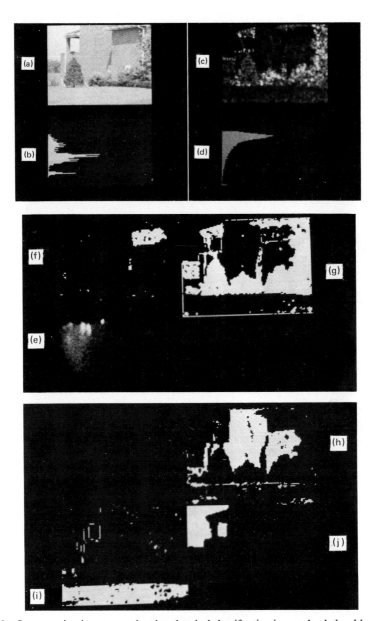

Fig. 6. Segmentation into textured regions by pixel classification in gray-level – local-busyness space, applied to a black-and-white version of the house image. (a) Image. (b) Histogram; note that there are only three main peaks. (c) Local busyness values, displayed as gray levels (busy = bright). (d) Histogram of the values in (c). (e) Scatter plot of gray-level – busyness values; note the two clusters at the upper left that overlap in gray level but differ in busyness. (f–j) Results of partitioning gray-level – busyness space to separate the five clusters in (e); the pixels in each class are displayed as white in parts (f–j).

One approach to reducing the computational cost of image matching is to use a cumulative measure of mismatch, such as the sum of absolute differences of gray levels, in place of the correlation measure of match. This allows the rapid elimination of positions for which a good match cannot possibly exist [3]. Another approach, which also reduces the sensitivity of the matching process to geometrical distortion, is to break up the template into small parts, find good matches to these parts, and then look for combinations of these matches in (approximately) the correct relative positions [5]. It should be pointed out that all of these matching schemes *detect* the presence of the desired pattern in the image, but they do not explicitly *extract* it; to do the latter, one can apply local segmentation techniques to the parts of the image where the pattern has been detected.

An interesting alternative approach to finding specific patterns in an image is to construct a transformed space in which each instance of the desired pattern maps into a point; peaks in this transform domain then correspond to occurrences of the pattern in the original image. The transforms used are known as *Hough transforms,* because a similar approach was first used by Hough to detect straight lines; see [14] for a review of such schemes. The following example shows how we can use this method to detect straight lines in arbitrary orientation: We apply local line detection operators to the image, and each detection, say, at orientation θ in position (x,y), is mapped into the point in transform space whose Cartesian coordinates are $[\theta, r \sin(\theta - \varphi)]$, where $r = \sqrt{x^2 + y^2}$ and $\varphi = \tan^{-1} (y/x)$. Here, $r \sin(\theta - \varphi)$ is just the perpendicular distance d from the origin of the line of slope θ through (x,y). If there are many detections that lie on the same straight line, say, of slope θ_0 at distance d_0 from the origin, there will be a peak in the transform space at position (θ_0, d_0). An example of this transform is shown in Fig. 7. The cost of this approach is comparable to that of

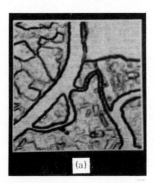

Fig. 7. Detecting straight lines using a Hough transform. (a) Image. Edge magnitudes (and directions, not shown) obtained from a portion of a LANDSAT frame. (b) Transform domain (see text). The peaks (underlined) correspond to the major edges in diagonal directions.

Distance from
center (pixels)

Slope

(b)

Fig. 7. (*Continued*)

0   2π

template matching, but it has the (possible) advantage of using a transform domain designed for the specific patterns being matched. For special classes of patterns, very simple transformations can be used; for example, if we want to detect horizontal lines in an image, we can project the image onto the y axis (i.e., sum its rows); horizontal lines map into peaks on this projection.

E. Partitioning into Homogeneous Regions

Earlier we discussed methods of segmenting an image into regions by classifying its pixels into homogeneous subpopulations based on peak or cluster separation in a histogram or scatter plot. This does often yield a segmentation of the image into homogeneous regions, but it is not guaranteed to do so, because it does not take into account the spatial arrangement of the pixels; if the pixels are arbitrarily permuted, the histogram or scatter plot remains the same, even though the image now looks like noise. In this section we describe methods of segmentation that deal with regions rather than with individual pixels. For a general treatment of such methods, see [9].

The most common homogeneity criterion used in segmenting images into homogeneous regions is low variability of gray level, for example, standard deviation below a threshold. Other criteria based on gray level can also be used, for example, closeness of fit to a polynomial of given degree. Another approach is based on similarity between a region and its subregions (or between the subregions themselves) with respect to given gray-level or local-property statistics; for example, we might call a square region homogeneous if the values of these statistics for (the square and) its quadrants differ by less than a threshold.

A partition of an image into homogeneous connected regions can be constructed by starting with an initial trivial partition, for example, into single pixels or into regions of constant gray level, and then repeatedly merging adjacent pairs of regions, as long as the larger regions resulting from these merges continue to satisfy the homogeneity criterion. An alternative approach is to start with the entire image and test it for homogeneity; if it is not homogeneous, we split it arbitrarily, say, into quadrants, and repeat the process for each quadrant. Note that this process may give rise to pairs of adjacent regions (obtained by splitting two neighboring inhomogeneous regions) that could be merged without violating homogeneity; thus, merging is generally necessary after splitting. An example of an image that has been segmented by recursive splitting into blocks (quadrants, subquadrants, etc.) and subsequent linking of pairs of similar blocks is shown in Fig. 8.

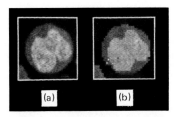

Fig. 8. Segmentation into homogeneous regions by recursive splitting into quadrants, subquadrants, and so on, until homogeneous blocks are obtained. (a) Image of a tank, obtained by an infrared sensor). (b) Results of splitting.

This class of segmentation methods requires the use of other data structures in addition to the original pixel array. For merging, we can represent the regions (at a given stage of the merging process) by the nodes of a graph, with pairs of nodes joined by arcs if the corresponding regions are adjacent. The statistics associated with each region can be stored at its node; and if it is possible to compute the statistics for a (tentatively) merged pair of regions direct from those for the individual regions, we can carry out the merging process directly on the graph, without the need to reaccess the original image. For splitting, we can represent the quadrants, subquadrants, and so on at a given stage of the process by the nodes of a tree of degree 4, where the root is the whole image and the sons of a node correspond to its quadrants. Here, it is necessary to refer to the original image in order to compute the statistics of the subregions each time a region is split.

The fact that region-based segmentation methods may require access to the image data in an arbitrary order is a potential disadvantage when it is necessary to access the image from peripheral storage. Thus, such methods are best applied to small images. On the other hand, region-based methods have the potential advantage that, in principle, they can be designed to incorporate information about the types of regions (sizes, shapes, colors, textures, etc.) that are expected to occur in images of the given class; thus, merging or splitting can be inhibited if it would violate restrictions on the expected types of regions. As a classical example, a region-based approach can be used to "grow" or "track" global edges (or curves) in an image, starting from pixels that have high edge magnitudes and accepting new pixels (i.e., merging them with the edge fragments already constructed) if they continue these edges.

Another advantage of pixel-based over region-based segmentation methods is that pixel-based schemes can be greatly speeded up if parallel hardware is available by dividing the image into parts and assigning a separate processor to segment each part, where the processors can share global information about segmentation criteria, if desired, and where they may also have to share neighbor information along the common borders of the parts. In principle, parallelism could also be used in region-based schemes by assigning processors to (sets of) regions, but this would require a very flexible interprocessor communication scheme to allow processors that contain information about adjacent regions to communicate. In pixel-based schemes, on the other hand, we can divide the image into, say, square blocks, so that the processor responsible for a block needs to communicate only with a limited number of processors that are responsible for neighboring blocks.

III. Region Description and Segmentation

A. *Topological Properties*

One of the most basic properties of a subset of an image is its *connectedness,* that is, the number of connected regions in the subset. (A precise definition of connectedness will be given in the next paragraph.) When we segment an image by pixel classification, the pixels in a given subpopulation may comprise many different connected regions (e.g., different printed characters, blood cells, or industrial parts), and it will usually be desirable to label each of these regions distinctively, so that it can be treated individually for description purposes. Labeling the individual regions also allows us to count them. The process of *connected component labeling* is a basic region segmentation task. Note that when we segment an image by partitioning it into homogeneous regions, each of the regions is automatically connected, because we are only allowed to merge adjacent pairs; thus, region-based methods immediately yield a decomposition into connected parts, each of which is represented by a node of the region graph.

Any subset S of an image can be specified by a one-bit overlay image in which the pixels belonging to S have value 1 and those not in S have value 0. (There are other ways of representing image subsets, using other types of data structures, that may yield representations that are more compact; we shall return to this topic in Section III.B.) Two pixels P,Q of S are said to be connected (in S) if there exists a path $P = P_0, \ldots, P_n = Q$ of pixels of S such that P_i is a neighbor of P_{i-1}, $1 \leq i \leq n$. Note that we have two possible definitions here, depending on whether or not we allow diagonal neighbors; these two versions are known as *4-connected* and *8-connected,* respectively. A *connected component* of S is a maximal set of mutually connected pixels of S.

To label the connected components of S, say, with positive integers, we examine the pixels of the overlay image in sequence, row by row. If the current pixel P has value 1, and all its previously examined neighbors (i.e., its north and west neighbors, and also its northwest and northeast neighbors if diagonal neighbors are allowed) are 0, we give P a new label. If any of them already have labels, and these labels are all the same, we give P that label; but if they are not all the same, we given P, say, the lowest of them and make note of the fact that these labels are all equivalent. When this phase of processing the overlay image is complete, we have created an array of labels in which pixels having the same label must all be connected, but pixels that are connected may still have different labels. To produce the final labeling, we sort out the label equivalences to determine the lowest label equivalent to each label; we then rescan the labeled image and replace each label by its lowest equivalent, which yields an array of labels in which pixels have the same label if and only if they are connected. The number of connected components is then the number of (unequivalent) labels used.

If the connected components of S do not have complex shapes, the algorithm is quite efficient, because it requires only two scans through the image, the first involving simple neighbor comparisons, the second involving a table look-up. For complex components, however, the algorithm may require us to sort out a large number of equivalent labels. Note that the algorithm involves more than just an input image (the overlay) and an output image (the array of labels). During the first scan, it requires a memory register to store the new label that was last used, so that the next available new label can be determined; and it also creates a list of equivalent label pairs, which is then sorted out to create a look-up table (of the least label equivalent to each label) for use in the second scan. A simple example of the operation of this algorithm is shown in Fig. 9. For a more detailed discussion of connectedness and related topics, see [13, Sections 11.1.7 and 11.3.1].

B. Borders and Skeletons

The border of an image subset S is the set of pixels in S that have neighbors not in S (in this definition, diagonal neighbors are usually not allowed.) It is trivial to extract this set in a single scan of the overlay image representing S: one discards 1s whose neighbors are all 1s.

Each connected component C of S is adjacent to various connected components of $\bar{S}$, the complement of S. (One of these components surrounds C, and the others, if any, are surrounded by it; the latter are called *holes* in C.) The set of border pixels of C that are adjacent to a given component D of $\bar{S}$ constitutes a closed curve. (This statement is true only if we use opposite

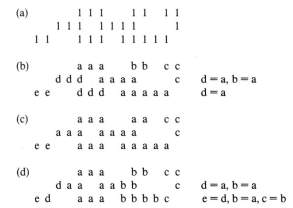

Fig. 9. Simple examples of connected component labeling. (a) Input overlay image; blanks are 0s. (b) Results of first scan when diagonal neighbors are not allowed. (c) Results of second scan. (d) Results of first scan when diagonal neighbors are allowed.

types of connectedness for S and $\bar{S}$, that is, if we allow diagonal neighbors in S, we do not allow them in $\bar{S}$, and vice versa.) It is possible to "follow" such a border, that is, to visit its pixels in sequence (note that it can pass through some pixels twice at places where C is only one pixel wide); this sequence can then be compactly encoded by specifying the moves from neighbor to neighbor, and S can be reconstructed if we know the codes of all its borders. Algorithms for border following and encoding and for reconstructing S from its borders are presented in the following paragraphs. For simplicity, we give these algorithms as if we were coding or reconstructing each border sequentially, which requires access to the overlay image in an arbitrary order. Encoding or reconstruction of all the borders can be done in a single scan of the overlay image, but the details are somewhat complicated and will not be given here.

Let P,Q be a pair of nondiagonally adjacent pixels belonging to C and D, respectively. Let A,B be the two pixels that form a 2×2 square with P and Q; and such that we are facing A,B when we stand between P and Q with P on the left, that is,

$$P \quad A$$
$$Q \quad B'$$

or a rotation of it by a multiple of $90°$. (Keeping P on the left implies that we will go around the border counterclockwise if it is the outer border of C and clockwise if it is a hole border.) If P,Q is the current (or initial) pixel pair in our border-following process, then the next pair P',Q' is given by the following rules:

If we allow diagonal neighbors in S				If not			
A	B	P'	Q'	A	B	P'	Q'
1	0	A	B	1	0	A	B
0	0	P	A	0	0,1	P	A
1,0	1	B	Q	1	1	B	Q

Note that P (or Q) may occur several consecutive times on the border as we follow the cracks between it and the pixels adjacent to it in D. The process stops when we get back to the initial pair P,Q; note that we may get back to P (with other Qs) before this happens if the border passes through P twice.

Successive (different) Ps that we visit as we follow the border are neighbors of each other (here, diagonal neighbors must be allowed, because we did not allow them in our definition of a border pixel). Thus, we can represent the border by a *chain code* of 3-bit numbers representing the successive moves from neighbor to neighbor, for example, denoting the directions to the eight neighbors of P by

$$\begin{matrix} 3 & 2 & 1 \\ 4 & P & 0. \\ 5 & 6 & 7 \end{matrix}$$

(Mnemonic: neighbor i makes an angle of $45i°$ with the positive x axis.) Alternatively, a 2-bit code could be used to represent the orientations of the successive pairs P,Q along the border; but this would not be more compact than the 3-bit code, because there are likely to be about 50% more P,Qs than Ps on the border. The chain code of the outer border of the large connected component in Fig. 9, counterclockwise starting from its upper left-hand corner, is 5407001700004431454344.

Given an initial pair P,Q and the chain code, we can reconstruct the entire sequence of pairs on the border. If the current (or initial) pair and its neighborhood are, for example,

$$\begin{matrix} A & B & C \\ D & P & E, \\ F & Q & G \end{matrix}$$

then the chain code tells us the values of some of the neighbors of P and also determines a new current pair according to the following rules:

If we allow diagonal neighbors in S			If not		
Chain code	Neighbors	New pair	Chain code	Neighbors	New pair
0	$E = 1, G = 0$	E,G	0	$E = 1, G = 0$	E,G
1	$C = 1, E = G = 0$	C,E	1	$B = C = 1, E = 0$	C,E
2	$B = 1, C = E = G = 0$	B,C	2	$B = 1, C = E = 0$	B,C
3	$A = 1, B = C = E = G = 0$	A,B	3	$D = A = 1, B = E = 0$	A,B
4	$D = 1, A = B = C = E = G = 0$	D,A	4	$D = 1, A = B = E = 0$	D,A
5	$F = 1, D = A = B = C = E = G = 0$	F,D	5	Impossible case	—
7	$G = 1$	G,Q	7	$G = E = 1$	Q,G

Analogus rules apply for the other orientations of P,Q. They allow us to "paint" all the successive pairs around the border with 1s and 0s on an initially blank array (we assume that 0 and blank are distinguishable). If all the borders of S have been painted in this way, we can reconstruct the entire overlay array of S by letting 1s and 0s propagate into blanks: on each row, from left to right, turn the first pixel into a 0 if it is blank, and turn a blank into 0 or 1 if its predecessor is 0 or 1. For further details about borders see [13, Sections 11.1.3, 11.1.7, and 11.2.2].

Another way of compactly representing a subset is as a union of maximal blocks of 1s. For example, for each pixel P in S, let S_P be the largest upright square (of odd side length) centered at P that is entirely contained in S. Call S_P maximal if it is not contaned in any other (larger) S_Q. Evidently, S is the union of the maximal S_Ps (every pixel of S is contaned in some maximal S_P, and they are all contained in S); thus, if we know the centers and radii of the maximal S_Ps, we can reconstruct S. The centers of these S_Ps constitute a kind of skeleton of S, sometimes called the medial axis, because they tend to be at (locally) maximal distances from the border of S. An algorithm can be given for constructing this skeleton in two scans of the overlay image of S; see [13, Sections 11.1.2 and 11.2.1] for further details. This representation of S is not as compact as the chain-code representation of the borders of S; the skeleton pixels do not form connected sequences, and we have to specify the coordinates of each of them separately, whereas for border pixels, we need only 3 bits to specify each move from neighbor to neighbor.

A different type of skeleton of S, which does consist of connected sequences of pixels, can be constructed by thinning S by repeatedly deleting border pixels in such a way as to preserve local connectedness. We call a border pixel of S *simple* if at least two of its neighbors are 1s and if in its 3 × 3 neighborhood, changing it from 1 to 0 would not disconnect those of its neighbors that are 1s. To thin S, we delete all the simple north border pixels (i.e., pixels whose north neighbors are 0); then we repeat the process for south, east, west, north, south, east, west, and so on until there is no further change. (We delete simple border pixels on only one side at a time to insure that connectedness is preserved and to produce a skeleton that is as centered as possible in S.) Note that we cannot reconstruct the original S from this skeleton without further information, for example, we can reconstruct S approximately if we know the maximal block radii at the pixels of the skeleton. If S is nowhere more than k pixels thick (i.e., if k thinning steps are sufficient), the thinning process can be performed in a single scan of the overlay image, keeping only k rows in main memory at a time. For further details see [13, Section 11.2.3]. An example of the effects of thinning, allowing diagonal neighbors, is shown in Fig. 10.

C. Size and Shape Properties

The *area* of an image subset or region S is simply the number of pixels in S; it is trivial to compute the area by scanning the overlay image of S and counting 1s. Area is not only an important property in its own right, but it also can be useful in resegmenting S; for example, one can eliminate all connected components whose areas are smaller than a given threshold. The *perimeter* of S can be defined as the number of border pixels in S or as the number of moves made by the border-following algorithm (Section III.B) in going around all the borders of S. By either definition, it can be computed in a single scan of the overlay image.

The height and width of S (or its extent in any given direction) are easy to compute, but they are of interest only if the orientation of S is known (e.g., in character recognition). A more interesting question is that of defining the *thickness* of S; note that different parts of S may have different thicknesses. Intuitively, the thickness of (a part of) S is equal to twice the radius of its maximal blocks (see Section III.B), and (a part of) S is *elongated* if its area is high relative to the square of its thickness. For example, characters are usually elongated.

Simple shrinking and expanding operations can be used to extract thin pieces of S. Let $S^{(k)}$ denote the result of expanding S k times, that is, changing 0s to 1s in the overlay image if they have 1s as neighbors and repeating this process k times. Similarly, let $S^{(-k)}$ denote the result of shrinking S k times, that is, repeatedly changing 1s to 0s if they have 0s as neighbors. (As in the case of thinning, these operations can be done in a single scan of the overlay image, keeping only k rows at a time in main

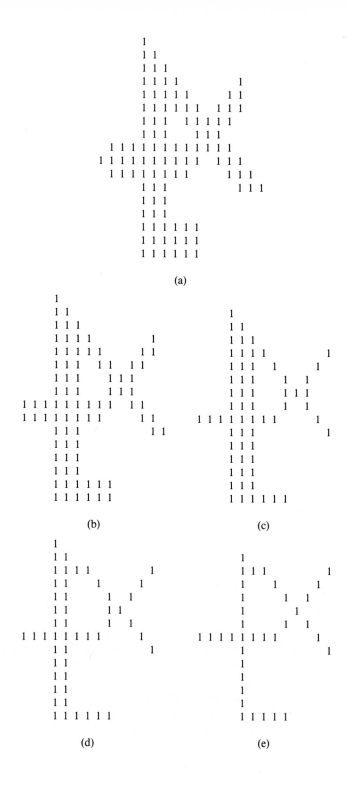

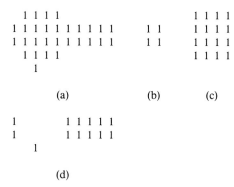

Fig. 11. Elongated part detection by shrinking and reexpanding. (a) S; (b) $S^{(-1)}$; (c) $[S^{(-1)}]^{(1)}$; (d) $S - [S^{(-1)}]^{(1)}$. The 10-point component in (d) is large and, hence, elongated.

memory.)   It can be shown that $[S^{(-k)}]^{(h)} \subseteq S^{(h-k)} \subseteq [S^{(h)}]^{(-k)}$.   If we shrink S and reexpand it by the same amount, that is, we construct $[S^{(-k)}]^{(k)}$, noisy pieces of S disappear when we shrink and do not reappear when we reexpand.   To detect elongated pieces of S, we can shrink, reexpand, and subtract from the original S, recalling that $[S^{(-k)}]^{(k)} \subseteq S^{(0)} \equiv S$; what remains are the parts of S that disappeared under the shrinking and reexpansion, and if such a part has large area (relative to k^2), it is elongated (of thickness $\leq 2k$), as illustrated in Fig. 11.

Shrinking and expanding can also be used to extract isolated pieces of S and to fuse clusters of pieces.   If we expand S and reshrink it by the same amount, that is, we construct $[S^{(k)}]^{(-k)}$, an isolated piece of S (at least $2k$ away from any other piece) simply expands and shrinks back; it does not become much bigger than it was originally.   But a cluster of pieces less than $2k$ apart from each other fuses into a large region when it expands, and this region shrinks back only at its borders, resulting in a component much bigger than the original pieces, as shown in Fig. 12.   Similar methods can be used to link broken edges or curves that have been detected in an image; but here, because we know the slope of each edge or curve pixel, we can use directional expansion (or search) to find nearby continuations, and we can allow pieces to be linked only if they smoothly continue one another.

The *curvature* of a border (or linked edge or curve) at a given pixel is the rate of change of its slope.   (Because the slope of a chain code is always a

Fig. 10.   Simple example of thinning, allowing diagonal neighbors.   (a) Input overlay image. (b – e) Results of successive deletions of simple pixels on the north, south, east, and west.   At the next north step, the uppermost 1 in (e) will be deleted; at the next south step, the leftmost of the lowermost 1s; at the next west step, the leftmost of the uppermost 1s.   There will be no further changes.

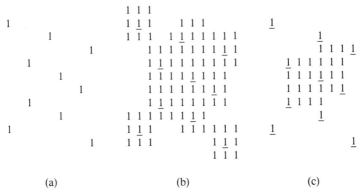

Fig. 12. Cluster detection by expanding and reshrinking. (a) S (b) $S^{(1)}$; (c) $[S^{(1)}]^{(-1)}$. In (b) and (c) the pixels in the original S are underlined.

multiple of 45° and because there are similar limitations on the possible slope values of edges or curves detected locally, it is often desirable to smooth the slope values by local averaging with the values at neighboring pixels or to estimate curvatures using vector sums of slopes on each side of the given pixel.) The *slope histogram* of a border (or edge or curve) gives gross information about its orientation, for example, peaks indicate orientation biases. Similarly, the curvature histogram gives information about wiggliness. *Angles* or *corners* are pixels at which the curvature is high; they are usually good places at which to put the vertices in constructing an approximating polygon. *Inflections* are pixels at which the curvature changes sign; they tend to separate the border into convex and concave pieces (peninsulas and bays). These concepts are illustrated in Fig. 13.

The ideas in the preceding paragraph are analogous to local feature detection concepts in image segmentation; slopes are analogous to gray levels, angles to edges, and wiggliness to texture. Global methods can also be used to segment borders (or edges or curves); parts that have given shapes can be detected by chain-code correlation (analogous to template matching); and region-based methods, involving merging or splitting, can be used to segment them into homogeneous arcs. Two particular global segmentation schemes that deserve mention are (1) the detection of sharp (convex or concave) spurs by comparing arcs with their chords (spurs are present if the arc gets far from the chord) and (2) the extraction of concavities by constructing the *convex hull,* the smallest convex set $H(R)$ containing the given region R, where the concavities are the connected components of $H(R) - R$. For methods of partitioning a region into convex pieces, see [9].

The chain code of a border (or edge or curve) is a kind of discrete slope intrinsic equation, giving slope as a function of arc length around the border

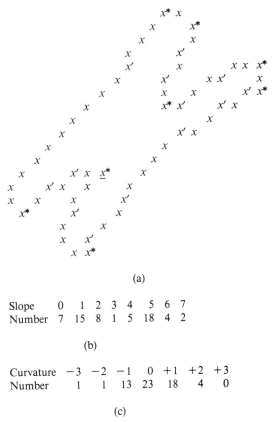

(a)

Slope	0	1	2	3	4	5	6	7
Number	7	15	8	1	5	18	4	2

(b)

Curvature	−3	−2	−1	0	+1	+2	+3
Number	1	1	13	23	18	4	0

(c)

Fig. 13. Slope, curvature, angles, and inflections. (a) Curve. The chain code, counterclockwise beginning at the underlined *x*; is 5565670211211111011010224454554221212345556555555555 67110100. Local maxima of curvature are starred, and inflections are primed. (b) The slope histogram, indicating at 45° orientation bias. (c) The curvature histogram; its concentration near 0 indicates that the curve is not wiggly.

(note, however, that diagonal steps are $\sqrt{2}$ as long as horizontal or vertical steps). The discrete Fourier transform of this one-dimensional representation can be used to process the border (e.g., to smooth it by supressing high frequencies) or to measure useful shape properties (e.g., if a given harmonic has an especially high value, the shape has a corresponding number of lobes). The same approach can be used with other types of equational representations, such as parametric equations (*x* and *y* as functions of an arbitrary parameter); here, we can still use one-dimensional Fourier analysis by combining *x* and *y* into a single complex variable $z = x + iy$.

All of these border-based concepts require relatively small amounts of computation, because they have chain codes or similar one-dimensional representations, rather than images, as input. For further details on all of the concepts discussed in this section, see [13, Sections 11.3.2–11.3.4]. For general reviews of shape analysis techniques, see [10,11].

D. Image Description

Many different types of image descriptions can be formulated in terms of the concepts defined in this chapter; the type of description actually constructed depends on the problem domain. For example, in character recognition we separate the characters from the background, usually by thresholding, and we segment the result into individual characters; if these do not correspond to connected components, some method of linking fragments or (more likely) splitting regions based on shape features must be used (perhaps, splitting into individual strokes and then reassembling these into individual characters). Each character can then be described by various shape properties or by using its relational structure description in terms of the strokes. Analogous remarks apply to other image analysis tasks. The methods described in this chapter provide a repertoire of techniques applicable to a wide variety of image analysis problems.

An important concept in defining image or region properties for description purposes is that these properties should be invariant with respect to transformations that do not affect the desired description. For example, the properties should usually be invariant to simple (e.g., monotonic) transformations of the gray scale (in the various spectral bands), at least if they are not too extreme. This is generally true for the segmentation methods described in Section II. When we segment an image by pixel classification based on separating peaks in a histogram or clusters in a scatter plot, the segmentation tends to be invariant to gray-scale transformations, because these will shift the positions of the peaks or clusters but should still leave them distinct. Local feature detection is relatively insensitive to gray-scale transformations, because it depends on local contrast, not on absolute gray levels. Similarly, partitioning into homogeneous regions depends on variability, not on particular values. Analogous remarks apply to segmentation based on texture; moreover, in texture classification, it is standard practice to normalize the gray scale, typically by histogram flattening, which yields a standard image that remains approximately the same if the original image is subjected to a monotonic gray-scale transformation. Histogram normalization is useful in template matching as well, because, otherwise, systematic gray-level differences can yield a large measured mismatch.

Invariance and normalization also play important roles in the measurement of geometric properties after an image has been segmented. One usually wants to measure properties that are invariant with respect to position and orientation (though this is not true in cases such as characters and chest x rays) and sometimes also with respect to perspective transformation. Most of the properties described in Section III are, in fact, essentially invariant to position and orientation, for example, connectivity, area, perimeter, width, and curvature. *Moments* (expressions of the form $\Sigma \Sigma x(n_1,n_2)n_1^i n_2^j$, $i,j \geq 0$, where x is gray level), which can be measured for images either before or after segmentation, can be combined in various ways to yield properties invariant under a variety of transformations and can also be used to normalize an image or region with respect to such transformations. Various generalizations of the autocorrelation or power spectrum can be used for similar purposes. Thinning can be regarded as a method of normalization with respect to width (e.g., of the strokes in characters). For a more detailed discussion of invariance and normalization, see [13, Section 12.1].

As mentioned in Section I, the description obtained as a result of segmentation and property measurement can often be represented by a relational structure specifying the parts, their properties, and their relationships, for example, characters can be described in terms of strokes having particular sizes and shapes and joined in particular ways. Such descriptions are sometimes hierarchial, that is, the parts are, in turn, composed of subparts, and so on. In structural pattern recognition [9], the input is identified by comparing its relational structure description with standard model descriptions. In syntactic pattern recognition [6] (see also [13, Section 12.2] for a brief introduction), the model is more implicit; a hierarchical structure is recognized as belonging to the desired class by verifying that it could have been generated by the operation of a given set of "grammatical" rules.

Models that describe real-world classes of images, whether explicitly or implicitly, are not easy to define. For example, it is very hard to characterize the particular combinations of stroke shapes and arrangements that define a particular alphanumeric character, and analogous remarks apply to many other classes of images. Nevertheless, in many domains, image analysis techniques can already outperform human analysts. Further advances in image analysis will play an important role in the development of perceptual systems for intelligent machines.

IV. Bibliographical Notes

Pixel classification is the standard method of segmenting multispectral imagery in remote sensing, using the pixel values in the various spectral bands as features. For an introduction to the subject see the book by Swain and Davis [15]. When there is only one band, the pixels must be classified on the basis of their gray levels, that is, by thresholding. The use of the gray-level histogram in selecting thresholds was introduced by Prewitt and Mendelsohn in [12] in connection with the classification of white blood cells. A brief survey of threshold selection techniques, including applications to character recognition and biomedical image analysis, is given by Weszka in [16]. An alternative to pixel classification is the split-and-merge approach, which attempts to partition the image into maximal homogeneous regions; this approach is reviewed by Pavlidis in [9].

Edge detection techniques are surveyed by Davis in [4]. The use of edge information combined with thresholding to yield convergent evidence for segmentation is discussed by Milgram in [8]. A detailed review of texture analysis techniques is given by Haralick in [7].

Template matching using (normalized) cross-correlation can be regarded as an application of matched filtering. A computationally cheaper approach, based on cumulative measurement of mismatch, was introduced by Barnea and Silverman [3], who also considered various methods of speeding up the matching process. In some cases, it is difficult to obtain sharply localized matches, but the sharpness can be improved by matching differentiated images (e.g., Arcese *et al.* [2]) or even by matching thresholded edge maps (Andrus *et al.* [1]). Conversely, if the matches are too sharp, they become sensitive to slight misregistrations between the images; but this can be alleviated by making the matching process hierarchical, that is, finding submatches and then looking for the proper combinations of submatches in the proper relative positions, as discussed by Fischler and Elschlager in [5]. The Hough-transform approach to template matching, based on specialized coordinate transformations, is reviewed by Shapiro in [14].

Shape analysis techniques are surveyed by Pavlidis in two review papers [10,11]. The structural and syntactic approaches to image analysis and recognition are reviewed in the books by Pavlidis [9] and Fu [6].

References

1. J. F. Andrus, C. W. Campbell, and R. R. Jayroe, Digital image registration using boundary maps, *IEEE Trans. Computers* **24**, 1975, 935–940.
2. A. Arcese, P. H. Mengert, and E. W. Trombini, Image detection through bipolar correlation, *IEEE Trans. Information Theory* **16**, 1970, 534–541.
3. D. I. Barnea and H. F. Silverman, A class of algorithms for fast digital image registration, *IEEE Trans. Computers* **21**, 1972, 179–186.
4. L. S. Davis, A survey of edge detection techniques, *Computer Graphics Image Processing* **4**, 1975, 248–270.
5. M. A. Fischler and R. A. Elschlager, The representation and matching of pictorial structures, *IEEE Trans. Computers* **22**, 1973, 67–92.

6. K. S. Fu, *"Syntactic Pattern Recognition and Applications,"* Wiley, New York, 1981.
7. R. M. Haralick, Statistical and structural approaches to texture, *Proc. IEEE* **67**, 1979, 786–804.
8. D. L. Milgram, Region extraction using convergent evidence, *Computer Graphics Image Processing* **11**, 1979, 1–12.
9. T. Pavlidis, *"Structural Pattern Recognition,"* Springer, New York, 1977.
10. T. Pavlidis, A review of algorithms for shape analysis, *Computer Graphics Image Processing* **7**, 1978, 243–258.
11. T. Pavlidis, Algorithms for shape analysis of contours and waveforms, *IEEE Trans. Pattern Analysis Machine Intelligence* **2**, 1980, 301–312.
12. J. M. S. Prewitt and M. L. Mendelsohn, The analysis of cell images, *Annals. N.Y. Acad. Sci.* **128**, 1966, 1035–1063.
13. A Rosenfeld and A. C. Kak, *"Digital Picture Processing,"* second edition, Academic Press, New York, 1982, volume 2.
14. S. D. Shapiro, Feature space transforms for curve detection, *Pattern Recognition* **10**, 1978, 129–143.
15. P. H. Swain and S. M. Davis, *"Remote Sensing: the Quantitative Approach,"* McGraw Hill, NY, 1978.
16. J. S. Weszka, A survey of threshold selection techniques, *Computer Graphics Image Processing* **7**, 1978, 259–265.

8 Image Processing Systems

John R. Adams
Edward C. Driscoll, Jr.
Cliff Reader

International Imaging Systems
Milpitas, California

I. Introduction

In the last ten years, digital image processing systems have seen steady advances in sophistication and orders of magnitude improvement in throughput. Many functions now operate at real-time rates that took hours

DIGITAL IMAGE PROCESSING TECHNIQUES

in earlier systems. Through the 1970s, image processing systems were used mostly for research into applications of digital image processing. However, several factors are now combining to produce a mature state of the art for production systems. We are experiencing a revolution in systems design and an explosive increase in the application of digital image processing to real-world problems.

This chapter is mostly devoted to the type of image processing system that is commercially available today. Such systems are designed for the analysis of image data. The text does not cover graphics data systems nor systems whose architecture is currently being developed. For example, artificial intelligence machines are being researched, and their progress has been chronicled elsewhere. Special systems such as compressors – expanders for image transmission are also excluded; however, the analysis of images forms the basis for most image processing that is being performed today.

The principal reason for special attention to image processing system design is that the volume of imagery data is incompatible with existing computer technology. One cannot afford enough memory and computer power to address the problem completely. If one uses conventional computer architectures for image processing, the system will be either I/O or compute bound for much of the time, depending on the algorithm being executed. Another reason is that the science of image processing is immature. Ideally, an image processing display should input image data and automatically output statements relative to the task at hand. This has been accomplished for only a few, simple, well-defined problems. For other problems, the human operator analyzes the data, and the image processor functions only as an interactive tool to maximize the efficiency and accuracy of the operator. This stresses the system design in another way, because it requires the real-time refresh of a display and, ideally, a real-time response, so that the hand-to-eye coordination of the human can be exploited. This need for speed is being met by special purpose hardware but has caused another constraint on performance: the system can become control bound. This problem has been further complicated by the recent introduction of the first multiuser systems.

We shall describe the hardware and software in some detail, because the performance of imagery systems can be strongly affected by constraints buried in the hardware or within the software operating system and also by subtleties of the design. By highlighting these issues, we hope to rationalize the sometimes uneven functional behavior of the systems and to guide newcomers around the pitfalls of building or buying their first system. In this first section we shall survey the history and the present state of the art in image processing. The next section discusses the distinct nature of image processing and the requirements of different application areas. The following sections then address the overall system architecture, image processing display hardware, and system software. We shall summarize with a discussion of some fundamental issues in image processing.

During the 1950s and 1960s a number of attempts were made to build analog image processing systems. Such systems were characterized by very large size, limited functional capability, poor image quality due to noise and temporal instability, and low reliability due in part to the use of mechanical devices such as drum memory; but these systems served to whet the appetite for interactive manipulation of images. Image processing really got started in the late 1960s with the development of economical mainframe computers and minicomputers. It was then within the financial grasp of research laboratories to establish a stand-alone minicomputer-based image processing facility and to gain access to bulk processing power. With the exception of special hardware for digitizing and printing film, the systems used standard hardware, and all the image processing capability was provided by software.

An early example of image processing software was the VICAR system [8], developed at the Jet Propulsion Laboratory in the late 1960s. It typified the mainframe, batch-oriented approach that was prevalent at the time. Images were viewed on primitive displays after a sequence of blind processing operations. Hardcopy image outputs typically were time exposure photographs of the CRT screen. These systems were either expensive commercial products, such as those from Link Singer and Information International, or one-of-a-kind laboratory devices. Later, a number of groups, such as the USC Image Processing Laboratory, gained some local processing power for the display by attaching it to a stand-alone minicomputer. After bulk processing, which was still performed on a batch-oriented mainframe, data were passed by means of tape to the minicomputer-based display system where it could be viewed and cosmetically enhanced.

The system architecture of the early 1970s established a model that is still prevalent today. At its heart is a conventional minicomputer with a variety of standard and specialized peripherals. Special-purpose peripherals provide the functions of digital image acquisition, interactive processing and display, and hardcopy (film) or videotape image output. Because such a system is not suited to intensive, high-precision computation, it is frequently augmented by the bulk computing power of a mainframe computer or, more recently, an array processor. Such systems have provided an excellent environment for research on the applicability of digital image processing to real-world problems. The limitations of these systems revolve around the large volume of image data and the incompatibility between the serial (one-dimensional) nature of the hardware and software and the two-dimensional nature of most image processing algorithms. Both single-function processing times and overall system throughput rates are incompatible with the requirements for production image processing systems.

The earliest commercial display system was the *Aerojet,* later to become the *Comtal.* It achieved real-time color refresh using disk memory; and from the outset it had the capability to perform radiometric manipulation, merging graphic overlays with image data, and interactively denoting points in the image by means of a trackball and cursor. The designers of these early displays were chiefly concerned with the then very difficult task of real-time color refresh. The displays were functionally limited to presenting the results of image processing experiments and global intensity or color manipulation. Though the designs were characterized by a simple, unidirectional flow of data over hardwired data paths, they were received enthusiastically, particularly because they allowed immediate display of processing results.

In the mid-1970s, the first, truly interactive processing systems with extensive real-time software were developed. For example, VICAR was transformed into Mini-VICAR, General Electric introduced the GE 100 for interactive processing of Landsat images, and Electromagnetic Systems Laboratory transformed their PECOS batch system into the interactive IDIMS system. While the design of the overall image processing system was evolving, image displays also advanced. A number of design innovations began a trend toward establishing the display as the central element of the image processing system. These were semiconductor random-access memory (RAM) for refresh memory, image addition and subtraction at video rates, hardware histogram computation, four-way split screen, hardware zoom, and the incorporation of a microprocessor for the control of interactive processing.

In 1976, International Imaging Systems introduced the first image processing display with a video rate feedback loop, the Model 70 [3]. It allowed the user to capture the video date after processing and to store it in another refresh memory. The image display was now capable of performing iterative processing tasks. Within a few years the feedback loop was common in a number of display systems, and many included arithmetic logic units (ALUs) in the processing path [21,26]. Today, feedback loops include increased precision (32 bits in the DeAnza IP8500) and video rate integer multiply (Vicom, Ikonas, and DeAnza). Designers of the evolving display architectures tackled both the I/O and the computational requirements of image processing [1]. The large volume of refresh memory was recognized to be an invaluable asset, and declining memory cost enabled multiple memories to be incorporated. Modular design provided the flexibility to use this memory for processes other than refreshing the screen. Feedback loop processing reflected a basic characteristic of many image processing algorithms: a sequence of simple operations that can be iteratively applied to every one of a large number of pixels.

Implementations of image processing displays that developed during the late 1970s and early 1980s make use of (i) special-purpose options for performing specific operations such as convolution and geometric warping at video rates and (ii) parallel architectures and feedback processing for implementing algorithms by iteration. Convolution and geometric correction are the most prominent examples of the use of special-purpose architectures to solve specific image processing problems. Video rate convolution was first introduced by Comtal in 1978 using a look-up table implementation. With the availability of video rate multipliers, the 3×3 convolution was implemented using a true multiply operation that increased the processing precision over the look-up table implementation (Vicom, 1981). The use of parallel architectures, such as DeAnza, International Imaging Systems, and Spatial Data, stresses the use of iterative processing using general-purpose components that solve image processing problems in nearreal time (usually less than 3 sec.). The parallel approach has been shown to be cost effective for a variety of applications, but it lacks the performance of the special-purpose architectures. Geometric processing, which is key to a large number of applications, was introduced in the early 1980s by Comtal and Vicom. At present these options are expensive and not widely used.

II. Current Context of Image Processing

In this section we shall discuss the characteristics of image processing that influence the way in which a system is best implemented and the constraints put on implementation by physical laws and current technology. Work in this field involves the transformation of input data by a library of algorithms to produce the desired results for a given application. We shall discuss the characteristics of image data and image processing algorithms and the requirements of various applications. Throughout this discussion we shall encounter the practical limitations of existing technology, and whenever possible we shall show the effects of these limitations on the realization of optimum data structures, algorithms, and processing scenarios.

A. Characteristics of Image Data

A set of image data is a discrete representation of a continuous field. Regular sampling of a continuous field, without regard to redundancy, produces a vast volume of data to store and handle. It would be ideal to have all these data in fast, easily accessible, cheap memory, but this is not practical. Consequently, image processing systems must resort to a memory hierarchy, in which relatively small amounts of data are stored in fast, accessible,

Device type	Transfer rate (Kbytes/sec.)	Online capacity (Mbytes)	Media cost ($/Mbyte)	Limitations
Tape	40–500	40–9000	2–5	Sequential access only
Disk	250–1500	5–600	7–25	Random line access only
Refresh Memory	1000–10000	1–16	200–1000	Random pixel access

Fig. 1.   Memory hierarchy characteristics.

comparatively expensive memory (RAM) and are supplemented by larger volumes of data stored in moderate-speed, limited-access, moderate-cost secondary memory (disk) and slow, sequentially accessed, low-cost tertiary memory (tape).   The characteristics of this hierarchy are summarized in Fig. 1.

The multidimensional nature of image data conflicts with serial (one-dimensional) devices, such as disks, and the inherently serial nature of conventional computer hardware and software.   The hardware is, therefore, particularly limited in its ability to perform spatial or geometric transformations.   Such algorithms use multidimensional coordinates to address image data, and the expense of mapping such coordinates onto a one-dimensional coordinate system can be substantial.   The geometric transformation problem is further compounded by the need to resample the continuous field of image data.   Unlike a transformation of graphics data, which means only the moving of discrete components to new locations, a transformation of image data requires the interpolation, into a new array, of pixels from the old.

Fortunately, image data usually have low precision, and this limited precision or dynamic range can be exploited to simplify implemention.   Although during intermediate steps of a processing scenario the dynamic range and sign of the data may vary, they can be calculated and bounded at each step, permitting the use of a fixed-point processor with intermediate scaling of the data.   Relatively few image processing algorithms need floating-point computation, although it is often used when algorithms are implemented with a conventional computer.

B. Characteristics of Image Processing Algorithms

The characteristic lack of complexity of most image processing algorithms must be viewed in the light of the fact that they may be applied to millions of data points.   In most cases, a long sequence of simple steps must be executed to output a single pixel. Conceptually, this sequence of steps must be repeated a large number of times to process an entire image.† The complexity of an algorithm determines the complexity of the system that is to solve it. The volume of data requires a cost–performance tradeoff between the extremes of a single processor that iteratively processes all the data points serially and a massively parallel system that processes all the data points simultaneously. In practice the optimum balance is sought, and these issues interact to make the system more complex, notably in system control.

In later discussions, algorithms will be considered to be one of two types, point processing or spatial processing. Point processing algorithms refer to operations that can be performed on each pixel of an image without knowledge of adjacent or nearby pixel values. As display technology has developed, operations such as image combination and image statistics have been included within the category of point processing, because they are performed on a pixel-by-pixel basis. On the other hand, in spatial processing, the output intensity is a function of the intensities of the input pixel and its neighboring pixels. The definition of the neighborhood varies with the algorithm. Geometric processing actually represents a third category, because it involves address transformation followed by image resampling; but because resampling is a spatial process, geometric processing will be considered to be a spatial process in the following discussions. Spatial processes may be divided into two subclasses: Algorithms that are based on *signal processing* execute in a predetermined number of steps, whereas those based on *image understanding* execute in a number of steps that depend on the specific problem and data.

The inherent dimensionality of image processing algorithms strongly affects the cost and performance of a system design. Point processes can be easily implemented in one-dimensional systems using simple I/O, but spatial processes require more sophistication. Fortunately, the data-access requirements of the algorithms possess structure that can be used to design efficient and cost-effective systems. The possible computational parallelisms of an algorithm are defined in Fig. 2. Note that a point algorithm can be expressed in a kernel or image form, although perhaps less efficiently, but a kernel algorithm cannot be reduced to a point form. In other words, these parallelisms are shown in an ascending order of complexity. Within the

† In fact, some image processing display architecture implement algorithms by effectively turning them inside-out. Rather than performing all the steps of an algorithm sequentially on each pixel in an image, recent implementations perform each step of an algorithm on all the pixels in the image before moving to the next step. This approach has proved successful on deterministic algorithms running in pipelined image processing displays with a memory capacity that can store interim results [12,13,33].

1. Pixel parallel $(N \times N)$—Each pixel in the image has its own processor.
2. Kernel parallel $(n \times n)$—A limited size subset of pixels will be processed by a single processor to generate each output point.
3. Image parallel (l)—An entire image will be sequentially processed by a single processing element or pipeline of elements.

Fig. 2. Definitions of parallelism.

definitions of parallelism, we have a hierarchy of useful I/O structures for various algorithms, as shown in Fig. 3. Again, these are ordered by complexity, and a simple I/O structure can be shown to be a simplified case of a more complex structure. Also, some algorithms can be reduced to combinations of simpler operators. For example, when an algorithm is spatially separable, the computation can be accomplished by two one-dimensional operations rather than by one two-dimensional operation.

C. Requirements of Various Applications

The diverse applications of image processing require operational scenarios that differ from each other in a number of respects. Most scenarios will be composed of a concatenation of individual operations, successfully

1. Point (1 pixel)—Single pixels must be successively accessed and they may be ordered as a raster scan. This I/O structure is used for radiometric operators and for image combination.
2. Raster Scan $(n \times 1$ pixels)—Successive lines of data must be accessed. This I/O structure is very prevalent for output data and many algorithms are implemented so that this is the input data format. An example of an algorithm that naturally requires this structure is DPCM image coding.
3. Sliding Block $(n \times n$ pixels)—A window must be successively centered on every pixel in the input image. This is the I/O structure for the large class of algorithms that are based on spatial convolution.
4. Orthogonal Stepping Block $(n \times n$ pixels)—A window must be successively located on adjacent $n \times n$ regions of the input image. The definition may be generalized to include limited overlap $(<< n)$ of the regions. This is the I/O structure for image transformations such as Fourier or cosine.
5. Random Sliding Block $(n \times n$ pixels)—A window must be successively located on adjacent pixels in the input image with a scan direction that is not orthogonal with the image axes and with sufficient variation in location to defy description by a simple structure. This is the I/O structure required by many geometric manipulations.
6. Blob $(n$ pixels)—An arbitrarily shaped region of pixels must be used. The set of all such blobs will not necessarily exhaust all the pixels in the input image. The blob may be interactively defined, or it may have resulted from a previous data adaptive process. This is an I/O structure for image understanding algorithms.
7. Random $(n$ pixels)—Scattered data are required from the entire image space. An example is ground truth for a supervised classifier.

Fig. 3. Definition of I/O structure.

reducing or transforming the input data to produce an output result. Within each step of a scenario, a particular application will have requirements that stress system performance to differing degrees. For example, noninteractive algorithms can often be executed as background processes, in which speed of response is not critical, because the user will be occupied by other tasks in the scenario. Similarly, the automatic process can be overlapped with other operations and is not time critical unless it is the gating item in a scenario.

Interactive algorithms can be the most demanding of system performance, because rapid feedback to the user is essential. We can define three time intervals of interest to image processing:

1. Real time. When the user operates a control, the displayed image changes smoothly, with no hysteresis. This fully exploits human hand-to-eye coordination for the fastest optimization of a process parameter.

2. Near-real time. The image changes within 3 sec in response to user input. This forces conscious comparison, which is less effective than the subconscious human feedback of real-time response but is fast enough to avoid user frustration or boredom. A class of function exists for which response times of more than 3 sec can be considered near-real time, because they represent a 10–100-fold improvement over processing times obtainable in a traditional sequential processor. Examples of these functions include two-dimensional Tukey median filtering [13,33], unsupervised classification (clustering) [12,13], minimum distance classification [12,13], spatial convolution, and geometric rubber-sheeting.

3. Nonreal time. These are processes that cannot be implemented to execute in 3 sec or do not represent a 10–100-fold performance improvement over sequential processing techniques. Typically, these processes are relegated to a batch mode.

At each intermediate stage of a scenario, the format of the data can change. Precision and dynamic range requirements will usually necessitate an increase of word length from that of the input pixels. Ancillary files and process parameters can be generated at one point in the scenario and then carried with the data for use by a later process. Such data can be very numerous and in extreme cases can even outweigh the volume of image data.

To characterize the performance requirements for a processing scenario, two terms are necessary. For a single process within the scenario, the response time is a key design parameter; for the whole scenario, the throughput is the important metric. These will be governed by the volume of data to be processed and the concatenated time of all the process response times. Each application of image processing has a different mix of these requirements. In Fig. 4 we have summarized the importance of different

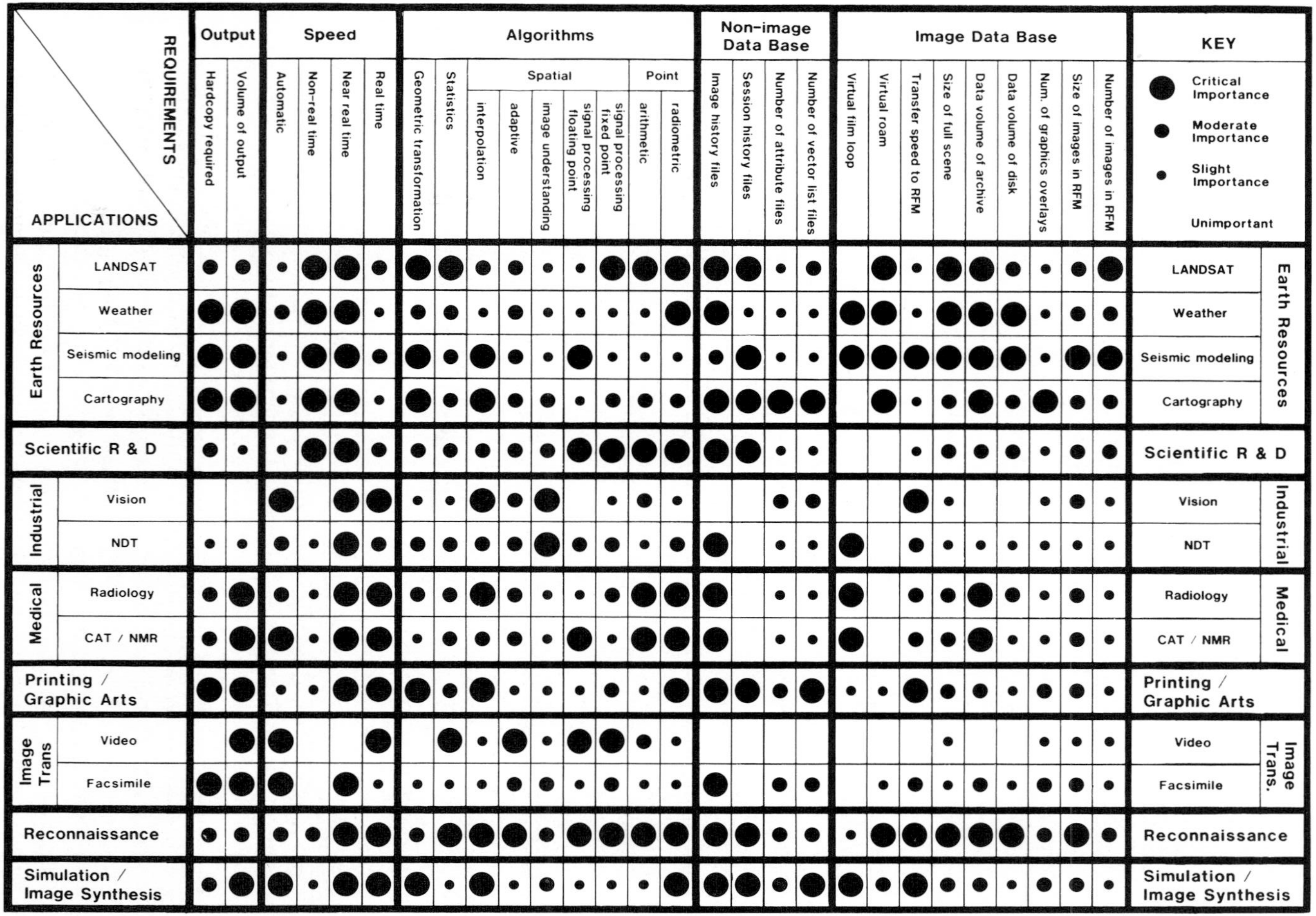

Fig. 4. Requirements for various image processing applications.

KEY
- ● Critical Importance
- ● Moderate Importance
- • Slight Importance
- Unimportant

Applications: Earth Resources (LANDSAT, Weather, Seismic modeling, Cartography) · Scientific R & D · Industrial (Vision, NDT) · Medical (Radiology, CAT / NMR) · Printing / Graphic Arts · Image Trans. (Video, Facsimile) · Reconnaissance · Simulation / Image Synthesis

REQUIREMENTS

Image Data Base
- Number of images in RFM
- Size of images in RFM
- Num. of graphics overlays
- Data volume of disk
- Data volume of archive
- Size of full scene
- Transfer speed to RFM
- Virtual roam
- Virtual film loop

Non-image Data Base
- Number of vector list files
- Number of attribute files
- Session history files
- Image history files

Algorithms
- Point: radiometric
- Point: arithmetic
- Spatial: signal processing fixed point
- Spatial: signal processing floating point
- Spatial: image understanding
- Spatial: adaptive
- Spatial: interpolation
- Statistics
- Geometric transformation

Speed
- Real time
- Near real time
- Non-real time
- Automatic

Output
- Volume of output
- Hardcopy required

algorithms or data characteristics when arrayed against a variety of image processing applications. From this matrix it can be seen that each application area requires a different mix of inputs and algorithms. In succeeding sections we shall describe the components and architectures that are available for meeting these applications needs.

III. System Hardware Architecture

An image processing system is a collection of hardware that is designed to process two-dimensional data digitally. Architectures are continually being presented for the arrangement of hardware components for image processing applications. In many cases, the optimization of an architecture is a matter of common sense, and image processing systems are optimized by means of the same reasoning as for other applications. In other words, when defining a system architecture, one should concentrate on the generic demands of image processing, such as high data volumes, two-dimensional structure, and (often) low precision. Conventional systems design can be used to address these demands within the broad spectrum of computer industry techniques. We do not have the space in this chapter to detail these techniques, so we shall only survey the tradeoffs associated with different architectures and the limitations presented by even the best architecture in an image processing environment. We shall break this discussion into two categories: those architectural issues that affect data transfer speed and those that contribute to processing speed.

A. Data Transfer

Data transfer concerns the movement of data through the classic memory hierarchy. The high demands placed on data transfer by the volume of image data demand a high-performance implementation of hardware components. At present, the data base is almost always controlled from the host CPU. For a single transfer within the memory hierarchy, the hardware components are well matched for speed; however, in a multitasking or multiuser environment, contention can be a severe problem.

Recently, specialized system architectures have dealt with these problems by use of multiporting devices such as disks and by use of multiple buses [23]. Although multiported devices, open up the potential for high throughput rates, they present nontrivial control problems. If the host is not controlling all data transfers from device to device, it is unable to control device availability. Thus, high speed by means of a direct device-to-device connection can adversely affect the performance of other areas of the system. The implementation of dual-porting also does not eliminate contention problems. If a disk and an array processor are actively interchanging large volumes of data in tight loops, can a host computer intersperse random disk accesses so it can go about its other business? The system designer must also watch for bottlenecks. Configuring multiple buses can merely shift contention problems back to the hardware that interconnects the buses. As a result, specialized system architectures that make use of multiporting and multibusing are usually appropriate only where the throughput requirements of a system are high and clearly defined, such as in a production-oriented environment.

The current limitation in image display technology to 512×512 or 1024×1024 resolution requires the system designer to provide for the review of larger images. A capability that allows the user to roam through an image larger than the display format has been implemented by a number of vendors. To obtain satisfactory roam speed while minimizing the impact on the overall system performance, it is desirable to provide the ability to transfer data directly from the image disk across the bus to the image display. Implementation of this capability requires special consideration in the design of the disk and display interface as well as modification of the device handler for both devices. The parallel transfer disk (PTD) has been used to provide a rapid update capability for image display systems. The PTD allows conventional access by the CPU while allowing the display to read out data using parallel heads to obtain rapid data transfer to the image display. The evolution of applications that require the review of large three-dimensional data bases, such as those used in seismic processing or nuclear magnetic resonance imaging, will increase the need for rapid transfer of image data.

B. Data Processing

The display has been given an increasing share of the overall computational load. The processing capability of today's image processor – display provides a true fixed-point array processing capability for image data. Although its capabilities are still largely restricted to the processing of limited windows of data, and it has limited computational precision, we are beginning to see implementations in which processing displays are used for their processing capability [1,13].

It may not be an efficient use of system resources to allow the display to iterate through the processing of a large image or to execute algorithms whose complexity would preoccupy the display for extended periods. It has been commonplace to allocate such processing to an array processor that operates in a background mode [32]. Array processors have tended to be oriented to one-dimensional signal processing problems and consequently have lacked the intrinsic ability to perform two-dimensional data access. This has restricted efficient processing to those functions that are separable. Fortunately, many useful algorithms possess this attribute. If double buffering of adequate size is used in the internal memory of the array processor, then the ideal balance can be struck, and the system will not be compute or I/O bound [18]. Another problem has been the complexity of efficient software coding. It has usually been necessary to program in low-level language, frequently with intimate knowledge of the array processor hardware.

The competing demands on processor time include high-level complex computation and lower-level interrupt servicing, resource management, and system supervision. It is advantageous to allocate only the required processing power to these tasks. The many ways in which this issue can be addressed have been referred to as *distributed processing.* In short, a mainframe or supermini can accomplish more if lower-level demands on its time are handled by local, unsophisticated microprocessors. The concepts of distributed processing and network architectures have been widely discussed in the literature [10], and their advantages clearly apply to image processing as well.

IV. Image Processing Display Hardware

The development of image processing displays over the last decade has been extensively reported [4,5,11,29]. This discussion summarizes the current state of the art and gives an overview of this complex subject in two major subsections. First, the major components of a typical image processing display will be individually described, and second, we shall discuss typical hardware architectures that tie these components together.

A. Functional Components of Display Subsystems

Figure 5 shows the functional components of a sophisticated image processing display system. These components are segregated on functional grounds, although in many cases such segregation also reflects a typical breakdown into physical circuit board assemblies. We shall isolate each component and indicate its purpose, the ranges of size and performance

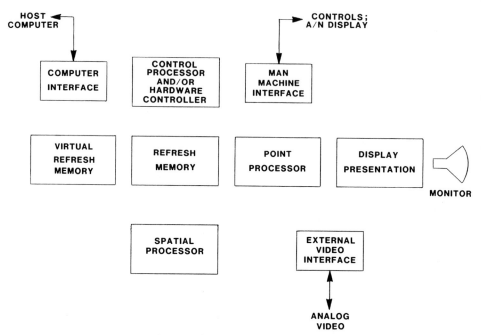

Fig. 5.   Functional components of an image processing display.

available, the implementation alternatives that are available, the tradeoffs involved, and the immediate effects these implementations may have on neighboring components.

i. Refresh Memory

Refresh memory rightfully belongs at the center of the image display system, because its design is critical to almost all the features of the display. In modern displays, refresh memory both refreshes the screen and provides memory for processing, including the capability to zoom and scroll the displayed image, to create split-screen presentations, and to roam the image on the screen over a larger image area.   These functions are nondestructive of the data in memory.   In most systems, refresh memory is designed in channels.   A typical channel can hold a single $512 \times 512 \times 8$ bit image and can thus refresh a monochrome image on the screen.   Three channels are used in parallel to display a color image.   Systems are available that provide between 1 and 64 channels.   There are systems with memories that are 1024×1024 pixels in size and others that are up to 16 bits in precision.

The addressing and control for refresh memory provide for the independent use of each channel within the system.   Typically, the channels can be programmatically associated to hold spatially larger areas of data and to hold radiometrically higher-precision data.   This allows display manipulations such as roam and sophisticated video rate processing of multiple images, with greater than 8-bit precision if necessary.   Recently, some systems have been designed that have a single, large memory.   They offer the advantage that images with sizes other than the size of a channel can be stored.   Internal design constraints, however, still restrict image size to a multiple of some fundamental size (typically eight pixels) and limit the flexibility of multiple simultaneous accesses.

Binary-valued graphics data are stored in a channel of memory that is physically similar to image memory.   In contrast to displays designed exclusively for graphics, these displays often have restrictions on independent access to single graphics planes, particularly in that they cannot be independently scrolled or zoomed for display.

Refresh memory is dual-ported, which allows its use for two operations simultaneously.   This provides the host computer with access that is independent of video refresh operations and permits use of the memory for intermediate storage during iterative feedback loop processing.   Control and addressing and handled differently for each port.   The control and addressing of the video port must meet the stringent requirements of the video refresh timing, but the control and addressing of the system port can range between the rigid video standards and random pixel addressing, depending on the operation being performed.   When the memory is addressed randomly, the access speed drops, due to internal design constraints.

Memory is addressed by a simple raster sequence for refresh of the screen.   This sequence can be trivially modified on a single-channel basis to provide integer zoom and scroll capability.   Zoom is provided by dividing the address clock and outputting the same pixel more than once.   Scroll is effected by adding an offset to the address that redefines the origin of the data on the screen.   Current systems allow some flexibility in defining the wraparound conditions that occur when scrolling an image, for instance, wraparound with black rather than image or wraparound with data from an adjacent memory.   Split-screen effects are accomplished by switching between memory channels during the refresh of the screen.

Refresh memory has usually been accessed by the host computer in one of two modes: first, by means of an interface that allows the programmer to access a random or rectilinear volume of data with a DMA transfer; second, by mapping a rectangular subregion of the channel into the host computer address space, so that it can be addressed directly as if it were main memory. Recent devices have allowed extension of the latter technique to include the entire refresh memory in the host computer address space. When binary graphics data are stored in an image channel, a bit mask capability is provided to allow access to single bit planes.

Since the mid-1970s, refresh memory has been constructed from semiconductor RAM. This technology replaced CCD shift registers and has become increasingly cost effective, especially because it allows auxiliary use of the memory for processing. Trends in RAM technology were summarized in a 1981 paper [26], in which it was shown that the technology has advanced by generation along with the capacity of the chip (i.e., 4K bits, 16K bits, 64Kbits). Performance curves shows that there has been a steady improvement in characteristics during the lifespan of each generation and a quantum improvement with each new generation. The curves also show an inverse correlation between the time required for access and the power consumption, error rate, and cost. Consequently, refresh memories have been constructed from relatively slow devices. It is typical to use a memory chip cycle time of 400 nsec, far slower than the rate required for video refresh. To obtain the data rates necessary for video refresh, memory systems are pixel multiplexed, with a factor of eight typically being used for the 512×512 display format. In addition, most systems multiplex each bit of intensity resolution into a different chip. Therefore, in a typical design, 64 RAM chips are accessed during a single memory cycle. The 8 bits/pixel are output on parallel lines, and the eight pixels are serialized for transmission to the rest of the system.

It may now be apparent that refresh memory is structured and not wholly random. Dual-porting is supported, because the time to refresh 8 pixels is 800 nsec, but the memory provides them in its 400-nsec cycle time. Therefore, alternate cycles are available for nonrefresh activity. The zoom and scroll capabilities interact with the memory multiplexing, and thus special additional hardware is required to obtain smooth control in both the horizontal and vertical directions. This is normally accomplished by outputting eight pixels into high-speed registers, and the required subset of them is then output to the system. Cycles are stolen during video retrace periods to refresh the dynamic RAM to assure data retention. The basically simple design of these memories and a conservatism with respect to operating the RAM chips well within their ratings makes refresh memory a very reliable resource. It is easy to diagnose faults, because diagnostics can be written to identify the individual chip location.

Most of the limitations of refresh memories result from the use of multiplexing. Due to the cost of the hardware involved, some display systems have not provided all of the integer values of zoom ratio. The binary values of 2:1, 4:1, and 8:1 are much easier to implement. The multiplexing also constrains column access, because the eight pixels that are simultaneously accessed are row (raster) oriented. Column access is eight times slower as a result. Because of the fixed structure of the RAM chip and the multiplexed architecture, channels are constrained to be specific sizes (invariably binary). This can cause operational limitations that software can only mitigate. In this context, note that in most image systems, the graphics overlays are stored in a refresh memory channel. Thus there is no ability to manipulate individual overlays simultaneously and independently. The design of most memories prohibits the "minification" of images for display. The refresh rate is designed to access, say, a 512×512 pixel region in a frame time. Although for a 1:2 minification of a 1024×1024 pixel image it would be possible to access only every other line in the column direction, it is beyond the speed of the memory to access 1024 points in the row direction and subsample them for display. Furthermore, the inherently orthogonal structure of the memory and the internal multiplexing of pixels prevent video rate access to the data other than in a row or column raster scan; it is not possible to display a rotated version of the image, for example. The discrete nature of the data prevents scrolling by a fraction of a pixel or zooming by an arbitrary factor. The preceding limitations make digital displays less flexible than film-based image analysis systems. Overcoming them requires the spatial operation of interpolation, which will be discussed later.

ii. Virtual Refresh Memory

Virtual refresh memory, a recent addition to display capability, provides a locally integrated magnetic disk to expand the capacity of refresh memory. This is an important addition to the classical memory hierarchy, particularly if its performance in exchange data with refresh memory is optimized. A key functional capability is virtual roam. The usual ability to roam the display screen over a wider area of imagery in refresh memory is extended by roaming the contents of refresh memory over a much larger image on disk. Performance ranges from (sustained) rates of 100–1000 pixels/sec depending on the degree of special purpose hardware involved. An alternate use of the capability is virtual film loop. A slow motion review of data frames occurs as fast as the disk can load them. This is typically up to six 256×256 frames/sec for an implementation that is not hardware enhanced. However, a combination of special disk to refresh memory interfacing and image compression for the data on the disk can produce a real-time capability (i.e., 30, 512×512 frames/sec). The capacity of virtual refresh memory may range up to 600 megabytes.

This capability has been implemented with conventional magnetic disks, Winchester disks, and parallel-transfer disks. The difference between the first two is a matter of capacity and cost. Parallel-transfer disks are used for speed and must be interfaced directly to refresh memory. Although this would be the ideal interface for any disk, the high cost of providing exotic hardware and software has led designers to use conventional techniques, especially the microcomputer bus, which is common in neoteric displays. Nevertheless, to achieve a reasonable performance some special integration is necessary, because ordinary computer disk operation is not set up for the continuous transmission of large files. In fact, in a conventional environment, interruptions will follow from a number of sources including noncontiguous file formatting, the need to seize the computer bus, disk rotational latency, and head seeks. The removal of these impediments is a matter for both hardware and software. The image files can be formatted to be contiguous and to be written on the disk so that track-to-track and cylinder-to-cylinder seeks are minimized. Roaming applications favor block-oriented data formats on the disk. To improve performance further, a special purpose disk controller is often used. This can provide local control to ensure that the computer bus is not released during transfer and that there is enough buffering to mask the latency and seek times.

It should be noted that in today's implementations the user is allowed to view images larger than available refresh memory only when point operators are applied to the visible window. More complex operators, such as iterative convolution, classification, and median filtering, are not always possible in real time on the window and are presently impossible on the data in virtual memory.

iii. Point Processing

Point processing can be subdivided into three categories: radiometric manipulation, arithmetic and Boolean image combination, and statistics. Radiometric manipulation can modify image grey levels by means of linear and nonlinear (including random) functions, within the precision ranges of the hardware. This is the basis for many of the most frequently used image processing operations, including scaling. Scaling is an important enhancement operation in its own right and also may be required at interim points during more complex processing to maintain precision. Image combination can involve multiple images or images and constants, although the latter can also be accomplished by radiometric manipulation. Finally, typical hardware assistance for the compilation of image statistics includes the detection of maximum and minimum pixel values within an image area and the collection of histograms and statistical moments. The use of the hardware statistics generation allows the processing display to perform many processing algorithms in near-real time that would not be possible otherwise; for example, the generation of a two-dimensional histogram of an 8-bit image can be completed in less than 20 sec. This capability also speeds recursive, statistically based processes such as classification. It should be possible to execute these functions, over arbitrary subregions of the image.

Radiometric mappings are implemented using look-up tables (LUTs), which are interposed in the path between refresh memory and the monitor. The LUT is implemented with fast RAM (less than 100 nsec access), using the pixel intensity as an address into the LUTs from which the output value is obtained. This event occurs at video rates, allowing the pixel gray values that are stored in refresh memory to be transformed to new values on the screen according to a function that is loaded into the LUT from the control processor. For example, the grey-scale range may be inverted, or perhaps the radiometric range may be modified by a logarithmic function to display a greater contrast in the darker end of the range. In some implementations for 1024×1024 displays or 60-Hz noninterlaced displays, the designer has chosen to interleave two LUTs, each processing alternate pixels. This provides the required speed without using chips, which require excessive power. In early systems, eight 256-bit RAMs were used, providing an 8-bit-in, 8-bit-out capability. In current (8 bits/pixel) systems, 1024-bit RAMs are frequently used. The extra two bits of input address are used to select among four 256-level transforms. These two control bits can be set programmatically for the whole image or can be set using a feature called *region of interest* (ROI). In this technique, graphics planes are used to store the control bits on a pixel-by-pixel basis. The graphics planes are then used concurrently with the image data. Establishing the ROI bits may have been accomplished by having the user define them interactively with a cursor, or they may have resulted from an algorithm such as classification.

The cost of LUTs to process 16-bit data has been prohibitive, so a hybrid approach of arithmetically reducing the dynamic range to some extent before using LUTs has been taken. In the DeAnza IP1172 display, which has 16-bit refresh memories, a 16-bit threshold value can first be subtracted, and then any contiguous 12 bits can be input to a LUT. In the 64K grey levels of data, therefore, it is possible to select any subset of 4K grey levels or any subset of 8K levels, quantized 2:1, and so on.

In most applications, changes to the video paths must be made during retrace periods to avoid noise patterns on the screen. Some systems attack this problem by duplicating all registers and tables. Asynchronous loading can be followed by synchronous switching to the updated register or table during the vertical interval. The objection is cost, especially because the logic is expensive. Another approach is to restrict all transfers that can affect video display to blanking intervals or, preferably, to provide a programmable control bit for the restriction of a given transfer to blanking intervals. The loading of LUTs for high-precision data presents problems due to the volume of data involved. A 4096-level LUT (for 12-bit data) requires a DMA rate of approximately 2 megawords/sec to write it within the approximately 1 msec duration of a single vertical interval. Some systems mitigate this problem by also making use of the horizontal blanking interval. It should be noted, however, that although this does not produce visible noise on the screen, it does change the radiometric mapping as the image is refreshed, which can introduce temporary visual artifacts. An alternative implementation, which provides the benefits of both of the above approaches, is to load a front and FIFO buffer and control the output by a vertical interval bit. This buffer can be built to handle low-speed input from the host and high-speed output to display subunits, thereby allowing complete tables to be loaded during the vertical interval.

Point-by-point image arithmetic is the other important form of point processing. This is conveniently implemented using arithmetic logic units (ALUs) that can perform a dozen or so arithmetic and Boolean algebra functions at real-time video rates. Eight- or 16-bit capability can be provided, plus a carry or overflow bit. The introduction of the TRW multiplier accumulator chip (MAC) has made true multiplication possible at video rates. Earlier, a quantized capability was implemented using LUTs. Multiplication of an image frame by a constant value (as required for spatial convolution) can be performed with a scaled LUT. In systems that have concatenated LUTs and adders or ALUs, multiplication has been implemented by use of a logarithm table in the first LUT, followed by an add or subtract, and then an antilogarithm in a second LUT. This approach provides multiplication suitable for simple image display applications but is not usable for true arithmetic processing. It was noted earlier that rescaling of the data is a required capability. During a concatenated set of operations, adds or accumulates expand the data dynamic range. Rescaling can be performed by bit shifters to exchange dynamic range with precision.

With the introduction of processing capability in image displays, the need for image statistics such as the minimum or maximum value and the intensity distribution (histogram) became immediately apparent. Several display systems can provide statistical information about the image that is being processed by the point processor. The min – max and image histograms are typically computed in one or two video frame times. When these are used in conjunction with ROI graphics planes, it is possible to obtain image statistics from arbitrary subregions of the display. For applications that perform image combination, image convolution, Tukey median filtering, and soon, the existence of image statistics is mandatory.

Restrictions in the point processor include the limited precision–dynamic range of the necessarily fixed-point arithmetic. The real-time capability may not extend to 2's-complement arithmetic, which commonly takes two frame times instead of one. Currently, there is no real-time pixel divide capability other than by the quantized methods mentioned.

iv. Spatial Processing

Spatial processing was introduced in Section II.B. The subdivision of functions into classes is relevant to the hardware design constraints for video rate processing. A study of the signal-processing-based class of algorithms [25] has revealed that they can all be expressed with spatial convolution or superposition as their kernel operation. Examples include high-pass and low-pass filtering, Sobel edge detection, and Wallis adaptive filtering. Many of these functions can be implemented using a small kernel, often only 3×3 in size, and many larger kernels can be decomposed to a sequence of small operators. Some systems execute a 3×3 kernel operation in real time. Thus, by the definition in Section II.B, these systems use a kernel parallel approach to implementation. For reasons of cost, image processing displays more often execute spatial functions iteratively, processing one kernel point per frame time and thus achieving a near real-time response. Designers and users of contemporary processing displays have learned that algorithms can be decomposed in unconventional ways to expose underlying structures that can be mimicked by low-cost, video rate hardware.

The other class of functions, image understanding algorithms, is not so well established as that based on signal processing. Considerable effort is being expended within the research community to develop special purpose array processors [14,24]. Functionally, these devices perform neighborhood operations on an adaptive inter-pixel basis and are usually limited to binary operations of a Boolean-algebra or conditional-testing nature. They can also be characterized by their ability to exchange value and positional data. The outcome of an operation may not be a value, but rather the spatial coordinates of points at which a specified condition was satisfied. The architectures tend to be based on pixel parallelism [15], with experimental implementations that take this to an economically reasonable level, somewhere in size between the kernel operators and a full frame.

Direct implementation of spatial processing capability has been infrequent until now due to cost. For the systems that have implemented a real-time 3×3 convolver, various techniques have been used. The Comtal Vision One/20 uses an array of LUTs to perform the multiplication, whereas the Vicom uses the TRW MACs. Both systems feed the array with three parallel data paths that contain the three lines of data required during the process. A more elegant alternative allows the use of only one line of data input by providing a "snake" of video line length delay lines at the input to the convolver [25,31]. However, in both the Vicom and the Comtal displays, the parallel paths are used for various other display and processing features.

A second approach to the implementation of spatial processing is the use of iterative processing techniques. Three mechanisms are involved in systems that iteratively perform spatial processing. A video rate processing pipeline implements the fundamental mathematical capability that can be applied to a frame of data in a single frame time. The width in bits of this pipeline determines the arithmetic precision that can be maintained. Refresh memory is used as interim storage for each iteration. The programmable ability to stack channels is used to provide a sufficient depth in bits per pixel to match the pipeline precision. Scroll offsets that are changed between each iteration are used to introduce the spatial relationship between the data and the algorithm operator. The advantage of the iterative technique is that it can also be used to perform local adaptive rank filters.

An example is the architecture used by DeAnza Systems [27]. A multi-stage pipeline processor organizes the basic operations of multiply, add, shift, and look-up within a feedback loop. This type of system does not attempt real-time spatial processing but provides an inexpensive fundamental processing power that can be used iteratively to yield a near real-time response. In the DeAnza IP 8500 display, the level of intrinsis parallelism is four. The 32-bit wide feedback path can be used for four 8-bit operations, two 16-bit operations, or one 32-bit operation. A simpler implementation is that of the Ikonas display [16], which provides a single processing path; consequently, an algorithm will probably take longer to run, but the cost is low.

A different class of processing systems combines point and spatial processing in a highly parallel refresh data path. The Comtal display provides up to 64 parallel paths, and the International Imaging Systems display provides up to 24 parallel paths between refresh memory and the display screen. The spatial dimension is introduced by using the ability to scroll many of the parallel processing paths independently. A feedback path allows retention of real-time processed results or iterative processing. With the exception of the Comtal convolver, the processing power on each path is very simple; but the large degree of parallelism, combined with ingenious software, allows these structures to perform most of the signal processing type of algorithms at near real-time rates.

Aside from the use of special pipeline processors to perform spatial processing, a number of displays have, for some years, allowed local image processing in the internal control processor. In the Vicom system, the

MC68000 was specifically chosen to do this effectively. Nevertheless, the inherent limitations of conventional processors constrain throughput. Another manufacturer, Aydin, which for some years relied on its internal 8080 processor, recently took advantage of a new technology to provide near real-time performance at low cost. A bit slice microprocessor is intimately interfaced with refresh memory. With its fast cycle time, it can randomly access memory and fetch, process, and store pixels in a pipelined flow, running at refresh cycle rates.

Geometric operators are required in many image processing applications, yet very little is available commercially to provide this capability. COM-TAL provided an option for the Vision 120 that allowed image translation, rotation, scaling, and skewing of source memory with bilinear interpolation. The geometric processor is capable of processing 512×512 image in 0.2 sec. This implementation provided a highly interactive capability for geometric processing of imagery but was never distributed widely, because it was very expensive. Other vendors such as International Imaging Systems and DeAnza have implemented translation, rotation, scaling, and skewing operations using an iterative feedback approach with nearest-neighbor resampling. The speed varied from 5 to 60 sec as a function of the operation being performed. Both the speed and the accuracy of this approach proved to be undesirable for most applications; but the cost was right, because no special hardware was required, only the judicious use of existing features such as scroll and feedback.

In 1982, International Imaging Systems introduced a hybrid approach that used the host processor to compute the addresses but performed the actual I/O and resampling operations within the architecture of the image processing display. This approach could perform a generalized warp on a color image in less than 60 sec. Again, however, speed and accuracy, although improved over that of iterative feedback techniques, were not sufficient for most applications. Some manufacturers are currently developing a compromise approach that will perform general geometric correction in near real time, less than 3 sec, with $\frac{1}{10}$ of a pixel accuracy using bilinear interpolation. In a few cases provision is being made for expansion of the interpolation technique to include cubic spline.

For broadcast video, several vendors today provide real-time geometric processing using up to second-order warping. Examples of this technology can be seen on the news as the image of the next news item flys into the field of view, or during the highlights of Monday-night football as the image of Earl Campbell unravels on your screen. This technology, although sufficient for the broadcast industry, is still too expensive for the general image processing marketplace and lacks the precision required in most digital applications. There should be a great deal of activity in the area of geometric processing during the next five years as the availability of low-power high-speed arithmetic processors increases.

Mention must be made of the efforts to build spatial processors that perform the image understanding algorithms. Recent progress has been covered in [14]. A typical example is the CLIP processor from the University of London, which stores each pixel in a processing cell and is programmed so that pixels retain or change their internal state according to the states of their neighbors. Iterative processing allows the propagation of selected conditions across the image space until a steady state is reached. (This example, which has little in common with a spatial convolution, demonstrates why the distinction between classes of image algorithms.)

Spatial processing is restricted by the limitations of current semiconductor technology. The fast multiplier chips are large, expensive, and, more importantly, consume a lot of power that must be dissipated as heat. It is not physically possible to build a large kernel convolver on a single printed circuit card, and a multicard implementation is very expensive. Other restrictions on spatial processing are a matter of specific display architecture or implementation. For example, although iteration is the basis for spatial processing in most displays, some displays do not allow feedback of enough bits to retain precision – dynamic range during the interim passes.

v. Display Presentation

Many video rate functions are required to present a composite image on the screen. The point in the system at which these functions are implemented varies among different designs, but it is convenient to discuss them collectively. Sophisticated displays provide many refresh memories, for which a means of selection must be provided. Some manufacturers now provide for multiple screens and, potentially, for multiple users. Programmable selection of channels allows any of the images stored in refresh memory to be output for display. For a monochrome image, the output of the appropriate channel can be routed to all three guns of a color monitor. For color images, any channel can be directed to any of the three guns. This allows both true color and false color capability. Instantaneous switching permits convenient reassignment of channels and colors; it also allows the display to flicker between two or more images, a capability that is useful in comparative analysis.

Pseudocolor allows the image from a single channel to be displayed in color, with an arbitrary (programmable) color assigned to each of the original grey levels. Pseudocolor capability is usually provided by look-up tables that input the 8-bit data from a monochrome image but output 24 bits, 8 each for red, green, and blue. Depending on the architecture, this can be implemented by multiplexing a single channel through three LUTs that are otherwise used as monochrome tables for separate channels. Alternatively, a dedicated pseudo-color LUT will be concatenated with the monochrome LUTs.

Most displays now provide a flexible four-way split-screen capability. A vertical split position can be programmably selected anywhere across the screen, with one image displayed to the left and another to the right. Similarly, a horizontal split can be created, either by itself or with the vertical split, for simultaneous display of portions of four images. In conjunction with scrolling the four components, an efficient comparative analysis can be performed. A variant of split screen allows arbitrarily shaped regions to define the split. Split-screen capability is achieved by a pixel rate switching from one refresh channel to another for the vertical split and by a switch during horizontal retrace for the horizontal split. The ROI split is simply accomplished by placing these switches under the control of the bits in the graphic plane defining the region(s).

A major function of the display presentation subsystem is the merging of graphics overlays with images. Each graphic can be assigned one of several colors. The graphic is merged either by replacing the image grey values with the solid color or by additively combining them. On the screen, the former seems to imbed the graphic in the image, whereas the latter appears to lay the graphic as a color transparency over the image. In some systems, a mode called color contrast is implemented. The graphic is caused always to assume the complementary color of the local image region. In systems providing display of multiple overlays, a means is provided to handle intersections. The coloring of graphic overlays and the merging of them with the image used to be implemented in hard logic, but it is now common to provide a programmable capability, usually by passing both the image data bits and the graphics bits through another LUT. The LUT can be loaded to color the graphics and select the required merge techniques. This approach allows the full range of colors and a diversity of merge techniques that is limited only by the imagination of the user.

Graphics generation has been a neglected capability within image processing systems. Few displays have a hardware vector generator and are equipped only to draw straight lines or conics in software. In many applications this is adequate, especially in systems in which the memory is mapped. Some displays include hardware character generators, but these are unsophisticated, usually limited to a single font, a few sizes, and a fixed output format.

The video output stage is also the point at which cursors and hardware-generated test patterns are merged. Today, the curson is frequently implemented as a graphic overlay. It is composed of a single programmable bit plane, but its size is limited to about 64 × 64 pixels. Cursor color and merge technique are programmable. Test patterns include a grey-scale wedge and a grid to facilitate the convergence of color monitors. The basic simplicity of the test patterns means that the whole screen can be filled by the repetition of a few lines; hence, they are stored in PROM and inserted into the video when selected.

If an intelligent implementation is used, there is little restriction in video output. One might note that all operations are performed on an integer pixel basis and only with respect to the orthogonal axes of refresh memory. Operations such as registering two nonaligned images require spatial processing.

vi. Display Monitor

The image display itself is a high-quality industrial monitor. It is normally calibrated to display a square frame of data with a one-to-one aspect ratio. Standard display sizes are 512 × 512 or 1024 × 1024 pixels, but 640 × 480, 1024 × 1280, and other sizes can be found. The broadcast TV industry rates of 30 (or 25) frames/sec with 2:1 interlace are standard. There is increasing use, however, of sequential scan displays, which refresh at 60 (or 50) frames/sec, without interlace. (These are also known by the misnomer repeat-field displays.) This provides a flicker-free display that minimizes viewer fatigue.

The analog part of the system is often its weakest point. Though there has been a steady improvement in monitors, the near perfection of the digital portion of the display is lost if the analog outputs are inferior. Unfortunately, the issue of displayed image quality is clouded by the use of ambiguous terminology. *Resolution* is a much abused term. A manufacturer will claim a raster scan rate as the resolution of the monitor, but there may be less data resolution than this indicates, because it is easier to design and implement the hardware to paint the raster on the screen than it is to put grey-scale information on that raster. The only acceptable test is to put test chart images on the screen and measure the resolution optically with a photometer. This test will measure the modulation transfer function (MTF) of the device. However, confusion does not end there, because there are several ways to specify MTF. One way is to specify the 30% MTF, the spatial resolution at which output response has fallen to only 30% of its value in the passband. It is more common to specify the limiting resolution at which the monitor has reduced a black-and-white test pattern to a uniform grey. There is dubious value in this, but as an example, it would be typical of a 1024 × 1024 color monitor to have a limiting resolution of around 800 lines.

Another concern is flicker. The classic broadcast industry solution has been 2:1 interlace. This fails for high-quality monitors, because individual scan lines are sharply defined. The location of a high-contrast horizontal feature in the image will appear to shift by one scan line at only 30 times/sec, and an irritating flicker results. Fortunately, it is possible to build noninterlaced displays with 512 × 512 resolution, and efforts are in progress to extend this capability to 1024 × 1024 displays. In another approach, advances in semiconductor technology have allowed a formerly 30 frame/sec interlaced display to be speeded up to 40 frames/sec (interlaced). The result is visually pleasing.

The restrictions on display monitors are unfortunately based on physical laws. Divergence of the electron beam and scattering of photons in the faceplate and phosphor cannot be avoided. There are also engineering limitations: 80–100 MHz bandwidth in the video amplifier is difficult to provide. And, in the case of color, there are basic difficulties in implementing the tricolor dots.

vii. External Video Devices

It is often necessary to interface displays with an external video environment. Most systems have the option of selectable, broadcast TV compatible video interfaces and can digitize from a TV camera or video tape machine. An advantage of digitizing displays that include a feedback loop is that they can integrate a number of frames of data. This considerably improves the signal-to-noise ratio (SNR) of typical video data. The capability to record data from the display system is also provided. In the United States and Japan, the 512 × 512 pixel digital format conflicts with the broadcast standard in which only 486 lines are visible; but in Europe, 512 × 512 pixels fit within the standard, and a special I/O is unnecessary.

The ability to interface to external video devices is provided by switching the whole timing of the display for the duration of this task. If this is the only action taken, then the aspect ratio of the image is modified; but this is usually tolerated. When one is interfacing to video tape machines, a time base corrector is frequently required. When working with color data, one also employs an encoder (NTSC or PAL) to convert from the digital display's RGB format.

viii. Display Control and Control Processor

As the sophistication of image processing displays has increased, control has become more important. The modern display contains up to tens of megabytes of RAM refresh memory, up to tens of LUTs, and up to hundreds of registers for processing or configuration control. Moreover, all these resources are operating at state-of-the-art speeds and must be synchronized to meet the requirements of frame synchronous processing. Display control operations can be divided into four categories: reading and writing data between refresh memory and the host computer, loading LUTs and parameters for processing and reading tabular results such as histograms, loading the programmable switches that configure the system for display or processing, and managing the interactive inputs and outputs for the hardware man–machine interface.

The various approaches to display control involve different combinations of basic techniques. In one, exemplified by Lexidate, a custom microprocessor is dedicated to the task. Several manufacturers have divided the responsibility between a conventional microprocessor and a special-purpose hardware display controller, whereas others have relied on the latter alone in the display itself, but with more responsibility in the host computer software. Depending on the design approach, the control processor may be intimately involved in real-time system operation, such as updating the cursor coordinate register. If this approach is taken, the control processor can become preoccupied.

The techniques that have been used to implement the handshake between the display interface and the hardware register or memory in the subunit are well established and include microsequencers and direct hardward handshaking. To handle frame synchronous operation, the display maintains a busy state until the retrace period begins. This should occur before the control processor or host computer times out. The transfer occurs on a single-word basis at DMA rates through a single-word register in the display interface or hardware controller.

Problems can arise when the computer system tries to access multiple subunits in rapid succession or when more accesses are required than can fit in the retrace interval. In the first case, the lowest-level handshake may still be in progress when the next request is made, and the resultant busy state can hang the interface. The latter case occurs during a sophisticated feedback processing operation, when whole frame times must be injected into the algorithm to allow all the registers to be set for succeeding steps. This truly control-band condition is magnified in multiuser systems, in which control contention can severely degrade response.

ix. Hardware Man–Machine Interface

The hardware implementation of the man–machine interface (HMMI) should be responsive and easily understood by the nontechnical user. It should provide a natural control for a given function. In a production environment, a real-time capability should be controlled by a technique that allows eye–hand coordination. As an example, image rotation should be controlled by a knob. The interactive environment can also allow certain processes to be concurrently active, for example, roam and radiometric manipulation. Hardware function buttons with visible indication of state can provide this capability. In production-oriented environments, with their limited or fixed scenarios, the HMMI can become very important.

The HMMI has been a rather neglected part of the display. In most cases, the user is left to fend with an alphanumeric terminal, a trackball, maybe a joystick or graphics tablet, and a few function keys. Interpretation systems have provided a comprehensive HMMI from the outset with many fixed function and programmable keys in a well-laid-out console. The designers have recognized that it is natural to push a key to select zoom and to push it further to select a zoom ratio. This approach has also been taken by Aydin. To some extent, the lack of interest in this area has been due to the predominate use of image processing systems in research and development, where the scientists have considered flexibility to be of primary importance and convenience secondary. In production systems, an HMMI that is designed for a specific task and is easy to use is the more important consideration. An excellent example is the HMMI for the Scitex electronic pagination system. A set of keys, trackball, and rotating knob have been engineered into a console for maximum correspondence between hand movements and iterative functions on the screen.

x. Computer Interface

The interface of the display to the control host computer is a critical element of the display system. In a classic system design this must handle all the data I/O for the display, and the typical rate of one megahertz is an obvious constriction in the system. The interface defines the control parameters that are available to the programmer and establishes the maximum data transfer rate. This interface has been neglected by most vendors and needs industry-wide enhancement.

There are two distinct categories of display interfaces: memory mapped and DMA controlled. Memory mapped systems map the internal memory, registers, and RAMs into the host address space. On the surface, this approach appears to provide the ultimate flexibility to the programmer. Memory mapped interfaces, however, have several drawbacks: They are not transportable, they do not support buffered I/O operations, and they are not easily usable in high-level languages. Principally because of their transportability, DMA controlled interfaces are prevalent. Memory mapped interfaces are, however, used frequently in internal control processors.

The design of the display interface must address the fundamental communications requirements: control I/O, data I/O, interrupt support, device status, device initialization, error recovery, and I/O bus protocol. These functions are usually separated into two categories. The first is the CPU interface, which handles the handshake with the host I/O bus. This is normally a 16-bit parallel interface that supports the ability to send limited control information (6 to 8 bits) and supports the device dependent status and interrupt conditions. The second is the display interface, which handles handshakes within the display.

The image display requires control information to establish data transfer conditions such as direction (read or write), packing, subunit address (e.g., LUT and refresh memory), DMA word count, system address (e.g., line 100, sample 25, or LUT word 350). The most common approach in sending this information has been to use the control signal lines that are provided in most general purpose interfaces. This approach has been unsatisfactory for three reasons. First, the interface can be set into an incorrect state because of bit errors. These can result from bad chips, faulty cables, or programming errors and can cause the wrong operation to be performed or the device to hang without explanation. Second, because each DMA packet generally requires separate control settings, multiple operations cannot be buffered into a single DMA operation. Finally, to minimize the overhead introduced in changing control conditions between DMA transfers, the device handler is coded to handle frequent control sequences with a single driver call. As a result, the device handler is typically a significant undertaking that must be repeated as the operating system changes or the device is interfaced to a new processor.

An alternative approach, introduced in 1975 [3], used a fixed-length header with each transfer to define all control information affecting that transfer. Because the header was presented as data to the interface, the programmer could buffer any number of operations into a single DMA operation. The device handler was significantly simplified, because the only I/O operations required were read and write of a data buffer. To protect against erroneous control sequences, the header included a checksum. The most significant drawback of the header approach was in situations in which many short transfers were required. This condition occurred most frequently when drawing vectors, characters, and so on.

The principal function of the hardware interface is the transfer of data to and from the display device to the computer I/O bus. Because most image data files are stored in a packed format, most image displays provide the ability to perform packed data transfers (two 8-bit pixels per word). As sensor technology evolves, we are seeing the need to pack and unpack 12-bit data into multiple 16-bit words. In the use of the image display for graphics applications such as character or vector generation, it has proven useful to provide a special operations mode for DMA transfers, in which one constant value is transferred and then is repeated until the DMA is exhausted.

The use of interrupts is essential in an image processing system. Typical interrupts available to the programmer include button pushed, cursor moved, feedback done, min/max complete, histogram complete, and vertical interval. In systems that support multiple computer interfaces, interrupts that allow the processors to coordinate interprocessor communication must be included. Each of these interrupts must be maskable so that its use can be controlled programmatically. In addition to the maskable interrupts, there are a number of unmaskable interrupts such as transfer complete and timeout.

A device status bit is generally available for each of the possible interrupt conditions. When an interrupt occurs, the programmer can read the device status word and determine what caused the interrupt. One useful status bit that is not associated with interrupts is a power-on status. The power-on status can be tested prior to each transfer to assure that the device is present.

The interface must provide the programmer with the ability to reset the device to a known state. This initialization is required for power-up and after error conditions. It should not modify any internal registers or RAMs other than which is required to return to a known control state. The interface should return an error status when invalid requests are issued. Most interfaces include a timeout interrupt that will generate an interrupt and error status if the device does not respond within a prescribed period of time.

To obtain a maximum burst transfer rate, the interface must be designed to transfer data to refresh memory in a format that is related to the refresh memory multiplexing. In a typical case of eight pixels being multiplexed in a single memory access for video output, the interface will usually request eight words from the bus in a single bus request. The eight words are buffered on the interface card and transferred to the display device en masse.

B. Architectures of Image Processing Display Systems

A dozen or so manufacturers make image processing display systems, and each uses a different approach to the design. Inspection of the architectures of these systems reveals some areas of commonality, and it is natural to try to classify these systems into distinct types. One would hope to find that each type is best suited to particular functions or that a given architecture is suited to a particular balance between performance and cost. However, attempts to do this become forced, because most of the designs use some combination of a number of architectural features. We shall discuss the overall architecture in four categories. These are (Fig. 6) the refresh data paths, including

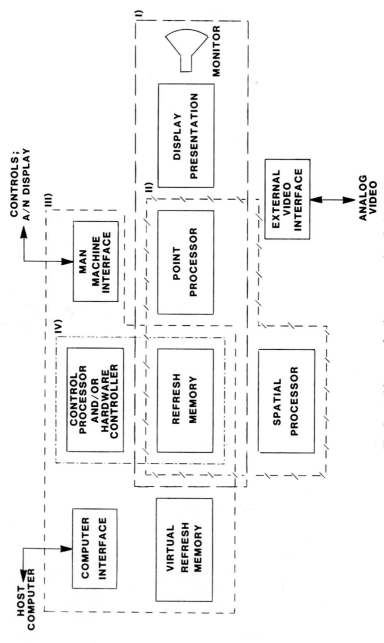

Fig. 6. Architecture of an image processing display.

point processing; feedback-loop, video-synchronous spatial processing data paths; DMA rate, control processor or host computer data paths (henceforth referred to as *control system* paths); and non-video-synchronous image processing data paths. The alternate approaches in each category will be discussed, highlighting strengths and cost impacts and using specific implementations as examples.

i. *Video Display and Point Processor Architectures*

The architecture at this point in the system is concerned with video rate data flows between the refresh memory and the monitor. Point processing has always been performed along this path. Variation of architectural approach is possible, because several refresh memories can be multiplexed into the paths, and a number of display monitors can be serviced. Also, the length of the point processing pipelines, that is, the number of discrete operations that can be concatenated, is a variable, as is the sophistication of each operation.

The two basic approaches to this area are illustrated in Fig. 7. In (a), the architecture is driven by the three color outputs, to each of which a processing pipeline is devoted. Multiple refresh memories are multiplexed into the three pipes. In (b), which might be described as refresh memory driven, a processor is described to each channel of refresh memory, and the multiplexing for display outputs follows. When multiple monitors are used, the multiplexer for either case is enlarged on its output side to provide the necessary data paths. In the usual situation, in which there are more memories than video outputs, there is cost advantage to the first architecture, and for many applications this is the most effective approach. The disadvantage is inflexibility when interacting between several alternate input images. Then, any point operators that are specific to individual images must be reloaded each time an image is selected. This places a burden on software and may overload the control or host processor. As an example, consider an application in which advertising copy is being prepared in a digital page makeup system. It is desired to cut and paste a page together from a number of subimages. Each constituent may have a different color balance as applied by pseudocolor LUTs. It will be much easier to use the second architecture, because as the task progresses, the artist may wish to go back to a previously included subimage and adjust color balance to blend with a recent addition.

With regard to the length of the pipeline, a common arrangement is to concatenate the operations of LUT mapping on a single image (i.e., monochrome basis), image combination, LUT mapping on a color basis, and merge of imagery data with graphics data, and so on. The architectures of

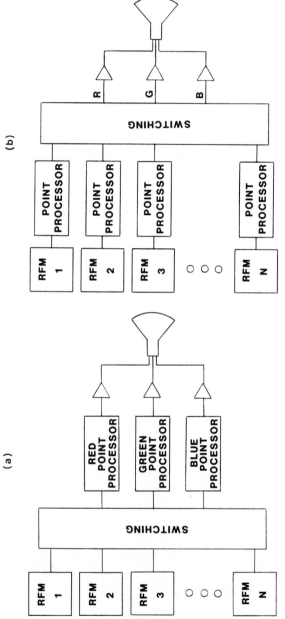

Fig. 7. Two basic approaches to video display and point processor architecture.

the International Imaging Systems' Model 75 and the Comtal Vision One/20 follow this arrangement and employ a large number of parallel pipes. Examples of displays that use the first architecture (see Fig. 7), include the Aydin 5216 and Vicom systems. DeAnza Systems displays use an architecture that is a hybrid of the two extremes, providing the flexibility of LUTs devoted to each channel of memory and the economy of dedicating all succeeding operations to color output pipes. The sophistication of the pipeline functions is strongly correlated with the cost of the system. It ranges from the minimal configuration of just three LUTs in the AED display to real-time convolvers in the Comtal display.

ii. Video Rate Spatial Processor Architectures

Since the mid-1970s, considerable processing power has been provided by the feedback loop. The key characteristics are that it is frame synchronous, runs at video rate, and aside from the refresh memory itself, is memoryless. Two basic architectures have been implemented, as illustrated in Fig. 8. In the first, the point processors are used as the processing elements, and the feedback path simply allows the result to be stored in refresh memory. In the second, there is a feedback path with processing elements, which is independent of the refresh path. From a purely topological point of view, there is just one difference between the two: The second architecture allows independent display and processing operations. This can be operationally useful when an iterative algorithm is being performed, because the video display portion of the hardware is still available for use. More importantly, this approach is more cost effective for multiuser systems wherein a single, powerful feedback loop processor can be shared without contention for display and point processing functions.

The actual implementations of the architecture have proven to be more significant than the topology. The International Imaging Systems Model 75 and Comtal Vision One/20 displays are both based on the first architecture, and both implement a considerable degree of parallelism in the point pro-

(a)   (b)

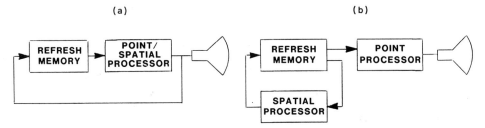

Fig. 8.   Two basic video rate spatial processor architectures.

cessor. As noted in Section IV.A.iv, an unconventional decomposition of algorithms can be used effectively to exploit this parallelism. For a given complexity of processor in each path, the number of iterations (i.e., the execution time) is inversely related to the number of paths. The Comtal display is a pure example of the architecture. The International Imaging Systems architecture, however, is an example of hybridization. The pure concept has been modified by including a processor in the feedback path with one input by means of the point processing path and the other directly from refresh memory. Another hybrid is the architecture of DeAnza Systems displays, in which the inputs to the independent feedback path (i.e., second architecture type) can nevertheless be switched to originate after the first stage of point processing in the refresh path.

There are advantages of flexibility in hybrid approaches and, in the latter case, a clear cost saving by not replicating the point processor within the feedback processor. For an iterative implementation of an algorithm to be successful, adequate computational precision must be maintained. The independent feedback loop architectures have been implemented with 16-bit or 32-bit precision and can economically accomplish this, because 16-bit ALUs and MACs are available. Problems arise when it is necessary to use LUTs, because their cost becomes prohibitive at larger sizes (greater than 12 bits, i.e., 4096 words). Current implementations of highly parallel architectures have, therefore, restricted their precision to 12 bits or less.

iii. Control System Architectures

The control processor or host computer side of the overall architecture is invariably organized with a bus structure, as illustrated generically in Fig. 9. Variations on this theme surround the issue of whether the bus is a special design or whether it is a commercial computer bus. There is also variation in control responsibility between hardware display control and local microprocessor control. Use of a commercial bus brings the considerable advantages of support by standard software and hardware interfaces. DeAnza Systems displays contain an internal emulator of the DEC UNIBUS and may be interfaced to a DEC host by extension of the host UNIBUS. Internal display addresses for memories, tables, and registers are mapped into a conventional UNIBUS address window. Because there is inadequate space for all refresh memory addresses, a hardware display controller acts as an indirect addressing scheme, mapping a window of refresh memory into the UNIBUS address space.

The Vicom display contains an internal MC68000 and VERSABUS, with all display addresses being mapped as main memory. Addition of peripheral devices such as disks is trivial for such systems. Other displays contain

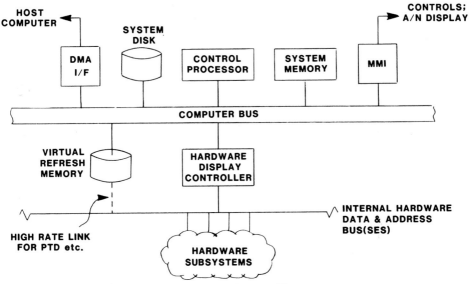

Fig. 9.   Control system architecture.

specially designed busses that follow conventional protocol but have been tailored for the specific use, notably that they run at a much higher rate. Ikonas and Aydin have taken this approach in combination with a dedicated control microprocessor. The third type of internal control bus is less flexible than a computer bus and serves as a specialized interconnect network between the display interface or hardware controller and the various subsystems.

iv. Non-Video-Synchronous Processing

Two approaches have been taken to processing within the display without use of the refresh or video feedback paths. The first is to process within an internal microprocessor. This is architecturally identical to using a host computer but has the advantage that all the data are in main memory, because it is making use of the refresh memory resource. Until recently, this approach had dubious value other than low cost, because the power of microprocessors was so limited (in both hardware and software). The advent of the MC68000 changed that, and a significant movement in this direction has begun, led by Vicom. The alternate approach is to incorporate bit slice microprocessors in an arrangement with refresh memory that is so intimate that it is realistic to refer to it as intelligent memory. Coupled with innovative memory architectures, this approach allows image processing to proceed much faster than DMA rates. In fact, it is becoming possible to process at the same rate as video, meaning that a distinction from feedback loop processing is moot. Ramtek has used this approach for their graphics displays for many years. Aydin has now implemented this approach for full grey-scale imagery data and has the distinction of covering all bases, because it also has an internal microprocessor. We can anticipate that rapid advances will cause bit slice technology to play a prominent role in image processing displays, with the caution that writing the microcode will be burdensome.

V. Image Processing Software

In this section, we shall discuss the major components of an image processing software system, their functions, and their interactions. In Section VI, we shall show different design and performance tradeoffs by discussing a series of competing issues in image processing, which will tend to stratify the range of image processing systems and show the reasons for making one tradeoff or another.

Image processing systems have a specific design intent that is summarized in Fig. 10. Around the outside of the system we see three major external factors that govern an image processing system: users, input data, and output results. Such a system should control the transformation of raw input data into meaningful output information under the user's supervision. We have expanded this diagram to give it some functional structure in Fig. 11.

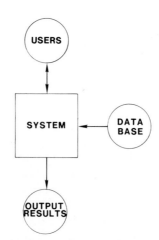

Fig. 10. Summary of the overall design of image processing systems.

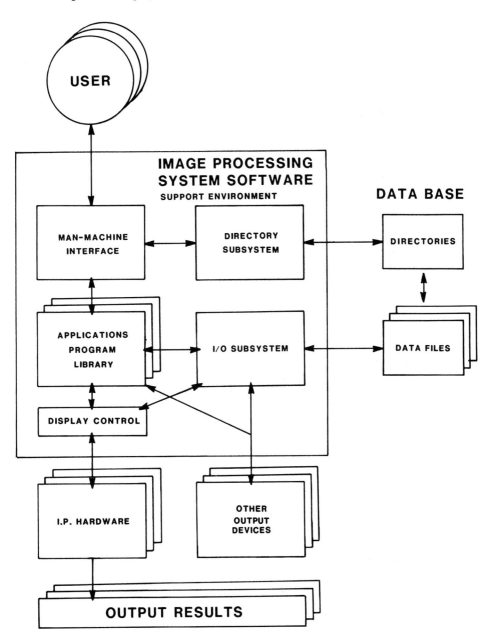

Fig. 11. Design plan for image processing systems.

Such a system can interface to multiple users, data bases, and output devices and can control a library of applications programs. In the diagram we signify this plurality by showing multiple layers where appropriate.

We can see that on a fundamental level the system has components that control the interface with each of the three externals. The user talks to the system through the software man–machine interface (SMMI). The applications library is accessed and controlled by a standard protocol from the SMMI. The image data base is controlled by directory and I/O subsystems. The outputs are created by a specific display control subsystem and a general purpose I/O subsystem. This categorization also has functional meaning in that it represents a good structural organization for image processing software.

A. Software Man–Machine Interface

A major component of any interactive system is the software man–machine interface (SMMI). It is responsible for managing the two-way flow of information between the user and the applications software in the system. In highly sophisticated image processing systems, this function may be accomplished by a separate program, but less-sophisticated software may rely on the host operating system. It can be argued that the principal difference between an image processing *system* and an image processing *software package* is the presence of this SMMI, which unifies the various applications programs with a common syntax and directory system. In this section we shall discuss four major topics: what an appropriate command language is and what bits of information it must provide, how SMMI accomplishes command entry, how the SMMI controls applications program prompting, and how the SMMI implements message and error reporting.

i. Definition of a Command Language

In many sophisticated systems, the SMMI appears similar to an interpretive programming langauge. The user passes command language instructions to the interpreter, and these are decoded and converted into a set of explicit instructions, a machine language. Then the proper modules are scheduled to execute those machine language instructions. The form of the command language can vary widely, from wordy, natural-language expressions to dense, assembly-language-like instructions.

A command line must provide certain minimum information to specify a task sufficiently, typically the input and output descriptions, the desired function, and any user-specified controlling constants or options. Inputs and outputs can refer to full or partial datasets on disk. The functions are usually fundamental types of image manipulation or transformation, with their varying input constants (e.g., convolve with kernel $= x, y, \ldots, z$) or options (e.g., histogram with option $=$ cumulative). A productive way of analyzing the information content and syntax of a typical command is to use the familiar structure of English language sentences [6,9]. Datasets and operations can be thought of as nouns and verbs; input and output datasets correspond to subjects and objects, with their subimage descriptions equivalent to adjectives. A fundamental function is the predicate of the sentence, and controlling constants and options are its accompanying adverbs and prepositional phrases. We can take a natural-language command and decompose it into its constituent parts as in Fig. 12. In this case we can see the difficulties of natural-language expressions: They are wordy and redundant with many colloquial phrases necessary only to blend the essential information.

From this example it is evident that a typical image processing operation can be specified by a set of basic building blocks: subjects, predicates, objects, and their modifiers. The subjects are inputs to the predicate function, with output results named as objects in this sentence-like analogy. Most systems avoid the complications of natural-language commands and ask the user to type in only the basic building blocks. We can take the information in Fig. 12 and transform it into a typical command line, as shown in Fig. 13. The essential information is present. Subject, predicate, and object are separated by delimiters that allow multiple specifications in those fields; and modifiers are appended to the nouns and verbs.

Such a command format can still allow for further condensation. Key words, such as nouns and verbs, will be known to the system; therefore, they can be specified only as is necessary to eliminate ambiguity. In our example, PSEUDOCOLOR can be reduced to PSEU; and depending on the user's image-naming conventions, SECTION can be sufficiently specified by SEC. Typically, new output images will have to be fully named the first

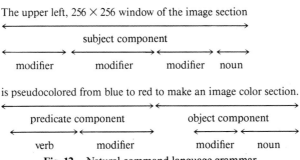

Fig. 12. Natural command language grammar.

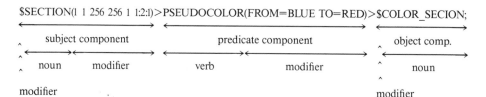

Fig. 13. Sample command language grammar.

time, but in the event that we are updating an existing image, COLORS may be sufficient. In addition, modifiers can be considered as always active, with default values to be used when they are not explicitly specified. Judicious selection of defaults can lead to great economy in a command language.

One of the most important command language features in an image processing system is the modifier for specifying image subsections when moving around a memory hierarchy. We can presume that a large parent image resides on disk or some other medium. It is common for a user to want to extract a portion of that image in either the spatial, spectral, or temporal dimensions. Normally, the spatial dimensions (i.e., samples and lines) are larger, and the spectral and temporal dimensions (i.e., bands and times) are more categorical in nature. The command langauge should, therefore, provide a simple means of extracting spatial subsections by windowing and spectral–temporal subbandings by listing. In the spatial dimension, there should be control of the physical subsection that the user desires, such as specifying the upper left pixal coordinates and the window size. Also, the user must be able to control any subsampling in both samples and lines.

The previous example shows a typical syntax. In the string (1 1 256 256 1 1:2:1) the first six numbers specify the starting sample and line coordinates of the window, its size in samples and lines, and the subsampling factors in samples and lines. The first colon indicates that the following numbers are a spectral band list; in this case, we are requesting just band 2. The second colon indicates that the next set of numbers is a temporal band list; in this case, we are requesting data only from the first temporal band, that is, the first date for which we have data. This type of list-based modifier with free field separators can be very powerful. The user learns to think in these explicit terms, and when a field is left blank, it simply reverts back either to the overall parent image size or to 1, depending on use.

Another method of condensing the command language involves the use of previously defined symbol macros and online string substitution. The user may be allowed to define a variable length library of symbol macros, for instance, the string UL (upper left) may be defined to mean the subarea specification (1 1 256 256). Then the user can type SECTION UL in a command line. The command interpreter program will scan its directory of symbol macros and substitute a new string where it finds a known macro, before it parses the line. This type of condensation is fairly simple to implement. It can lead to problems if the user is not required to flag the symbol macros; in such a case, an image named AWFUL might be transformed to AWF(1 1 256 256) unintentionally. Consequently, symbol macros are often implemented with a flagging mechanism that instructs the command interpreter to check its macro directory only when it encounters flags. Control characters, quotes, and case changes are among the flagging methods used.

So far, we have been examining the command language on the basis of individual commands. Interpreter-based programming languages, such as BASIC, are more than individual commands. Procedures can be created by combining individual commands and creating a local database of variables. The same should be possible for image processing command languages. An enhancement procedure can be specified as a series of discrete steps, with fundamental image processing operations as building blocks. Clearly, the previously defined command lines can be strung together in a sequence to do some procedures, but other procedures may require constructs such as looping, conditional branching, recursion, and shared data variables.

Conditional branching explicitly requires a data variable to test. This data variable typically comes from a global data base, that is, its value can be manually assigned in the command language or returned from a function. The previous discussion of symbol macros and substitution can be expanded to provide us with this construct. We can look at symbol macros as shared variables with character string values. The user can then be given commands to create, modify, pass, test, calculate with, or loop on these variables with numeric or logical values. With such a global database, available to all functions, the remaining constructs are easily implemented. Commands can be stored in disk files or remembered when first typed and then reexecuted singly or in a loop at a later stage. A group of commands can be defined as a command procedure, stored temporarily or permanently; then, with proper stacking of command queues and local variables, that procedure can be executed recursively. The resulting command language is very powerful as a single-function investigative tool and as a set of building blocks for creating more complex procedures.

Some production-oriented systems provide different ways of typing commands. Though an alphanumeric keyboard has been assumed in this discussion, there are simpler ways of specifying the same basic information, albeit with loss of detail. These issues will be covered in Section VI.C.

However, it seems that a combination of menu-driven and command-driven SMMIs provides the most flexible command interface, allowing the user to select a comfortable mode of operation.

ii. Command Entry and Interpretation

The command interpreter must perform a sequence of tasks, checking for error conditions at each step, in order to process and execute a user command. These tasks include prompting for command line, accepting command line, symbol macro substitution, abbreviation expansion, command syntax check, input validity check(s), function validity check(s), output validity check(s), prompting to correct problems, logging of commands, and queueing desired program.

The first contact that a user usually has with the system is with its prompt, which tells the user that the system is ready to accept commands. In sophisticated systems it also indicates what subsystem the user is presently in, what subdirectories are open, or what database is available. In addition, it will usually prompt in a manner that effectively labels the command the user will type, so that the line can later be recalled. Therefore, the prompt can have a great range of sophistication, from saying "I'm ready" to functioning as a programming aid.

The command interpreter must then accept the command line from the user. This may involve decoding command button pushes in menu-oriented systems or simply reading an ASCII string from a keyboard. Again, it can become more sophisticated. For example, if the user wants to type in a long command that exceeds the line length of the terminal, the command interpreter must be smart enough to wait until it has a complete command. This is usually accomplished by defining special characters to mean continuation and terminator. If a user wants to type multiple commands in a single line, the command interpreter must recognize this and parse those commands independently.

The first step in processing a command from the user usually will be to check the string for any explicit or implicit abbreviations and substitutions. If the user typed any symbol macros that need to be replaced by their full meanings, these must be found and substituted. If the user referred to any global database variables, these must be detected and their respective values substituted for their symbols. If the user used any incomplete but unambiguous names, these can be expanded. If there are any unrecognizable elements in the command or invalid uses of continuation, termination, or delimiting characters, basic command syntax can be checked at this stage, even though the interpreter has no real idea what the user means by this command. This type of check will detect most typing errors, for example.

The process of checking the validity of such a command line usually involves breaking the line into discrete *tokens* [19], which are logical elements, such as the previously discussed nouns, verbs and modifiers. They are then cross-referenced with the known database and directories. Are the inputs in this command line to be found in the available directory(s)? Are they online and valid for this function? Are their modifiers consistent with their size? Are they presently being modified by another user or protected by security? The command interpreter tries to answer these questions and then prepares a list of inputs and their characteristics, sizes, and locations for the function.

The function(s) specified in the command line are checked in a similar manner. Multiple functions may be specified in some systems, and they must be analyzed for compatability. Sophisticated systems will also perform a check on the validity of function modifiers at this stage. This is usually more difficult, because modifiers vary in name and value from one function to another, whereas input and output data files are relatively consistent in their valid modifiers. Unsophisticated systems usually leave function modifier checks to the applications program itself at a later time, which wastes time on scheduling and execution if an invalid condition is then detected. The applications program may then be forced to jump to an error condition or temporarily return to a reprompting subsystem to resolve the difficulty.

Outputs are checked in a manner similar to that for inputs, with some added provisions. If the output is a new file, is there enough disk space, for example? If it is an old file, will the new data fit? Again, a list is created of the valid outputs. At any point in the checking, the command interpreter may encounter problems. It can simply print a diagnostic error message and reprompt, or it can try to resolve the problem interactively. The latter is more user-friendly and, therefore, preferable; but it is not easy. Usually, if a token is unrecognizable, an error condition is generated. However, if the problem is with an inconsistent or out-of-range value, sophisticated command interpreters will prompt the user for a correction. Furthermore, these systems often make provisions to guide the user through the entire palette of possible modifiers. This can be done at the command interpreter level or later.

At this point the command line has been effectively comprehended, and the interpetation task is complete. Sophisticated systems will now record this command in an external file. The user can, therefore, review the history of a given working session, of an individual image, of the use of an individual function, and so on. Previous commands can be reexcuted singly or in procedures. They can be recalled and edited for resubmission.

Finally, the desired applications program is identified and queued for execution. This can be overseen by the command interpreter directly, in single-user sequential systems, or it can be supervised by a specialized controller, allowing multiuser, multitasking, parallel-processing systems. Single-user systems typically wait for applications program completion before issuing a new prompt. A more sophisticated, multitasking system may immediately prompt for a new command after queueing the last one, effectively running in parallel with its own applications programs.

iii. Applications Program Prompting

Situations can arise in which there is a need for later communication with the user, outside of graphic interaction devices such as cursors and function buttons. Many applications programs require typed interaction, for example, when annotating images on the display or adding comments to a documentation file. In addition, some functions may involve options or parameters that are difficult to check at the command interpreter level, and so inconsistencies must be resolved by the applications program. Also, the user may want a guided tour of the available options, with default values shown. This is easier to handle at the applications level; it simplifies the command interpreter's task and allows more complex, programmatic checking of parameter interactions and values. On the other hand, reprompting at this level presents difficulties for the logging of command histories and creation of procedures. A tradeoff must be made between the earlier and later modes of reprompting, and the usual solution is a little of each.

Reprompting at the applications level is often handled by a subprocess scheduled by the applications program. The user should be reprompted in a manner that clearly states the name of the required value(s), the number and nature of responses expected, and the default value to be used in the case of no response. Sophisticated systems may have different prompting levels, from brief to fully documented, with beginning users distinguished from experienced users.

Parameters and options are usually organized hierarchically, so that the specification of one option will automatically bring into play new defaults for other, unspecified options. When a function is created, default values will be established by a preliminary investigation of the different ways a function may typically be used, an ordering of options to separate the different possible scenarios, and finally the specification of a treelike set of default values, some of which may be conditional on other responses. In some cases, a parameter can be mandatory. For example, the user may want to rotate an image, in which case one can have defaults for the axis of rotation (the image center) and the resampling method (bilinear interpolation); but the angle of rotation is fundamental to the command, so it *must* be provided.

iv. *Message Reporting*

The final major area of man–machine interaction concerns the reporting of messages from all levels of the system. Such messages may be merely advisory, or they may report some unrecoverable situation. Most systems keep the user informed of system status. Examples include *percent complete, elapsed time,* and *program-did-this* messages. Perhaps the most important message output is error reporting. Without a succinct report of an error condition, including its source, magnitude, and outcome, a user can be faced with extreme frustration by a simple oversight.

Some errors are detected by the host computer at a very low level and cannot be trapped by the applications program. They are usually caused by programmer rather than user mistakes. The remaining errors should be trappable in the applications program and specifically concern image processing. For example, to produce a ratio requires two inputs; if the user has provided only one input file or has provided two incompatible files (i.e., one color and one monochromatic), then an error condition exists, and the user must be correctly informed. Ideally, this message should take the form of "Image COLOR is an inappropriate input to the function RATIO." Some systems may provide briefer, faster error messages as an option for the experienced user, for instance, "Invalid inputs." These are faster because they are printed immediately. When a longer, more explicit message is needed, it usually must be retrieved from an error directory, often by an independent, clean-up program; and all this complexity takes time.

v. *Relationships between Command Entry, Prompting, and Message Reporting*

These three areas of man–machine interaction represent all the distinct ways such interactions can take place. Notice that command entry involves user input, prompting involves user input *and* machine output, and message reporting involves machine output only.

In normal use, these three types of information exchange may themselves interact. For instance, if the entered command does not satisfy the applications program, it may resort to prompting: If an error condition is found and an error message reported, the applications program may prompt the user to correct the problem. If either command entry or prompting encounters an unexpected situation, they will print an error message. Consequently, in the design of a man–machine interface, these three interactions should be considered together. When program size is a limiting factor, it is advisable to try to segregate man–machine interactions in an independent supervisor program, so that each applications program does not have to pay a price for message input/output. Character strings take surprisingly little space to store, for example, one character per byte; so a 200-word paragraph takes only 100 integer words of space. This should be contrasted with the typical increase of 5 to 7 *thousand* words in program size caused by using Fortran formatted terminal I/O only to print a few messages.

B. Directory Subsystem

Systems are distinguished from mere software packages by the sophistication of their shared elements. A system manages diverse collections of data, documentation, and programs. In addition, the internal operation of a system may require shared access and management of a variety of resources to customize the system and to foster intercommunication among components. These tasks are usually accomplished with the help of directories that contain descriptive information about those resources being managed. In sophisticated systems the following types of directories will usually be found:

1. Image database directory
2. Function directory
3. Symbol macro directory
4. Procedure directory
5. Online help directory
6. Device configuration directory
7. Message directory

Some of these directories can be characterized as variable length lists, which the user updates and modifies in the course of normal system operation. Others are relatively fixed in content and are used to allow simplified initial configuration or global control.

The directory system must be able to organize the user's environment into hierarchies if called for. In other words, the user may want to define a local environment of immediately accessible, private, or incomplete working files that is a subset of a larger, global environment. Directory references in command lines would refer only to the local environment, with global references requiring more explicit specification. Local data might be restricted to a dedicated storage area, whereas global data resides on shared devices at a distant, larger-scale facility. In another application area, the user may want to segregate the data base by project or geographic location. This hierarchical structure can apply equally to all directories in a system. For example, individuals may want to customize their error messages or procedures but still be able to refer explicitly to those of another project or user. Such constructs are already being used in sophisticated host operating systems. Whereas a hierarchy of directories can be reduced to a single local directory, the ability to subdivide directories can lead to greater efficiency and security in managing database searches and access.

In general, designing a system to make wide use of directories is a good practice from a structured programming point of view. Access to directories is usually constrained to one-time searches and is not done in computational loops, where it might substantially slow a system's performance. Rather, it is performed principally during command interpretation and applications program initiation and completion. Furthermore, it is good practice to define and adhere to a consistent directory structure for all directories, even though they may have very different uses. Consistent structure allows the coding of only one set of directory access routines, which simplifies support, directory maintenance, and editing and permits optimization of that limited set of routines.

i. Image Database Directory

The most important directory in the system concerns the image database. This database will be made up of discrete images with varying sizes, precisions, and contents on varying devices and media. The image directory links a description of each of these images to a user-selected name, thereby allowing the user to manipulate parts or all of an image without needing to know the details about its physical characteristics. The database can then be stored in an internal format that optimizes system performance with the directory providing a user interface. The directory can be structured to permit database management (DBM) capabilities. For example, this can include renaming, deleting, archiving, transfering, locking, and searching by partial or full name, device, owner, history, project, or size. In addition, this type of directory structure can permit the user to define logical images, that is, images that are logical subsections of other, physical image data files [17]. In this way, the user can enter an image of a large area as a disk file and then define windows containing individual cities as logical images. When these logical images are referenced, they will direct the system to use that window of the parent image. This can be taken to a very complex level by sophisticated systems.

ii. Function Name Directory

A directory can also be used to help manage the processing environment. Sophisticated systems provide a directory of user-defined names that point to the available applications programs in the system. The user is then free from specifying actual program names and can customize the specification of processing tasks without regard for the peculiarities of the host operating system. This directory can also contain an options list and a prompting scenario, so that the SMMI can process the commands it receives, verify that it has sufficient data to proceed, and if not, prompt the user for necessary and optional parameters.

In addition to this simplification of the user interface, such a directory can provide the ability to customize the performance of applications programs. For example, the allotment and structure of working buffer space can be controlled by user-defined values in this directory. A change in the memory size or partitioning of the host computer can be globally updated in the whole system by adjusting values in the directory.

iii. Symbol Macro Directory

As previously discussed, it is useful to permit the user to define symbol macros to simplify the input of command lines to the SMMI. Like variables in a programming language, each symbol macro has a character-string name and a content that can be characters or values. When that name is encountered in the processing of a command line, it is replaced by its content. One example might be to define a macro named SOBEL-GRA-DIENT, which contains a string that defines the size and kernel values of a Sobel convolution operator. Then, the user can effectively run a SOBEL-GRADIENT function, and the command interpreter will decode this to mean the convolution function with options specifying the proper kernel. In another example, a user might extract summary statistics for an image and assign the resulting mean, 137.3, to a symbol macro MU. Later commands would instruct the system to normalize another image's histogram using the parameter MEAN = MU. The command interpreter would decode this to MEAN = 137.3.

iv. Procedure Directory

The user may want to define a collection of commands as a procedure, analogous to a subroutine in our image processing programming language. This can be done in an editor, or it can be done in the course of a processing session. The user may like the results of a series of commands just executed and choose to save them as a procedure. It will be saved in this directory under a user-defined name for later recall and execution.

v. Online Help Directory

User-friendliness requires a hierarchical system of assistance for the user. The novice should be able to type HELP at any point in the system and receive an appropriate response. A first level of help should provide a quick description of where the user is and what options are available. In most cases, this will be sufficient; but in some cases, the user may want more detailed information about a specific function or error. A typical online help directory will contain first-level information about general topics and a pointer scheme to a separate, detailed database of second-level help. In most systems, this second level of help consists of an online copy of the users manual.

vi. Device Configuration Directory

Large image processing systems will often consist of a pool of devices including image processing displays, image hardcopy devices, floating-point array processors, and tape drives. In this environment it is essential to manage both shared and devoted access to these devices. In addition, it is often necessary to store the unique configurations of each device, for instance, the amount of image memory available in image processing display B or what apertures are installed in film scanner D.

In any system there are different levels of temporal sharing. Some devices must be devoted to one user for long periods of time, whereas other devices can be shared on an instruction basis. Device allocation is usually accomplished by the host operating system through its file system. However, access management for multiuser devices is much more complex, and the host operating system has little to contribute. Managing configuration information about available devices is related to the issue of shared access. In effect, when one user acquires a portion of an image processing display resource, it is equivalent to reducing that device's available resources, that is, changing its configuration. Therefore, configuration management can be static in single-user image processors and dynamic when multiple users are sharing the architecture of a multiuser image display.

vii. Message Directory

When the applications program detects an error condition, it must notify the user (see Section V.A.iv). Sophisticated systems will maintain a directory of messages that can be accessed by a standard routine. The applications program will avoid the costs of doing direct terminal I/O by calling this recovery routine with a pointer to the appropriate message in the directory. In the case of an error condition, a recovery routine will fetch that message from the directory, fill in the function and image names necessary to describe the error, and queue the message to be printed. Although such a system seems unnecessarily complex, it does provide useful features. First, the messages in the system, like the image and function names, can be customized and can even be in a different language. Second, the system does not need repeatedly to allocate space for the same common message, such as *Invalid Input,* throughout the system; each error trap can simply point to the same message in the directory. Third, the system can isolate the cost of formatted terminal I/O in the SMMI. Finally, such a structure allows the message directory to interface easily with the other directories in the system; for example, a HELP response to an error message can be handled by a message directory pointer to the online help directory.

C. Applications Programs

An image processing system consists of systems components and applications programs. The latter are fundamental to the performance of an image processing system, whereas the former provide the efficient working environment. Each applications program embodies an algorithm, which is expressed in a structured, optimized manner that allows its efficient execution by the hardware resources of the system. However, working against this optimization is the need to control redundancy in the code of each applications program. Maximum speed can probably be obtained by coding a unique program for each distinct algorithmic option, but at the cost of much redundancy. On the other hand, the minimum amount of code and maximum amount of generality can be created by coding a whole family of algorithms together in one program, with conditional branches controlling the flow of the program to implement the requested algorithm. The resulting program will probably be enormous, require disk swapping, and run slowly. A functional categorization of algorithms must balance these two conflicting demands.

Image processing applications functions tend to fall into clusters of processing techniques. For example, the arithmetic functions of addition, subtraction, multiplication, and division have virtually identical program structure. Effectively, the programmer creates "boiler-plates" that can be copied and edited to create new functions. Functions are grouped according to their image access requirements (see Section II.B).

D. Input/Output Subsystem

One of the distinguishing characteristics of an image processing system is the processing of large volumes of multidimensional data. The user or programmer must be able to move data around the memory hierarchy mentioned in Section II. Consequently, the I/O subsystem is critical to system performance, ideally providing both flexibility and efficiency. Flexibility reduces the time required to write and debug new programs, improves internal documentation, and provides a consistent outward appearance to users and programmers. Efficiency of I/O significantly controls the overall throughput of the system. The I/O subsystem that provides the necessary flexibility while still retaining a structure that allows for maximum throughput must address requirements in the following areas:

1. Image data storage
   a. Image data types
   b. Internal storage formats
   c. Image file types
2. Image access
   a. Pixel addressing
   b. Image subarea definition
   c. Data type conversion
   d. Image organization – reorganization

i. Image Data Storage Formats

The data types supported by an I/O subsystem are determined by the input data precision, the required computational precision of an algorithm, the internal word length of the host CPU, and the desired output product. The radiometric precision of the input data or sensor is the principal factor in determining the smallest data type to be handled. Traditionally, this has been stored in 8-bit pixels, but recent sensor technology has led to image of 10-, 12-, and 14-bit precision. This can necessitate the storage of pixels at 16-bit precision to preserve I/O efficiency. The word length of the CPU often dictates the selection of data types to be used for processing the data. Systems implemented in the past on 16-bit minicomputers have generally supported at least 8- and 16-bit image data types. The 16-bit data type is necessary, because most processors do not allow arithmetic on 8-bit data.

The internal processing precision required by the processing algorithms determines any additional data type requirements. Although the input data may have only 8 or 16 bits of resolution, the applications algorithms often require additional computational accuracy to perform the desired operation. Applications involving two-dimensional Fourier filtering normally require the storage of a transform domain image with at least 64 bits of precision.

Because most image processing systems involve the use of graphics in some form, a complete I/O subsystem must include some capability to handle graphics (binary) data. Usually, these binary data are created at a precision that is lower than the minimum the system can flexibly manipulate, and this can lead to several problems. The most efficient storage method for binary data is to pack the pixels into 8-bit bytes or 16-bit words.

In less sophisticated systems, binary pixels are often stored as one bit per byte and thereby suffer size inefficiency, because such systems do not have the size and performance overhead necessary to pack and unpack bits. However, almost all systems resort to this inefficient packing when processing (particularly subsampling or subsectioning) this type of data.

To maximize I/O throughput, files should be structured so that data can be accessed with a minimum of disk head movement for most processing tasks. This is best accomplished by defining the physical record size to be an integer multiple of the disk sector size and allowing multirecord access during one physical I/O operation. Using physical record sizes that are not integer multiples of a sector will normally require that the programmer perform file I/O operations through higher-level routines, which are significantly slower.

The logical structure of the data within the files varies for different applications. One example is single-band (black and white) image processing systems, which historically have been implemented with a simple line-oriented file structure. This format has proved quite satisfactory and efficient for point operators but is lacking for spatial operators such as local adaptive enhancement and convolution. With these spatial algorithms, it is more efficient to store the data in a block-structured format, allowing the programmer to obtain a rectangular region from the image in a single I/O transfer. Point operators can usually be implemented in this format with the same efficiency.

In a multiband (color, multispectral, or multitemporal) image processing system, the system designer must be concerned with an additional dimension, such as time or frequency. Two organizational possibilities must be considered, band sequential and band interleaved. In a band sequential format, all the data from band 1 are stored on the disk prior to band 2, and soon. This structure is most efficient when the processing algorithms, such as spatial convolution or Fourier transform, require access to each of the image bands independently. In a band interleaved format, the multiband data are interleaved on a line-by-line basis, so that all the bands are stored together for each line of the image. This structure is most efficient when the processing algorithms require that all pixel values corresponding to a given location be present in memory at the same time, for example, spectral classification and principal components analysis. In systems in which only one format is supported, the band sequential method generally provides the best overall performance. Some systems, however, permit a variety of internal formats, allowing the programmer to store the data in the format that is most appropriate for the intended application.

The ability to categorize files within an image processing system by type or content has become increasingly important. Image processing systems are now being asked to combine varied data sources, for instance, thematic map data, topographic elevations, and geophysical data. In addition, files are generated during processing for storage of statistics, control points, and so on. The type of data contained in each file should be distinguished in the database. By assigning each different form of file an independent file type, the user and software can quickly differentiate between files. When functions require multiple files that have different file types, the software can distinguish between them when the files are opened.

ii. Image Access

There are a variety of requirements that must be met to provide the necessary image access capabilities within a system. The data must be accessible by pixel locations in image space. The system must be able to provide any precision and ordering of the data the user requests, by on-the-fly data type conversion and transposition, if necessary.

The ability to request an I/O operation based on absolute pixel coordinates within the image is an important feature. Because the I/O subsystem supports a variety of data types and possibly several internal organizations, it is important that the programmer be able to reference all I/O operations relative to the desired pixel rather than the record number. The I/O subsystem should take the burden of calculating physical record numbers and offsets, effectively mapping a virtual image onto a physical disk, for example. The programmer must, as a minimum, be able to request a line (or column), rectangular region, or image volume using image coordinates to define the size of the region. The use of pixel coordinates in all image I/O calls simplifies the coding to the programmer and makes the resulting code more readable as the system evolves over a long period of time.

A complete I/O subsystem allows the user to specify subsectioning, subsampling, and subbanding parameters at the time of program execution. The I/O subsystem then resolves the parameters and presents the subsampled, subsectioned, and subbanded picture to the applications program. In addition to an explicit subarea definition specified by the program, the I/O subsystem is expected also to perform transparently any subarea operation requested in the command line. In affect, we are dealing with a hierarchy of coordinate systems that must be sequentially resolved to provide the applications program with the requested data. An example is shown in Fig. 14.

At the top level, we have a disk or tape resident, parent image. The next level may represent a local window of the parent image that has been extracted for detailed processing. A third-level subset of the data may be explicitly called for in the command line. At the bottom level, we have the data subset that is actively being used by the applications program. When

Parent Image	BAY'AREA'CCT	3500 × 2500 × 4
Subject Image	PALO'ALTO'WINDOW	750 × 750 × 3
	(from parent with	(500 1000 1500 1500 2 2 : 4 3 2)
Command Specification	PALO'ALTO'WINDOW	(100 100 512 512 1 1 : 2)

	Samples			Lines			Bands	
	Start	Number	Inc.	Start	Number	Inc.	Start	Number
Program Request								
(in local command coordinates)	1	512	1	1	3	1	1	1
(in subject image coordinates)	100	512	1	100	3	1	2	1
(in parent image coordinates)	600	1024	2	1100	6	2	3	1

Fig. 14. A hierarchy of coodinate systems.

this bottom-level data subset is being requested, the program may choose to define it in terms of any of the higher-level coordinate systems. A powerful I/O subsystem will permit this flexibility, transparent to the program.

During an I/O operation, the program should have the capability of specifying the data type of the internal buffer that is being used for the requested operation. The I/O subsystem should provide any conversion necessary between the actual image data type and the requested data type. This capability is used in algorithms that require that the processing be done at higher precision than that of the input data, for instance, Fourier transform, spatial convolution, or spectral classification. For data that do not require conversion, there should be no loss of efficiency. The I/O subsystem should also provide the program with the ability to reorganize the order of the data during an I/O operation. In systems that allow a variety of internal image organizations, this permits the applications programs to be written so that they are independent of the organization of the data on the disk. The reorganization of the image data during the I/O operation introduces a significant overhead, so it is important that each application program be written to handle the data organization that is most applicable.

iii. Flexibility versus Speed

An image processing system is generally evaluated first on how fast it handles those functions that are important to the specific application area for which it was designed. However, its longevity and success will be determined by its flexibility. This determines a system's ability to grow and evolve as the requirements of the application field change. Unfortunately, speed and flexibility are often in conflict during system design and implementation. Systems that address only one or the other tend to be short-lived, even if they achieved success early in their development. The systems that have survived the test of time have paid close attention to both criteria and have found a balance.

One approach to a successful implementation is the identification of the cases in which speed is paramount. These functions can be specially treated by lower-level I/O routines that provide reduced flexibility. Fast low-level I/O routines interface to the existing data structures and can be used by the slower, more-flexible, higher-level routines as well. They by-pass the overhead required to provide flexibility. This approach allows the system to perform critical tasks efficiently without affecting the overall flexibility of the system. It has been used effectively in optimizing the transfer of imagery from the host computer disk to an interactive display device when the user does not require the general capabilities of the I/O subsystem.

iv. Ancillary Information

I/O subsystems frequently are required to maintain global information relative to the image, such as statistics and processing history. Some systems have allocated directory space to store statistical information and fill these fields the first time that the user requests that statistics be computed. Any subsequent requests for statistical information are satisfied by access-ing the directory records, significantly improving overall system perform-ance. The information recorded might include the minimum and maxi-mum value, the mean and the standard deviation, and the image histogram. For multispectral or multitemporal image files, this informa-tion can be stored on a band-by-band basis, that is, one set of statistics for each band.

Image histories have been incorporated in a number of systems. This information can be used to reconstruct at a later date a processing sequence that led to a given image. To implement a complete image history system, all interactions between the user and the system must be logged. A variety of techniques are used to store the history information for the user. The simplest implementation is to store the history of all previous images, plus the command and parameter information used to generate the current image, in label records that are kept with the data. This results in large history records for images from iterative processing and duplicates previous history information at every iteration, but it also results in a format that is easy to retrieve and present to the user. An alternative approach is to store only the history of the current processing with each image and to store a pointer to previous images, from which additional history can be obtained.

This approach reduces the storage required for history labeling but increases the overhead in retrieving the history labels to be presented to the user. In addition, it introduces the problem of how to reconstruct the image history when previous images representing interim results have been deleted from the system. Finally, there are advantages and disadvantages to storing history data in the image file or in separate files. The first approach is conceptually dirty, because it mixes image data with documentation. However, the second approach can result in a plethora of small history files that can become a management problem. A hybrid solution stores history information in a directorylike file that allows both rapid access and simplified management.

E. Display Control Subsystem

A specialized image processing display has become an integral part of most systems. In this context, the structure and efficiency of the software that controls the display determines the interactive potential of the overall system as well as its overall flexibility and efficiency. This subsystem can be broken into four distinct pieces: device handler, hardware control routines, buffered I/O, and refresh memory management.

The device handler provides the basic software interface between the image display and the host operating system. It must provide three basic functions: device initialization, data I/O (read and write), and interrupt processing. The device handler should be restricted to these functions only, leaving the device-dependent control to routines at a higher level. This will simplify the implementation and support of the device handler and also simplify the changes required as the hardware evolves.

Each of the functional components of an image display is usually controlled by a specific routine. A typical image display might require 20 to 30 of these hardware control routines to manage all aspects of display operation. Above this, there is usually a level of routines that combines the individual hardware components or loads them with fundamental operations. These two levels might be represented by the routines LUT and LINEAR LUT, for example. The first controls the LUTs, whereas the second loads a selected LUT with a linear ramp. The use of modular hardware control routines simplifies the coding of image processing algorithms, making them easier to document and support. Figure 15 demonstrates how a two-dimensional convolution function could be written using an image processing display controlled by such routines.

Buffering techniques allow a program to maximize the performance of the image processing display by grouping multiple command transfers into a

```
          DO   100 COL = 1, N
                DO 90 ROW = 1, M
C
C                Compute and load Lookup Table Weights
C
                DO 80 I = 1, 256
                   WEIGHTS(I) = function (kernel, (ROW, COL))
    80          CONTINUE
C
                CALL LUT (FCB, WEIGHTS, 0, 256, CHANL, WRITE)
C
C                Setup ALU Command
C
                CALL ALUCMD (FCB, FUNC, WRITE)
C
C                Compute and output X and Y scroll values
C
                SCROLX = function (COL)
                SCROLY = function (ROW)
                CALL SCROLL (FCB, SCROLX, SCROLY, CHANL, WRITE)
C
C                Issue Feedback Control Word
C
                CALL FEEDBACK (FCB, SOURCE, DEST, WRITE)
C
    90          CONTINUE
   100      CONTINUE
```

Fig. 15.   Sample program code.

single DMA transfer. When a buffered I/O approach is used, display commands are queued in a buffer until either they are explicitly flushed, the buffer is full, or a read command is issued. For graphics applications such as vector generation and histogram plotting, or iterative algorithms such as convolution, median filtering, or unsupervised classification, the buffering provides a performance improvement of 2 to 15 times depending on the exact nature of the algorithm.

The significance of buffering can be demonstrated by analyzing the timing requirements of the convolution example used in the previous section. At the completion of each video frame, the programmer has approximately 1 msec to prepare the processor for the next frame of data. If the data are buffered into a single DMA operation, there is ample time to send the necessary data and command sequences. If buffering is not used, it would normally require more than 1 msec to transfer the necessary information to the display, thereby missing every other frame. In the convolution example used in the previous section, an entire convolution sequence could potentially be contained in a single DMA operation. Furthermore, if double buffering is implemented, not only does the programmer gain the performance improvements discussed previously but also the next sequence of operations can be buffered and queued for operation while the last set are still being performed. A common problem in systems implemented using the new multiuser, multitasking image processing displays is their tendency to become control bound. The use of a double buffered I/O approach significantly reduces the impact of controlling the processor. Implementations that use memory-mapped interfaces are not able to take advantage of buffered I/O approaches, because each of the registers and RAMs is accessed by performing a move instruction to the corresponding address.

The management of images stored in the refresh memory is a necessity in a complete system implementation. The cost of implementing refresh memory for the image processing display has fallen steadily over the last decade; as a result, the number of refresh memories available in an image processing display has steadily increased. In the early 1970s, a typical system provided three images of 512×512 by 8-bit resolution, allowing the display of a single-color image. At that time, refresh memory management was not widely utilized, because the user usually only had the ability to store a single image within the image display. Today, systems are available with up to 64 refresh memories, with systems containing 12 or more memories being quite common. In these systems, the user can store several images concurrently in refresh memory; therefore, some form of refresh memory management capability must be provided to keep track of and control the contents of the refresh memory. The internal processing capabilities of an image processing display require the generation of interim and final results, which are stored in refresh memory. Refresh memory management allows the program to obtain available refresh memory to store these results. Because the displays are currently capable of generating 8-, 16-, 24-, and 32-bit results, the refresh memory management capability must allow the programmer to obtain images of varying size and intensity resolution.

F. System Support Environment

Structured programming generally leads to a hierarchical organization of subroutines in an image processing software system. In other words, the system structure should identify and segregate logical subtasks, in effect, breaking down the program into a tree of routines. This segregation into subroutines can be done along two lines in particular. First, it is advisable to isolate tasks that most functions must perform, such as open and close datasets, read and write logical blocks of data, initialize hardware, and conclude normally and abnormally. This reduces redundancy in the system code and makes the program easier to understand later. Second, it is often advisable to segregate code that is dependent on the host computer environment, that is, so-called machine-dependent code. Examples include calls to operating system primitives, host specific programming features, and specialized routines for processing bits, bytes, and characters. Experience has vividly shown that even the most stable systems must weather operating system updates, at least, and typically are transported to new, often unrelated hosts every five years or so. It is essential in systems that are designed to be transportable, and it can lead to efficiencies even in systems that do not transport. In those cases, isolation can simplify later realization of specific enhancements. If you discover a powerful new operating system feature for comparing strings, it would be helpful to know that this string comparison was isolated in one routine, rather than scattered over 200,000 lines of Fortran code.

Image processing system software can be divided into four distinct categories, as in Fig. 16. During system implementation, each routine should be put into one of these categories and designed accordingly. Even in systems

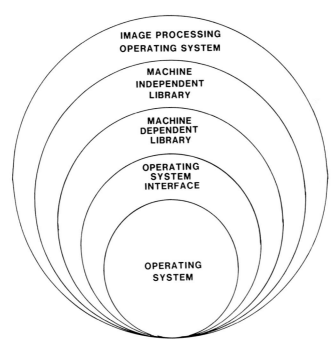

Fig. 16. Categories of image processing system software.

that are intended for internal use only in a single operating systems environment, one should categorize the routines into the four categories; but one need not be as careful when assigning a specific task into one of the four categories.

i. The Operating System

The operating system (OS) is generally supplied by the computer manufacturer. Device handlers to provide a software interface between the host computer and nonstandard computer peripherals must be added to the operating system. The OS normally provides ample support for the standard peripherals such as disk, tape, and line printers. Peripherals that may require additional device handlers include image processors, plotters, image digitizers, and image recorders. In situations in which the OS does not provide the necessary features, special-purpose devices should be integrated using additions to the operating system rather than changes to existing code. This often leads to some code duplication, but it simplifies changes as the OS evolves.

There is an important tradeoff between a fixed address space OS and a virtual OS. In fixed-address-space OSs, a significant portion of the development and maintenance of the system can be consumed merely in attempting to make the program fit. In the minicomputer environment, this can consume as much as 60% of the software cost. In addition, if the programmer must resort to overlays to fit into a fixed address space, the resulting code is usually slower and more demanding of the system disk.

ii. The Operating System Interface

The OS interface library consists of a variety of subroutines that allow the systems programmer to use features such as device I/O, devide handler linkage, and multitasking. The type and number of routines that fall into this category depend on the size and purpose of the system. Because image processing requires the handling of large volumes of image data, the I/O that is performed within the image processing system must be done efficiently. As a minimum, a set of routines must exist to read, write, open, and close tape, disk, printer, and terminal files. Generally, these routines will access low-level operating system intrinsics.

The use of multitasking is commonly found in image processing systems. Routines are provided within the OS interface library that allow any system or applications program to programmatically start a subtask. By isolating these functions into a small set of routines, the system and application programs requiring this capability remain independent of the specific OS environment. The use of multitasking allows the designer to isolate independent portions of the system, so that they can be individually modified without affecting the overall system. Multitasking is also used in some systems to allow the user to perform multiple tasks simultaneously to increase overall system throughput.

iii. The Machine-Dependent Library

The machine-dependent library consists of a set of functions that provide data handling capabilities not normally found in high-level programming languages such as Fortran. In transportable systems, routines must be provided for all bit, byte, or character handling. In the smaller fixed-address-space operating systems such as RSX-11M (DEC) and RTE VI (HP), message formatting routines must also be provided to save space. By implementation of a simple set of formatting functions, each task can link only the capabilities required for that particular function. This library also includes specialized routines that are optimized to maximize performance of specific functions. In any system there exists a subset of tasks, critical to the overall performance of the system, that are best handled in assembly language or the native system programming language of the host computer.

iv. The Machine-Independent Library

Most code within the image processing system is contained in the machine-independent library. All of the systems and applications functions reside within this library. For the purpose of control and documentation, these routines are often split into a variety of distinct categories, such as math library, display utility library, and parameter interface. In a transportable system, over 90% of the code should reside in the machine-independent category.

VI. Issues Involved in Evaluating Image Processing Systems

Certain issues are necessarily addressed in the design of most image processing systems, and the decisions and tradeoffs that are made have a major determinate impact on system architecture. We present here a discussion of a sampling of these issues, without implying that any one issue is more or less important than any other.

A. Transportable versus Host Specific Systems

The creators of an image processing system are faced with a distinct choice: should the system be designed to suit today's ideal host or to accom-

modate a variety of different hosts? This choice is partly a marketing decision, because a transportable system will appeal to more potential cus-
tomers. On the other hand, identifying and standardizing on one host (or
family of similar hosts) can increase performance and decrease support
costs. Also, transportable software is one thing; transportable hardware is
quite another.

Most image processing display implementations can be divided into two
categories. First, there are devices that actually plug into the host computer
bus, which makes it very difficult to change to another host manufacturer.
Fortunately, host computer manufacturers have shown a tendency to continue supporting their old buses on updated hardware (e.g., the DEC
Unibus). Second, there are devices that are connected by means of direct
memory access (DMA) interfaces. Generally, these devices are more easily
transported, but this is affected by the generality and simplicity of the DMA
interface chosen, the necessary control sophistication required of the interface, and the amount of internal intelligence in the image processor itself.

Software transportability is also linked to complexity and sophistication
[7,20]. Simple subroutine libraries can be expected to transport easily, as
long as they have been coded according to ANSI language standards (e.g.,
Fortran, 77). However, highly sophisticated image processing operating
systems are harder to transport, because they make wider use of the exotic
features in programming languages, which often are not standard, and
because they try to optimize their performance by taking advantage of
nontransportable features of the host operating system [2,28]. However, as
discussed in Section V.F, the machine-dependent and operating system
interface libraries isolate these functions, making it easier to make necessary changes when transporting or upgrading the system.

Host-devoted systems can optimize for their environment. For example,
they can integrate their command language with that of the host operating
system, thereby simplifying user learning. When they depend on a very
sophisticated host operating system, they can often take advantage of the
host's command interpreter directly. Access to the full range of peripheral
devices can be much simpler. On the negative side, host-devoted systems
can so thoroughly integrate themselves with their local operating system that
long-term support problems result. New versions of the operating system
from the host manufacturer may require substantial modifications to the
image processing code. This problem can be compounded by local modifications to the operating system itself, for example, to accommodate higher-
performance disk interfaces. Although performance may be improved,
support problems are magnified, and the future lifetime of the product may
be reduced. Experience shows that a given family of host computers has a
market lifetime of four to eight years. Too thorough an integration can trap
a system and doom it to obsolescence.

Transportable design usually requires highly structured programming to segregate the transportable from the untransportable code and to minimize the more complex support problem. When a system is designed to be transportable, over 90% of the code will need to be produced only once in a transportable programming language. Proper segregation of the tasks in the system will still allow a high degree of optimization, capitalizing on exotic operating system features or assembly language efficiency. Nevertheless, a price is paid for transportability. Speed is usually slightly reduced, and the command language of the system bears little relation to the host command language. With a proper investment of design skill, both of these objections can be minimized. The speed of a system is more controlled by intelligent programming than it is by exotic software tricks. And a properly designed system should allow provision for more than one command language, if it is deemed necessary.

There has been some discussion in the last few years about passing the transportability problem up to the host operating system. The widespread popularity of Bell Labs' UNIX operating system has prompted many manufacturers to offer it on a half dozen hosts [30]. Would it make sense to offer an image processing system under UNIX, thereby allowing simple transportability to those hosts? First, it does not necessarily follow that because the operating system transports, we can assume that any complex scientific software package will transport along with it. (Systems software experts have nightmares about the meaning of words such as UNIX-like.) There will still be a great deal of dependence on programming language transportability and instruction set compatability.

The principal objection will probably be performance. Letting the operating system cope with the whole transportability problem reduces the programmer's ability to optimize for specific host features. The price paid for generality, conceptual elegance, and transportability in an operating system is often speed. Such operating systems are more vulnerable to inefficiency than high-level scientific programs, because they must deal at the most fundamental level with the CPU architecture itself. For example, UNIX is often very slow; benchmark comparisons of VAX VMS versus VAX UNIX have generally shown the latter to have a factor of two or three in performance degradation [22]. Handling the transportability issue at the interface between the operating and the image processing system allows a measure of optimization for the particular demands of image processing.

B. Single-User versus Multiple-User Systems

Multiple-user image processing systems require that special attention be paid to a number of issues that are not of concern in the single-user environment. The principal issue is the necessity to share a variety of directories, files, and resources among the users. In addition, the SMMI must keep track of files that are being created but are not yet in the directory in order to prevent multiple users from creating files using the same name. The selection of a computer hardware configuration and operating system also assumes increased importance in the multiuser environment. Because the image processing system will operate as a subsystem to the computer operating system, the performance and features of the operating system in a multiuser environment will be a major influence on the overall system performance.

In the multiuser environment, most directories and data files are designed to allow (although not require) shared access among the system users. This means that the directory access routines must provide an internal locking mechanism that protects the integrity of the file as it is being updated by more than one user. Specific checks must be made to insure that one user does not delete files that are open and in use by other users. Of course, multiuser systems must also address the security problem. Most systems rely on the operating system to provide this security, but the image processing directory subsystem may augment it.

A variety of global hardware resources within an image processing system must be allocated in a multi-user environment. These resources fall into three basic categories. First, there are resources that will be needed by a user for a sequence of operations. These resources might include the image display, map digitizer workstation, and so on. Access to these resources can be controlled by acquire and release commands. Second, there are resources that are required for a specific function such as a magnetic tape, image digitizer, or plotter. The user obtains access to these resources implicitly by selecting the function for which they are required. Third, there are resources that are shared on an instruction-by-instruction basis. The best example is the programmable array processor. The array processor is allocated to a program or user only for the execution of a specific array procedure (such as perform an FFT on a data buffer). Note that the resource is locked for the time required to transfer the data to the processor, apply the selected array instruction, and return the data. Other users may then use the array processor, even though the first user's program may want it again in a few milliseconds. This type of contention is usually managed by the device's I/O driver and the host operating system.

Some image processing display manufacturers have developed architectures that can support multiple users. Although they may provide all the required physical resources to support multiple users, these devices can easily become control-bound. Bottlenecks can develop when command requests destined for each user's resources are forced to be queued to pass through a single component. In such environments, resource management can be very difficult and the factor that most limits performance.

C. Research-Oriented versus Production-Oriented Systems

An important aspect of any system design is that of selecting the goals and priorities of the final system. Image processing systems can be categorized into production-oriented and research-oriented groups. The design criteria and constraints of each are distinctly different.

The production-oriented system generally concentrates on a specific application area and involves the application of a fixed set of processing scenarios. The production system can be characterized by the following:

1. Limited application
2. Fixed processing scenarios
3. Emphasis on processing efficiency
4. Fixed data formats
5. Fixed output products
6. User friendliness
7. Rigid data base control

In digital radiography, for example, the primary functions to be performed are data acquisition and data review, in which the user can perform limited image processing tasks to enhance the interpretability of the results. The user is typically involved in repeating fixed sequences of processing steps. Because the user is not usually familiar with image processing or computer technology, the selection of processing steps and parameters is usually presented in a multiple-choice format. A menu orientation is typically found in production environments because of its self-documenting nature and ease of customization to fixed processing scenarios.

Processing speed is critical in most production processing systems. Because the production system can be tailored to the specific requirements of the application, critical computational and I/O tasks are implemented for optimum performance. Production-oriented systems often are sold in high volumes of identical configuration. This presents an opportunity for customized hardware design that can lead to maximized performance at minimum cost. The output products of a production system normally are produced with a consistent format and scale, allowing for standardization of output hardware, photo processing, and control software. The cataloguing and control of output data products are normally an important aspect of a production system. In medical applications such as digital radiology, the system must provide the ability to store and retrieve the patient records for future reference and must guarantee that patient records cannot be interchanged.

Although the production system provides many challenges to the system designer, it has the distinct advantage of having a singleness of purpose that allows the designer a framework on which to make critical decisions. The research environment, however, does not provide the designer with clear priorities. The research-oriented system is characterized by the following:

1. Diverse application areas
2. Diverse function repertoire
3. Multiple data formats
4. Significant user interaction
5. Modifiable programs
6. Varied I/O products
7. Transportability

The application of a research and development (R&D) system can cover the entire field of image processing. In a university environment, one system can be used to perform medical, remote sensing and industrial control research. Consequently, it will contain a large library of image processing applications programs designed to handle a wide variety of image data types and internal storage formats. The I/O subsystem must reflect this variety and maximize system throughput.

The R&D image processing system must be designed to handle a variety of image inputs, including various tape formats, video digitizers, and image scanning devices. Unlike the production user, the R&D user produces varied output products requiring additional flexibility in the hardware and software used in generating output products. The R&D users often are experienced in image processing and computer technology and will demand significant flexibility in the command language of the system. Unlike production users, who need to be presented with multiple-choice menu-oriented operation, R&D users typically know what they wish to do and want to be able to specify the processing sequence directly and explicitly.

The R&D environment demands the constant addition of new processing functions and scenarios. This requirement places emphasis on the modularity of the system design so that new capabilities can be added without significant change to the remainder of the system. Furthermore, this type of system typically spans many generations of a computer operating system and often is moved to a new computer system entirely. As such, transportability becomes very desirable.

Unlike the production users, who require efficient performance on a known set of computational and I/O tasks, the R&D user expects the system to perform efficiently over a diverse set of functions and data formats. Because the R&D environment requires that flexibility and modularity be maintained at all costs, the performance is generally obtained by developing specialized routines to handle the key processing tasks. Routines selected for specialization are implemented once the feasibility of a particular algorithm is established, using the existing system structure. Nevertheless, production-oriented systems will usually be faster than research-oriented systems on a well-defined problem.

D. Video Display versus Image Processing Display

The image display has evolved from a simple frame buffer in the early 1970s to a significant processing element in the overall system. Image processing system designs of the 1970s typically used the image display as a simple frame buffer, performing all significant processing in the host computer or programmable array processor. These systems quickly become I/O limited as they pass data to and from the various peripheral devices. To minimize the movement of image data across the CPU I/O bus, system integrators have implemented dual-ported disks to allow direct access of disk databases from the array processor or the image display (ESL, floating point systems, CSPI), parallel transfer disks to provide rapid loading of the image display (Ford Aerospace), and direct transfer of image data to and from display refresh memory to the array processor memory (see Section III). These approaches minimized the transfer of data across the CPU I/O bus and generally enhanced the overall performance of image processing systems.

During this time, the cost of RAM was falling rapidly. In the early 1970s, a system with three channels of refresh memory and simple video rate LUT processing would cost about $70K. Today, for the same $70K a user can purchase a system containing 16 channels of memory (4 megabytes). As the memory costs fell, suppliers began to implement processing components as discussed in Section IV, using the refresh memory to hold interim processing results. The display was now able to perform a variety of processing algorithms with integer precision and recursive processing. Since the image display processed an entire frame in $\frac{1}{30}$ sec, the design of algorithms was tailored to the particular hardware architecture. Algorithms such as Tukey median filtering were implemented using the parallel processing capability of the image processing display, which executed between 2 to 100 times faster than conventional approaches implemented in sequential processors [6]. In many cases, the actual algorithmic designs use an approach that is counter to the traditional approach of sequential machines due to the parallel processing attributes of today's image display.

The majority of the display manufacturers today provide image displays with significant processing capability. These processors stress LUT transformations, image arithmetic, logical operations, image convolution, and geometric correction. The existing designs represent an evolution of processing technology from the original frame buffer display of the early 1970s and as such live with the constraints imposed by video addressing and data rates. We are seeing a trend toward the design of processing displays in which the processing elements are logically separated from the burden of the actual video refreshing of the monitor, allowing the image processing display to operate at different speeds and with different addressing constraints.

In some ways, this issue can be reduced to a hardware versus software dichotomy. Is it better to process images with conventional computational resources and provide only a simple frame buffer display for viewing the output results? Or, is it better to expand the performance of an image display to fill the need for specialized, high-volume computation in image processing? These image processing displays sacrifice precision and algorithmic complexity to speed and interactivity. Whether this is an effective tradeoff, is determined by the application area and production requirements of the user.

VII. Conclusion

Several factors make the present a time of rapid evolution in image processing systems design. The principal driving force is the rapid advance of semiconductor technology. Both the performance and cost of RAM memory and of processing elements have improved significantly over the past decade [26], permitting first a practical demonstration of image processing capabilities and then a cost-effective implementation for everyday problems. A second factor, interrelated with the first, is a maturing of image processing science. Much effort during the last decade has been devoted to the research and development of image processing techniques and their application. Certain applications, notably airborne reconnaissance, have sponsored this work with an urgency that has overridden cost considerations. As a result, algorithms have been refined, data characteristics have been well defined, specific problems have been explicitly described in terms of solution by image processing techniques, and scenarios have been developed for routine solution of these problems. This has motivated the design of innovative system architectures that are dedicated to providing production image processing in a timely and cost-effective manner. A third factor has been the growth of digital remote sensing and communications technology. The volume of collected data has increased by orders of magnitude. Notably, much of the data is collected in digital form. Data are now generated by many techniques across a wide spectrum of wavelengths; synthetic aperture radar, x rays, and ultrasound are a few examples. Collectively, these factors have resulted in a commercial momentum for image processing. As awareness has spread, digital image processing is being considered as an approach to solving multifarious problems.

Image processing is just beginning to realize its potential. From its inception in the late 1950s and 1960s, it has been pursued largely as a research field. Only in the last few years has it become commercially profitable, which can be seen in the shift of ownership and investment among the major manufacturers. Large, high-technology companies have recently bought out or sought control of at least six of the early participants in the field. There has been an increase in the number of startups and spinoffs in the last two years, and some computer industry giants are entering (or reentering) the field. There has also been an explosion in the number of applications that are considering image processing. The combination of increased commercial investment and increased applications interest will accelerate the development of new capabilities and the configuration of specialized system architectures. No doubt, today's future trends will be covered in the next volume's historical overview.

References

1. J. Adams, Display or processor?, *Proc. SPIE* **301,** 1981, 48–53.
2. J. Adams, and E. Driscoll, "A Low Cost, Transportable Image Processing System," presented at First A.S.S.P. Workshop on Two-Dimensional Digital Signal Processing Oct. 1979, unpublished.
3. J. Adams, and R. Wallis, New concepts in display technology, *Comput. Mag.* August, 1977, pp. 61–69.
4. Advances in display technology., *Proc. SPIE* **199,** 1979.
5. Advances in display technology, II., *Proc. SPIE* **271,** 1981.
6. J. W. Backus, The Syntax and Semantics of the Proposed International Algebraic Language of the Zurich ACM-GAMM Conference, *Proc. Intl. Conf. Inf. Process.* UNESCO, 1959, pp. 125–132.
7. P. Brown, "Software Portability: An Advanced Course." Cambridge University Press, 1977.
8. K. Castleman, "Digital Image Processing." Prentice Hall, New Jersey, 1979.
9. N. Chomsky, Three models for the description of language, *IREE Trans. Inf. Theory,* **IT2,** 1956, 113–124.
10. *Computer Magazine,* 1981. (I.E.E.E. Computer Society, contains a number of relevant articles.)
11. Design of digital image processing systems, *Proc. SPIE* **301,** 1981.

12. E. Driscoll, High speed classification of multispectral imagery. *J. Appl. Photogr. Eng.* **8,** (3), 1982, 142–146.
13. E. Driscoll, and C. Walker, Evolution of image processing algorithms from software to hardware., *Proc. SPIE* **271,** 1981, 271.
14. M. J. B. Duff, and S. Levialdi (ed.), "Languages and Architectures for Image Processing." Academic Press, New York, 1981.
15. M. J. B. Duff, Propagation in cellular logic arrays, *Proc. of Workshop on Picture Data Description and Management,* (I.E.E.E. Catalog No. 80CH1530-5), August, 1980.
16. N. England, Advanced architecture for graphics and image processing, *Proc. SPIE* **301,** 1981.
17. K. Fant, Interactive Algorithm development system for tactical image exploitation, *Proc. SPIE* **301,** 1981.
18. B. Gordon, The search for the 10 second fast Fourier transform (FFT), *Proc. SPIE* **149,** 1978.
19. D. Gries, "Compiler Construction for Digital Computers." Wiley, New York, 1971.
20. S. Hague, and B. Ford, Portability—Prediction and correction, *Software Practice and Experience,* **6,** 1976, 61–69.
21. L. Hubble, and C. Reader, State of the art in image display systems, *Proc. SPIE* **199,** 1979, 2–8.
22. D. Kashtan, "UNIX versus VMS: Some Performance Comparisons." S.R.I. International survey for D.A.R.P.A., 1981.
23. J. Mannos, Powerful hardware/software architecture for a minicomputer-based interactive image processing system., *Proc. SPIE* **301,** 1981, 129–134.
24. *Proc. of Workshop on Computer Architecture for Pattern Analysis an Image Database Management* (I.E.E.E. Computer Soc. Catalog No. 378), Nov., 1981.
25. C. Reader, and S. Fralick, "The Interactive Image Processor," ESL Technical Report, 1981.
26. C. Reader, and L. Hubble, Trends in image display systems, *Proc. IEEE* **1981,** 606–614.
27. L. H. Roberts, and M. Shantz, Processing display system architectures, *Proc. SPIE* **301,** 1981.
28. B. Ryder, The PFORT Verifier, *Software Practice and Experience,* **4,** 1974, 359–377.
29. 1982 Society for Information Display International Symposium Digest of Technical Papers, Vol 13.
30. R. Thomas, and J. Yates, "A User Guide to the UNIX System, Osborne." McGraw-Hill, New York, 1982.
31. P. Wambacq, J. DeRoo, L. Van Eycken, A. Oosterlinck, and H. Van den Berghe, Real-time image computer configuration, *Proc. SPIE* **301,** 1981.
32. G. Wolfe, Use of array processors in image processing, *Proc. SPIE* **301,** 1981.
33. G. Wolfe, and J. Mannos, Fast median filter implementation, *Proc. SPIE* **207,** 1979, 207.

Author Index

Numbers in parentheses are reference numbers and indicate that an author's work is referred to, although his name is not cited in the text. Numbers in italics show the page on which the complete reference is listed.

A

Aaron, G., 216(53), 219(53), *225*
Aatre, V. K., 25(9), 47(9), *48*
Adams, J., 292(1, 3), 300(1), 318(3), 352(2), *359*
Aggarwal, J. K., 83(3), *108*
Ahmed, N., 189(1), *223*
Aho, A. V., 25(1), 28(1), 47, *47*
Algazi, V. R., 180(3), 184(3), 192(4), 200(6), 202(4), 220(2), 221(5), *223*
Allison, L. J., 43(54), *49*
Anderson, A. H., 115(1), 142(2), *167*
Anderson, B. D. O., 100(1), *108*
Anderson, G. B., *49*
Andrews, H. C., 3(5), 15(4), 16(5), *47, 47, 51,* 53(1), 58, 59(1), 60(1), 63, 70(1), 71(1), 73(1), 75(1), *76*
Andrus, J. F., 268(1) , 286, *286*
Arcese, A., 268(2), 286, *286*
Arendt, J. W., 166(15), *167*
Arguello, R. J., *48*
Arnold, J. F., *223*
Aron, J., 232(51), 247(51), *255*
Ascher, R. N., 218(8), *223*
Ataman, F., 25(9), 47(9), *48*
Au, S., 84, 86(2), *108*
Awtrey, J. D., *49*

B

Baba, N., 165(3), *167*
Backus, J. W., 329(6), *359*
Baggeroer, A. B., *253*
Barett, E. B., *51*

Barnea, D. I., 270(3), 286, *286*
Barrett, H. H., *49,* 166(14, 15), *167*
Bates, R. H. T., 165, 166(52), *167, 169*
Batson, B. H., 178(47), *225*
Bell, T. H., Jr., *253*
Bellman, S. H., 165, *167*
Bender, R., 165, 166(30, 31), *167*
Bergin, J. M., *253*
Berry, M. V., 165, 166(6), *167*
Bhaskaran, V., 187(40), *225*
Biberman, L. M., *48*
Billingsley, F. C., *48*
Blackman, R. B., 238(3), 252, *253*
Bosworth, R. H., 180(11), *223*
Boyd, D. P., 166, *168*
Bracewell, R. N., 165, 166(7–9), *167*
Breedlove, J., *49*
Brillinger, D. R., 248(4), 253, *253*
Brooks, R. A., 166(11), *167*
Brown, P., 352(7), *359*
Budinger, T. F., 166, *167*
Budrikis, Z. L., 47, *49*
Burg, J. P., 237(6), 253, *253*

C

Camana, P., 185, 222(9), *223*
Campbell, C. W., 268(1), 286(1), *286*
Campbell, K., *51*
Candy, J. C., 180(11), 206(10), 207(10), *223*
Cannon, T. M., 14(14), *48,* 60(2), 70(2), 75(2), *76*
Capon, J., 232(7), 237(7), 238(8), 243(8), 244(7), 252, *253*
Carter, G. C., 241, 252, *255*

Subject Index

A

Adaptive filtering
 examples, 37–38
 general scheme, 35
 using subimages, 34
Adaptive image coding, 181
Adaptive transform coding, 200
Additive image noise model, 55
Algebraic reconstruction algorithms, 136–147
 Numerical methods, 138
 Ray definition, 136
 Ray sum, 136
 Theory, 153–157
Algebraic reconstruction technique (ART)
 artifacts, 141
 computational algorithms, 141
 salt-and-pepper noise, 143
Aliasing distortion, in image reconstruction
 artifacts, 157–165
 criterion, 121
Analog image processing systems, 291
Apriori knowledge in, image enhancement, 36
Arithmetic logic unit (ALU), 307
Array, 233
Array processors, 301
Autoregressive modeling, 249–250

B

Backprojection
 definition, 122
 fan, 130
Bit allocation, 190
Blackman–Tukey method, 241
Block quantization, 188

Blurring (image), 3
Border curvature, 281
Border pixels, 275

C

Causal filtering, 91
Chain code, 277
Characteristics of image processing algorithms, 294
Coarray, 233–234
Color pigments, 43
Command interpreter, 332
Command language, 328–332
Connected component labeling, 274
Constant area quantization (CAQ), 185
Constrained least-squares restoration, 68
Constraints for image restoration, 67
Convex hull, 282
Correlation matrices, 237
Covariance, stationary Gaussian process, 94

D

Data adaptive spectral estimation (DASE), 243
Database management for images, 337
Delta modulation, 177–178
Detector bias field, 95
Differential pulse code modulation (DPCM)
 block diagram, 176
 open loop, 179
 vector DPCM, 202–203
Diffraction tomography, 116
Discrete Fourier transform, 15
Display-related image characteristics, 293–294